EDWARD SYLVESTER MORSE

LONDON HUMPHREY MILFORD
OXFORD UNIVERSITY PRESS

Edward Sylvester Morse

EDWARD SYLVESTER MORSE

A Biography

BY

DOROTHY G. WAYMAN

CAMBRIDGE, MASSACHUSETTS

HARVARD UNIVERSITY PRESS

1942

PRINTED AT THE HARVARD UNIVERSITY PRINTING OFFICE
CAMBRIDGE, MASSACHUSETTS, U.S.A.

See, what a world of ground
Lies westward from the midst of Cancer's line
Unto the rising of this earthly globe
And shall I die, and this unconquered?

—*Tamburlaine the Great*

PREFACE

Edward Sylvester Morse is worth reading about because his life is a case history in the American way of rugged independence, personal initiative, and democratic sharing with the community. He was a pioneer both in the laboratory method of applied science to replace canonical decree and in the building up of public museums instead of private collections in princes' palaces. Morse was no plaster-of-paris saint. He had faults of character that hampered his work, offended his friends, gave his enemies subjects for gossip. He paid tuition to no college; but Harvard, Yale, Bowdoin, and Tufts awarded him honorary degrees; foreign governments bestowed medals and decorations upon him. With no training in art, he illustrated his own books, made scientific drawings for the Smithsonian Institution, and became the world authority, consulted by museums on three continents, on Oriental ceramics. A scientist who loved learning, he maintained all his life that the ordinary man was entitled to measure common sense and ordinary reasoning against the theories of the specialist. His formula for scholarship was simple: keen observation, accurate memory, constant reflection and analysis.

To a great extent, this account of Morse is based upon manuscript materials in the possession of Morse's family and friends. Without the frank and generous coöperation of his daughter and son, Edith Morse Robb and John G. Morse of Concord, Massachusetts, and of Mr. Oliver C. Gould of South Portland,

Maine, this biography would have lacked valuable source material — the journal Morse kept between 1859 and 1863, numerous old photographs, and a series of letters, covering many years (1856–1925), written to his lifelong friend, John Gould. Morse's handwriting was so illegible, and his spelling and punctuation so erratic, that a few minor aids to reading have been supplied in quotations from manuscripts, though the journal's vagaries, for the sake of their flavor, have been left for the most part untouched. The original journal is in the possession of Mrs. Robb; the letters of John Gould are in the possession of Mr. Oliver Gould. Letters from other persons to Morse are cited from originals in the possession of the author, of Mr. John Morse and Mrs. Robb in Concord, the Portland Natural History Society, and the Peabody Museum of Salem. For permission to quote from these materials I am indebted to the persons and institutions mentioned and to the following: the late Dr. Glover M. Allen, Mr. George R. Agassiz, Mr. Talbot Aldrich, Mrs. C. E. Beecher, Mr. Poultney Bigelow, Mr. Morris Carter, Dr. E. G. Conklin, Dr. E. R. Cumings, Mr. Robert M. Dole, Mrs. E. F. Fenollosa, Dr. Simon Flexner, Mr. Cass Gilbert, Jr., Mr. Francis G. Goodale, Dr. L. O. Howard, Mr. F. M. Ives, Mrs. David Starr Jordan, Mr. John E. Lodge, Dr. A. Lawrence Lowell, Mrs. George H. Monks, Dr. G. H. Parker, Mr. J. D. Phillips, Mrs. D. P. Rhodes, Mr. Thorwald Ross, Mrs. Elihu Thomson, Mrs. William Morton Wheeler, and the authorities of the Boston Society of Natural History, the British Museum, the Gray Herbarium of Harvard University, the Royal College of Science, the Smithsonian Institution, the United States Navy Department, and the University Museum of the University of Pennsylvania.

Irreplaceable data on Morse's years in old Japan, 1877–1883, were fortunately gathered during my visit to that country in the summer of 1939, before war between Japan and the United States interrupted the peaceful pursuit of historical research. Japanese scholars gave both time and effort to help me gather records of Morse's day and preserve old memories.

Individuals connected with institutions in Europe, Japan, and the United States with which Morse was associated in his lifetime have been invariably helpful and cordial. In addition to material in the possession of the family, the chief collections of Morse papers or memorabilia are in the Peabody Museum, Salem, Massachusetts, and the Science Building, Imperial University, Tokyo. Some of the materials I used have been deposited in the Widener Library of Harvard University.

The words "Thank you" seem inadequate to express my appreciation of the help I have received during the past three years. I like to think that we all shared the pleasure of collaboration in the making of a permanent record of an American scientist's personality, and that the record may encourage youth of the future to find careers as valuable to the community as Morse's.

D. G. W.

Boston, Massachusetts
December 26, 1941

CONTENTS

ILLUSTRATIONS

FOREWORD

I WELL REMEMBER the first time I met Professor Morse. It was at a meeting of the Thursday Evening Club during the winter of 1907. After the meeting I stood talking to Mr. Alexander Agassiz, who had been interested in the subject of my informal address. I had just come back from the East Indies, and he had been considering going there for years and was interested in my firsthand and fresh impression. I was setting forth the beauties of Halmahera when Morse stepped up. Mr. Agassiz introduced us one to the other. Morse had been interested in a slide showing a Papuan native drawing a bow, a picture which my wife had taken at Humboldt's Bay in New Guinea. He was then much interested in the methods of arrow release, and the Papuan way of pulling the bow string showed particularly well in this picture. One thing led to another, and Morse came to the house frequently. Then I joined the Tavern Club and saw him there constantly.

One occasion stands out most vividly in my recollections, although the details are a bit hazy. I think it was on the occasion of the great Salem fire, when Dr. Samuel Mixter and I drove over to see if we could help get Morse moved in case the fire spread towards his house. It may have been on some other occasion, for Dr. Mixter and I called on him not infrequently, driving over from Swampscott, where Mixter lived, to Salem. We walked in quite casually, for I cannot remember that Morse ever locked his front door. The Professor on this occasion was sitting on a square white pine box with his feet on the table, playing a flute with his nose. He said the Solomon

Islanders did this regularly, and he was going to learn to do so too. Then he jumped up, so sprightly that anyone who knew him could see him do it in his mind's eye now, and said, "See what's in this box." It contained a beautiful crystal jar on which was engraved: "This contains the brain of Edward Sylvester Morse, born June 18, 1838, died.   He told us that he had debated whether he should leave his brain to Burt Wilder's collection at Cornell or to the Wistar Institute in Philadelphia. He tossed a coin to settle the matter. Wistar won and sent him the jar in appreciation of its good fortune. There was a special reason why Morse's brain was of more than usual interest. You will learn all about that in this book. I may say here that I remember on one occasion his taking my daughter Mary on his knee and drawing a dextral and a sinistral snail shell at the same time, one with each hand, and then drawing the two sides of a butterfly, or a dragonfly, with absolutely perfect symmetry, using a pencil in each hand. The children, of course, were fascinated, for, more than this, he could pat the top of his head with his right hand and rub his stomach with a circular motion with his left, which, if you will but try it, you will find to be an almost impossible feat.

It is hard to understand why there has been no biography of Morse until this late date, for he was certainly one of the most lovable, versatile, and erudite members of the scientific community hereabouts. That this biography has been at last prepared as a labor of love is all to the good. It is even more satisfactory that it is entirely adequate.

THOMAS BARBOUR

Director, Museum of Comparative Zoölogy

Cambridge, Massachusetts

EDWARD SYLVESTER MORSE

CHAPTER ONE

THE TWIG IS BENT
1838–1858

NEW ENGLAND in late spring knows sudden days of unseasonable heat, days that bring out the elm leaves in a rush and affect human beings with lassitude and dislike for the routine of life. Spring fever, say the New Englanders.

On such a day, May 17, 1856, two youngsters spent their Saturday half-holiday on the water front of Portland, Maine. John Gould was a bank clerk. Edward Sylvester Morse, who was approaching his eighteenth birthday, worked as a draughtsman over blueprints in the locomotive factory of the Portland Company. Morse wrote in his diary that night:

John came down. Played with the dog in the water. Rolled a great India rubber something at the dog. Played Physa Schologizeim to John. Acted like a pumpskewking.

He meant that they acted like fools, but the church of his father, Deacon Johnathan Kimball Morse, interpreted literally the Biblical injunction that threatened hell-fire to him who called his brother fool. From pumpkins, New England boys carved saw-toothed, grinning jack-o'-lanterns for Hallowe'en antics, and so, to describe laughing folly, Ned Morse coined his word "pumpskewking."

No one, on that May afternoon, could have persuaded Deacon Morse to envision his muddy hobbledehoy on the beach as one of America's most distinguished scientists and educators. The boy had been expelled from every school he had attended: the Portland village school, the academy at Conway in 1851, the Bridgton Academy in 1854. His teachers complained of his boisterous spirits, his impatience of authority, and his neglect of lessons for wandering in the woods or along the shore.

The rest of the family conformed more comfortably to the community pattern. Deacon Johnathan Morse — the curious spelling was common in New England — was a partner in the Byron & Greenough Company, dealers at wholesale and retail in beaver hats, caps, furs, and buffalo robes, at 148–150 Middle Street, on the corner of Free Street, Portland. He was a descendant of Anthony Morse who emigrated from England to Newbury, Massachusetts, in 1635. His wife, Jane Beckett, through her mother, Abigail Dyer of Portland, traced her ancestry back to Joseph White of Roxbury, Massachusetts, whose son, the Reverend John White, was graduated from Harvard College in 1698 and settled at Gloucester. The wife of the Reverend John White was said to be the daughter of an English earl, and had brought from England a Greek lexicon, highly valued. Jane Beckett Morse was romantic enough to repeat to her children a family legend which claimed Thomas à Becket, Archbishop of Canterbury in the twelfth century, as an ancestor. Her husband took more pride in the provable fact that his forebear, Thomas Morse of Haverhill, New Hampshire, was the first man in the countryside to own a chaise.

Johnathan K. Morse and Jane Beckett were married at Portland in September 1829. Their first child, Charles Beckett Morse, was born there in 1831. Between 1831 and 1836 the family lived for a time in Hallowell, Maine, where Johnathan Morse carried on his fur business; George Fred, 1834, and a girl, Charlotte, who died in infancy in 1836, were born there. Later the Morses returned to Portland, where Edward Sylvester Morse was born on June 18, 1838. Three younger sisters followed: Caroline, 1842; Mary Beecher, 1843; Nellie Kimball, 1849.

The eldest son, Charles, who was spoken of as lovable and sweet-natured, died of typhoid fever on January 21, 1850, in his nineteenth year. At his funeral a minister more doctrinal than Christian in spirit preached a sermon dooming the departed to hell-fire because he had not been saved by baptism. Ned Morse was twelve years old when his brother died; the effects of the sermon brought bitterness to his boyhood and lingered throughout his life. His father's reaction was to become deeply religious, fanatically determined that his remaining children should find salvation through the Baptist Church. Jane Morse, tight-lipped behind her black crape veil, declared as she turned from her eldest son's grave, "Never, never again will I step inside a church door." She probably encouraged the interest Ned soon manifested in spiritualism, spreading rapidly through America in the 1850's from the New York farmhouse where the Fox sisters had heard their mysterious rappings.

Fred Morse, four years older than Ned, "took after" his father, as the country people said, being serious, patient, and earnest. Ned's teasing ways tormented him often. Once, while he was chopping wood, the everlasting chore of New England boyhood, Fred's temper flared, and he hurled the hatchet at his mischievous little brother. His aim was not deliberate, and no injury was done, but Fred was so shocked at his own act that he begged from his mother two spools and a long piece of thread. Thereafter, for years, he carried the spools in his pocket. Whenever he felt his temper rising, he took out the spools and painstakingly reeled the thread from one onto the other before he trusted himself to speak. Sometimes it took more than one reeling to restore self-mastery, but Fred Morse never again lost control of his temper.

Ned had not the same quality of self-analysis and persistence. He gloried in mischievous methods of avoiding irksome routine. Grandmother Abigail Dyer Beckett, widowed and living with her son Charles near by, came often for supper with the Morse family at Number 40 Pearl Street. Ned's duty was to escort her home through the dark streets carrying a lantern. At the age of thirteen, he was already devoted to science, studying and classifying land shells, and he grudged the time spent on convoy duty. One evening, at the approach of each pedestrian, he held the lantern high, proclaiming shrilly, "My grandmother, gentlemen, my grandmother!" The indignant old lady specified Fred as her escort on future evenings.

Shells were the current craze of seaport towns. Every New England parlor had a few exotic shells in the whatnot. Seafaring sons brought back to New England farmhouses great pink Venus's-ears and placed them on each side of the doorstep. Linnaeus and Lamarck in Europe, Amos Binney and Dr. Augustus A. Gould in Boston, made a science of classifying families, genera, species of shells. Bright-hued lustrous shells like the *Olivas, Cyprinas,* and *Cythereas* from tropic seas brought from thirty to three hundred dollars apiece in London, Paris, or Boston. A rare Venus shell was sold for five thousand dollars. Lamarck named a turreted white shell that cost a hundred guineas *Scalaria pretiosa* for its market value.

Ned Morse, like that other Portland boy Henry Wadsworth Longfellow, knew "the bearded sailors on the wharves." While John Gould was at school at Bridgton Academy, Ned wrote to him:

March 6, 1854: . . . The Sarah Sands left Wednesday for Liverpool. There was quite a crowd on the wharf; they gave three cheers on board and went off on her voyage but for lack of commonsense in the Johnny Bulls she went aground on a bank in front of the Portland Company's works and there staid till morning. I got acquainted with the boatswain and a sailor on board the Sarah Sands and they promised to get me a lot of shells. Especially the sailor, for he had been a fisherman. He said that they would go a-dredging . . . and get them alive and perfect and sell them in the Shell stores in Liverpool. . . . He also stated that he got a dozen shells in Egypt . . . and had looked in all the Shell stores in Liverpool and had only seen two like them. . . . Find out whether the folks up there would like any of our shells. . . . Look around and see what chance there is for fossils and report.

Ned Morse had no money to purchase specimens in the "Shell stores"; he had had no education to direct or train his interest in science; yet from his twelfth year he had been collecting shells, and he had conceived the idea of making a complete collection of local shells. He was collecting the smallest, ugliest types, the land shells, ignored or overlooked up to that time. He had inspired John Gould in the same pursuit, as is shown in another letter to John at Bridgton in which he refers to "our Maine Collection":

October 29, 1854: . . . I have found the *Helix minuta* at the Portland Company's works. I suppose you know where the stable is.

On the bank running down to the water near the Stable is the *Helix minuta* by the 100,000,000. Yes, sir. The specimens are very nice and I have got a great many of them. I have bought a microscope for $3.00, a very nice one, I tell you. I have been looking at some *Pupas*. . . . I have discovered about half a dozen *Bulimas* in the box and one or two *H. minutas*. . . . We will begin that book as soon as you say, the sooner the better.

One of Ned Morse's brief interludes of schooling had been at Bridgton Academy with John Gould in the summer term of 1854. He always remembered a student there who asked him bluntly, "Is your father rich?" Morse answered, no, his father was not rich. "Neither is mine," said the boy, whose name was Hiram Ricker. "But he's got a fine spring of water, and he wants me to find out a man with money to make something of it." Because they were not rich, the Morses thus missed a chance to make a fortune by financing the Poland Spring House, popular and profitable summer resort of Maine for many years. At Bridgton, Ned and John Gould made collecting expeditions along the sparkling Maine streams or in the green forests. They climbed Barker's Mountain searching for *Physa,* a minute land shell, and had such a glorious day that sixty years later John Gould would write to Morse, "Remember when you said Phoza for Physa we tumbled down laughing?" It was the same land shell, their stock joke, to which Ned referred when he wrote in his diary that warm May evening of 1856 ". . . played Physa Schologizeim . . . acted like a pumpskewking."

The authorities at Bridgton Academy sent Ned home in the autumn of 1854. They resented his carving of desks, and he resented a stunning box on the ear from a master. Deacon

Johnathan thought Ned had better go to work. Fred, aged twenty, was employed at the Portland Company, of which he was ultimately to become general manager; he secured a job there for Ned, as draughtsman because of his knack for drawing. The pay was four dollars a week. Ned got on as badly with John Sparrow, the manager, as he had with his schoolteachers. Shells remained his major interest, the occupation of all his leisure time.

He spent some of his first earnings for a tattered, secondhand copy of *Report on the Invertebrata of Massachusetts* by Dr. A. A. Gould, published in 1841. He joined the new Portland Natural History Society and borrowed from Dr. William Wood a book on New York shells. Land shells were so little known at the time that Jackson's report for the *Geology of Maine* in 1836 listed only three species. Dr. J. W. Mighels of Portland listed twenty-two land species, thirty-four fresh-water species, in the *Catalogue of Shells of Maine* in 1842. J. W. Chickering, Jr., of Portland, an acquaintance of Ned Morse's, in 1854 had just published a *List of Marine, Freshwater and Land Shells Found in the Immediate Vicinity of Portland, Maine;* its title was almost as long as the list.

Morse's diary shows his continuing preoccupation with shells, shared now with John Gould since the latter's return from Bridgton to enter the bank as a clerk:

January 2 [1856]: Pleasant but cold. Talked quite a while in the [Portland Company's] shop on Spiritualism. Went to see Mr. Chickering's [shell] Cabinet; found that he had got five different species of Freshwater and Landshells of Maine that we have not got. Went down to John Gould's in the evening; carried some of my Shell books with me. Read to John on the Lymnidea.

March 3: John Gould and I hauled upon a sled a lot of fossils from wharf. This was ballast from Cuba.

A landslide at Bramhall's Hill in Portland that spring, as the frost worked out of the ground, exposed a rich deposit of fossils and forged one of the first links in Ned Morse's career.

May 4: . . . Introduced to Mr. Thomas T. Bouve of Boston, mineralogist. Went to landslide with him. He came up in forenoon to see my Cabinet. Spent evening with him.

Bouvé was a regular attendant at the meetings of the Boston Society of Natural History and had probably been advised to hunt up Ned Morse in Portland by a fellow-member, Charles J. Sprague. Ned had known Sprague when he was a scholar at Bridgton and had earned Sprague's friendship by his zeal in collecting fungi. Two weeks after Bouvé's visit to Portland, Ned "went to Boston in the cars. Wrapped up fossils that Mr. Bouve gave me. Went to the Boston Society of Natural History. Called on Mr. Whittemore, talked on shells for a long time. Came back to Portland in boat *Atlantic*."

The natural inference, when we find Ned Morse a month later resuming school, which he had abandoned two years earlier, is that the Boston men had counseled him to enlarge his education. He quit his job at the Portland Company, spent the first day of June "drawing shells; packing my trunk," and the next day went "with Lydia Berry to have my picture taken; went to Bethel in the afternoon."

At Bethel, Morse was exposed to the contagious influence of Dr. Nathaniel T. True, a born teacher and ardent scientist. In 1835 the little town of Bethel opened a high school, with thirty-five pupils and a two-and-a-half story brick building surmounted by a cupola and bell. Nathaniel T. True, teaching school to earn money for a medical education, was its first principal. Later he received his medical degree from Maine Medical School and practised briefly before returning in 1847 as principal of Gould's Academy at Bethel. "He liked the study of chemistry, botany, geology and mineralogy . . . ," said William B. Lapham, who knew him. "It was his delight to take his spring and summer classes in botany, through the fields, pastures and woods . . . or his pupils who were interested in geology and mineralogy up to Paradise Hill, and sometimes even to the mountain tops. . . ."[1]

Short of stature, black-haired, with blue eyes, quick in his movements and his judgments, this teacher understood and fostered the quick mind, the insatiable interest in nature, of young Ned Morse. Earlier teachers in other schools had disciplined Morse's restlessness and inattention to books by a box on the ear. True encouraged the boy's interest in collecting, appreciated the scientific spirit that concentrated on land shells. Instead of scolding when Morse played truant down at the Androscoggin's banks or up in the green mountains that stand like a western wall beyond Bethel Hill, he assigned the boy as assistant to visiting biologists or mineralogists. It was through True that Morse met, and accompanied on field expeditions, Charles J. Sprague and Thomas Bouvé of the Boston Society of Natural History. Thus was forged the first link in the chain of his future career.

Bethel was sixty miles north of Portland, on the Androscoggin River, where the old route of Indians and fur-traders

[1] W. B. Lapham, *History of Bethel* (Augusta, Me., 1891), pp. 144–145.

wound towards Canada. Near by, at Paris Hill, were mines of tourmalines, amethysts, beryls, garnet, and rose quartz. Ned Morse was not a mineralogist; he wrote in his diary, "Every day I go in swimming and collect shells." At first the return to school was interesting, as he wrote to John Gould.

June 10 [1856]: . . . I study and feel like study. I kinder like Latin Grammar. . . . My studies are Latin Grammar, Boyd's Rhetoric, Arithmetic and Grammar. . . . I room with Edwards; he is a first-rate fellow and a Free Thinker. McKusick is a sick-looking clown and of the Calvinist Hellfire. . . . [He] takes his Bible down into the kitchen every night and prays. . . . [He] brought me down a tract to read and I carried him up a Spiritualist *Telegraph*. . . . This A.M. I got into an argument with McKusick on religion. [He] got raving and says, "If you don't repent, if you aren't born of the Spirit, you will be Cast Out." He said the last in an exciting tone. I took it cool and told him about being born of the Spirit. . . .

Ned Morse had been attending spiritualist meetings and reading spiritualist tracts and periodicals for six years, since the death of his brother. The arguments with McKusick set him to examining the subject in the scientific way he had developed in researches in Mollusca. He wrote to John Gould arranging an experiment:

June 18: . . . I have seen Jordan and he says he will set Friday night from 7:30 to 8:30. Do you understand? I give below a list of the performances you are to go through; you can mix these up as much as you want to but don't do anything else but what I have set down. 1) Singing 10 minutes. 2) Whistling 10 minutes. 3) Writing 10 minutes. 4) Standing 10 minutes. 5) Reading 10 minutes. 6) Laying on Bed 10 minutes. I shall see that my time is railroad time and have a list before me and send you a list of the things as they come, like this: What is John doing now? You think this over and ask your Guardian Spirit to answer them direct as he can. If it succeeds, I shall want you to put it in form for the Spiritual *Telegraph*. Won't that be good, hey?

Five days later, disillusioned, Ned records the failure of the Guardian Spirit to answer direct.

June 23: Old Fellow, about the Circle, we could not expect much through an inharmonious circle and what is more, I would not put myself to the trouble of running back and forth to gratify the idle curiosity of a few darned fools.

September 7: Don't believe I shall stay up here long; come home or die. Sick these last two or three days; feel kinder sick now, weak as a rag! Raining like Old Scratch. I don't feel as if I was a-going to stay here the whole term. Don't believe I shall. Don't feel like it. Wont, I swear. . . .

November 6, 1856: We will have some good times, I tell you. The Natural History Society will have their first meeting in this month and we shall tend them, of course. I shall be proposed there by S. H. Beckett as a member; then I can get books out and we will read and write about our "Maine Collection" [of shells].

Before he quitted Bethel Academy for Portland, Ned Morse found, on September 28, 1856, an earth-colored land snail, one thirty-second of an inch in length. A smaller, less conspicuous object never determined the whole future course of a man's life. At the Boston Society of Natural History on November 19, 1856, "a communication was read from Mr. E. S. Morse of Portland, Maine, on *Helix astericus* and specimens of the shell were presented." A fortnight later the Society's most noted conchologist, Dr. Augustus Addison Gould, "reported on the communication read at the last meeting from Mr. E. S. Morse of Portland, Maine, on a species of *Helix,* the most minute of any yet observed . . . believed to be a distinct species." Probably Charles Sprague and Thomas Bouvé beamed and nodded approbation, told each other that there was a boy to watch, a bright lad.

Deacon Johnathan saw nothing to get excited about in a dingy speck of a snail shell. Here was his boy home again, having failed to stick the course through at a third school; high time he buckled down to earning his keep. A year of monotonous hours making blueprints and drawings at the Portland Company was as the "ten thousand million years" Ned had imagined. By January 1, 1858, nineteen years old, Ned Morse took refuge from frustration and conflict with his environment in a journal. In his first entry he took stock of his situation.

> My Cabinet is in disorder. How much shall I do towards putting it in order by the end of January? My draughting,—I hope I shall have some office with good pay before the year is finished. I am now getting only two hundred dollars a year and working under a man who *cannot* appreciate a good drawing.

He prefaced this with a New Year's resolution: "My endeavours will be to go through the year without smoking or drinking spirituous liquors. . . ."

His endeavors in this line were to give him a lifelong occupation. Frequent temperance pledges, backslidings, defiance, and remorse enliven the pages of his diary. His efforts probably were due to the exhortations of Deacon Johnathan at first and to a trace of the old Puritan conscience which imputed sin to anything pleasurable. Morse was drunk only once in his life, but the cigar or whiskey he enjoyed always tasted the better because of its flavor of something forbidden. The journal entries mirror the unsophistication of a poorly educated country boy, joined strangely with a mature devotion to his self-taught science. He had been elected secretary of the Portland Natural History Society and on January 18, 1858, read to the society his first "paper," on the "Progress of Conchology in Science" — "with much embarassment," he recorded. Three nights later, like any youth of twenty, he had an entirely unscientific good time.

January 22: Played three games of whist at Abbie Lowell's house. Went home with Lutie at 10:30. Coming home stopped . . . at Mechanics' Hall, Spiritual Levee. Got there 11:30. . . . Looked around, saw Miss Capen, danced with her two or three cotillions and contras. Glorious time. Went home with her and her sister. . . . What a nice time I have had this evening.

Virtually untutored as he was, Edward Morse had the faculty of vivid description and an enjoyment of expression. He scratched away in his journal with a quill pen for hours each evening, even though, his mind filled with zoölogical problems, he could not envision the row of books he was to write, beginning thirty years later. His description of a fire one night when he was house-bound with a severe cold has a degree of observation and unity of point of view unusual in an untrained youth.

February 12 [1858], Friday: My first remembrance of the day began thus. I think I was dreaming of the performance of a Chinese juggler. The ringing of the curtain bell woke me when instead of the Juggler's bell, I heard the Third Parish bell ringing all alone. I laid a long while hearing this bell ring, but heard no one cry Fire! It seemed as if the inhabitants was all dead. At last the bell stopped and I supposed the alarm was false.

Soon I heard a man run up the front steps in the other part [of the two-family house] and bang! bang!! violently on the door.

Then a fearful twitch of the bell wire and jingle went the bell in the back part of the house. Then came the bustle and noise of Mr. Beckett out of bed on the floor, thump, thump across the floor to the window and then the window was lifted.

"Stinson's clothing store is on fire," says the man on the steps, which was Dr. Robinson. . . . In the meantime the Third Parish bell, aided with the deep, clamorous tones of St. Stephen's began ringing again.

The window slammed down. Footsteps retreated down the steps and along the sidewalk. The whole mansion was awakened. The boys and Mr. Beckett in the other part each leaving the house most boisterously according to their degrees of smartness and spryness. Soon the [Morse] girls came up into the room with timid inquiries if it could be seen. Father, for I slept with him, being sick, hearing the intelligence imparted to the other part [of the house], hops out of bed and dresses quickly, inquiring of Mother who sleeps in the next room, for his fire bags.

Mother answers, "Now, J. K., you are not going to the fire this cold night!"

"Yes, yes, dear; I think I shall go. Do you know where the fire bags are?"

Mother answers, "Well, wrap up your head and neck warm." She tells him afterwards they [the canvas bags for salvaging the belongings of the fire victims] are up garret, and Father pokes off up in the attic. Grandmother shivers in her bed, positive that J. K. will ignite some combustible up garret. [The Deacon was carrying a candle or small lamp in his hand, since it was before the days of gas or electricity.] J. K. finds them, bundles out. Slam goes the door and all is still again. . . .

No! not so. The firemen are alive, for after laying on my pillow wishing I was well enough to go out, I hear the distant noise of an approaching engine. It is Young America and [as] they rattle down by I hear familiar voices shouting, "Wake her up, now, will yer, Boys? Hip! Hip! Hip!"

There were, of course, at this period in Portland, and in most American cities and towns, no water mains, no municipal fire departments. Wooden houses, heated by open fireplaces or iron stoves in the living rooms, lighted by candles or kerosene or whale-oil lamps, frequently caught fire, and shingled roofs spread the blaze rapidly to other structures from sparks and burning embers. On August 14, 1847, the house occupied by the Morse family at 4 Cotton Street, Portland, burned to the ground in a conflagration that destroyed a street of houses. After the ashes cooled, Deacon Morse found in the brick fireplace oven a pot of beans, "cooked to a turn," which the family ate. Ned, nine years old, was deeply impressed by the incident. He savored baked beans to the end of his life. "Baked beans for supper. Tip top!" is a weekly phrase of his diary. Fires, also, had almost the force of a compulsion for him. Wherever he was, whether at fourteen or forty years of age, at the sound of a fire bell, he would run for miles to watch a conflagration. Like all New England boys of his day, he was versed in the names, the uniform, the equipment, of the volunteer engine companies who turned out when the fire bells rang to haul the engine to the scene by ropes, to carry out furniture and fill their "fire bags" with the small belongings of the victims, to form a line to pass leather fire buckets from the nearest well or stream, and to pump the engine by man power to furnish water for the hose lines. Ned's rebellious disgust against the man in authority over him at the Portland Company, Mr. Sparrow, crops out in connection with his favorite topic of fire engines.

February 19 [1858]: . . . Went to work this afternoon for the first time since I have been sick. John Sparrow has made another *bull* or rather did not take advantage of a trade that was lucid as anything could be. Mr. Hitchens wanted to sell the Engine Penobscot, a little old engine, to Mr. Sparrow for two hundred and fifty dollars; or rather Mr. Hitchens owed Sparrow or the Portland Company a sum of money which was . . . a bad debt. . . . Well, Sparrow being deprived of commonsense refused to give that sum 250.00. Mr. Sparrow went over to the Engine House and examined it but would not take it at that price as part payment of the debt. The day after, Mr. Fred Pukey bought it at that price, 250, hired a man who took it all apart for about 25 dollars and the Portland Company were glad to buy the old copper pieces for .18 cents per pound (in Boston they would have paid .23) which amounted to the sum of 288.00 for the copper flues alone and this was paid in cash to Mr. Pukey who received his money back and enough more to pay all his expenses in taking it apart. He has now on hand a large lot of brass composition and cast and wrought iron which he can sell for about 300.00. A good bargain that.

Unconsciously Ned was translating his hatred of the office work that took him away from his shells into dislike of his taskmaster, the unfortunate Mr. Sparrow. The diary, without a break, continues:

I wish I had time to paint or draw every Shell found in our State, copying descriptions from the standard authors. I might sell such a work as that for a good sum. Our Society [of Natural History] talks of having publications and if they do it will be a convenience to me. Instead of sending my species up to Boston, I could find a Journal [to publish the articles] here.

Ned had neglected his shells many evenings to play whist and "walk home with Lutie." Her nineteenth birthday occurred on March 22, and young Morse gave her "a little workbag and a pretty knife." He also evidently asked her a question to which she said no, for the diary concludes that day's entry with "that she may secure a husband who shall be rich in noble attributes of the soul is my earnest prayer and wish. As I can be no other, I will be a brother to her." He took long, despondent walks in the spring evenings, with the piping of hylas shrill in the soft dusk, and in a week composed his soul to return to science.

March 28: Mr. Waterhouse and Mr. Stinchcourt of the Portland Company were up here today to see my Cabinet. They were much pleased. Labelled [shells] and moved a large table which is quite a luxury in my Cabinet. Have a new Journal for the purpose of writing out in full my Shell matters. . . . I wish I had more time on my Cabinet.

Lutie's face continued to come between him and the shells he was labeling. In the small community of Portland, naturally Ned heard the gossip of the neighborhood, and it inspired an unbrotherly jealousy.

April 10: Is it true that Lutie is interested in a young man who can boast of nothing, one who is vulgar and takes pride in thinking himself a libertine? Ah! Lutie, take care. He will not spare you, though you are one of God's frail flowers. . . . Dr. Burr and his lady, Miss Alba Rae and sister came to see my Cabinet this evening. How I enjoy showing my Cabinet to persons who appreciate looking at it. . . .

Evidently God's frail flower did not care for shells. Ned, trying to forget her, brought out a box of shells collected the previous summer on Mount Independence, Woodstock, Maine, and found among them another new species, which was to be christened *Helix milium* after it had been sent to the Boston Society of Natural History. Even this could not raise his spirits.

April 13: Arranged shells . . . *Cyclas rosacea, Pisidium altile* and *mirabile*. . . . Headache; bad cold. Blue. I wish I felt like writing. Wonder if I shall live the year out. . . . O dear me, I wish I had money. How much I might do towards developing the Shells of Maine.

April 17: Bought a new hat. John Dow found a lot of *Planorbis trivalvis Say* at Cape [Elizabeth]. He also found at Deering's Cove *Helix harpa* and also one species of new *Helix* like one I found out at Blackstrap. Now that the *Helix* is new, what shall I call it? I cannot think of a name appropriate.

April 18: Sunday. Wrote description of animal of *Bulimus neurpa Say*. This afternoon John and I went down to see the baptism of a large lot of converts to Casco Street Church. The shores and piles of lumber were crowded with people. The wind blew strong from the N.W. and it was anything but comfortable standing and waiting. John and I staid long enough to denounce it as foolish and ridiculous, a farce having nothing to do with our days whatever and then went over to Deering Oaks in hopes of finding the *Helix* that John Dow found.

Even John Gould was not immune to the contagion of springtime. Going down to Gould's house a few evenings later to talk of shells, Morse found John entertaining a young lady. Morse, in his character of misogynist, left them to their own devices, going out of the room "to a Firkin of dates and how I regaled myself. I eat over a pound I am positive." (Like other rural New Englanders of his day, Ned Morse, despite grammar at Bethel Academy, declined the verb "eat, eat, eaten.")

It is possible that, had Lutie consented to marry Ned Morse, he would have settled down to placid life in Portland. Certain museums in Cambridge, Boston, Salem, and Tokyo might have had a different history; certain books on science, ethnology, ceramics might never have been written. Lutie, however, was in love with someone else. Ned Morse concentrated on shells, defying his father, who thought science a waste of time and an impious prying into the ways of the Creator.

April 24: Saturday. Beautiful day. . . . I went down to Mrs. Barbour's and she gave me a *Cytherea*, very good specimen. Mr. Barbour lent me Sprague['s] list of fungi in the Proc. B.S.N.H. ult. I felt singular and yet tickled to see in print *Tympanis morsei*.[2]

Thus came recognition, the scientist's recompense and spur. They had named a species for Edward Sylvester Morse, aged twenty years.

[2] See the *Proceedings* of the Boston Society of Natural History, VI (1859), 320.

CHAPTER TWO

THIS UNWELCOME BODY
1858–1859

FRICTION, frequent argument, over religion was nearing a crisis between Deacon Johnathan Morse and his son Ned in 1858. A "revival" sweeping through New England fired the father to sterner disapproval of the spiritualism in which Jane Beckett Morse secretly found solace for the loss of her eldest son, and which Ned, with youth's argumentative turn, openly upheld.

March 6 [1858], Saturday: I had promised Father that I would go up to the morning prayer meeting. Though I had asked him in vain to go to Mechanics Hall [to a spiritualist "levee"]. So at 8 o'clock I started and in a short time I found myself in a . . . vestry in company with a crowd. The older ones having for the most part been brought up to it, while the young ones without exception were weak, nervous-looking persons, fit subjects for such an excitement, such a stampede that is rushing over the country. . . . Deacon Thwing from Boston, a most dark, forbidding-looking man with such doleful cadences, a voice as if from a coffin. . . . All their talk is about the saving of souls. Saving of Souls? Why, man, can you look at Nature and talk of soul-saving? . . . Every single boy I see here are [*sic*] weak senseless babies, pale gloomy-looking bugs. God has nothing to do with this, nothing whatever. God teaches us to smile. Nature, the flowers, all are smiling and beautiful, but here the gloominess, the desolation. They are so immured in this repulsive belief that even the joyous Sun beaming in at the windows cannot lead them to adore their creator in the openness of their hearts . . . pouring out their fiendish, hellish ideas they cause the young heart to quake and tremble and at last from very fear they "Join the Church."

March 13: . . . Met Charlie Clark who lives below us. He has just been converted to religion, Methodist. I talked with him almost an hour. His talk was that of a person greatly excited. Poor fellow.

March 14, Sunday: . . . At Mechanics Hall with John this evening . . . heard a discourse from Mr. Willis, the discharged Harvard student; most intellectual lecture and done me a lot of good. I wish these young revivalists could have heard it. It would be impossible for me to describe our feelings after hearing Edgar A. Poe speak through the medium in the beautiful manner in which he did. We resolved firmly to act better and more in accordance with spiritual teaching and to sit alone each day.

March 18, Thursday: It seems like Sunday. Bells ring morning, noon, and night, and the churches are opened two times in the day and in the evening. I meet at all times persons either pouring in or crowding out of some prayer meeting. But I will not talk with them. I have made better resolutions.

March 21, Sunday: Labelled an immense number [of shells] for Cabinet. . . . If I keep on at this rate I shall have my Cabinet in fine order before summer sets in.

Thus the scene was set, with clangor of church bells in the air, the revival spirit animating the father, Lutie's rejection making sore the heart of the son, and, deeper yet, his passionate desire for the scientific career at which his father scoffed. The sight of his own name in a catalogue, *Tympanis Morsei*, raised his frustrated yearning to the pitch where he defied his father on the two issues of church and work. It was Saturday night when a Portland neighbor showed him the list of fungi, the work of Charles J. Sprague of Boston, in which he saw that Sprague had attached his name to a new species.

April 25, Sunday: . . . Wrote to Sprague and sent species No. 80 fungi. This afternoon walked . . . at low tide. Got beautiful specimens. . . . Found a Crab which I carried down to Fuller's and dissected.

April 26. Monday: Back ached this morning. Staid at home today. Arranged shells. This evening went down to John's; eat dates.

That was the last straw. He had refused to go to church, mooning along the beach instead. He had neglected a job that paid four dollars a week to putter over shells, and the backache was a poor excuse if he could walk down to the Gould boy's house in the evening. Deacon Johnathan "had it out" with Ned. It must have been a bitter dialogue. At the end, Ned flung upstairs and poured his heart out in his diary, while his father in the adjoining room on his knees prayed aloud for the wayward sinner.

April 27: O, how blue I am. Blue! Blue! I wish I knew more about the law of suicide. What is the state of a self-murderer after he has left this earth? . . . Even now, as I write, the murmurings of a prayer come through the walls of my room from the other side. I can respect the author of the prayer and still pity [him]. . . . O, pity that your offering were not for the Spirits. . . . Two things cannot exist together. Our hearts simultaneously cannot contain love and hate. . . . So with God there is no room for Hell or Devil, as Christians of today hold up. . . . O, what folly, superstition. . . . And now I pray that my time may come shortly . . . , and as I lay in my deathbed I may recognize my dear sister Charlotte; [1] though I have never seen her, still she may be with me when I leave this Earth life. And, my God, I also pray that my dear Father may

[1] She had died before Ned was born.

recognize thy Spirit World and be devoted and refined. . . . Ah, Mother and sisters, as you watched those little chickens tonight and felt joy in seeing them fresh from the egg, so do Guardian Spirits watch and hail our birth into the spirit land. . . . And let us strive to be as pure and untainted when we enter our spirit life as the instinct that glistened from the little black eyes of that young bird. I am unhappy on this earth. . . . And still must I give up the only thing that I cared to live for? Must I indeed give up these beautiful things God has given us here solely to enjoy . . . ? Give it up, I say, and turn myself to the support of this unwelcome body, gross and low? If so, let it be of short duration. . . . Now for the work. . . . My Guardian Angel, keep me in my resolutions.

Meanwhile, Jane Morse came to the rescue. After a week, a painful week, with a father nursing self-righteous resentment while a son brooded darkly on suicide, she persuaded the father to let the boy visit her old friends, Captain and Mrs. Agry, in their former home, Hallowell. Perhaps she even plotted, as mothers do, having Mrs. Agry "spontaneously" write inviting Ned up.

May 11, Tuesday: Woke up (Persons generally do in the morning) and felt sleepy, tired and cross. While cogitating on my day's work it popped into my head that I was going down East. Not time, however, so I dropt to sleep again. Soon I was up and after much fussing I got ready . . . walked down to the Steamboat T. F. Secor. Two bundles, carpetbag and valise, a plant and a long pole. Truck enough for a small boy.

The plant must have been a gift Jane Morse was sending to Mrs. Agry, one of her own nurture, of course — perhaps a hybridized geranium or a begonia. The fact that Jane insisted, undoubtedly in spite of impatient protests, that Ned should carry a plant with all his other baggage is suspiciously like evidence that she had begged of Mrs. Agry the favor of taking Ned away from home just at that time. He goes on:

Clouds looked dark and foreboding. Found myself aboard and started at 7 A.M. at a merry rate down the harbor. Shall pass over this. Suffice it to say that the boat rocked, a few were sick and your uncle . . . felt a slight indisposition around the lower region rounding Seguin. Bath looked busy. Crafts on the river. Every now and then persons would board the boat with boxes of Shad. Large weirs, I think they are called, stretched out in the river.

In May on the New England coast, of course, the alewives would be running upstream, bound for inland ponds where they spawn. Alewife, shad, and herring are names applied indiscriminately to a kind of fish that appears off the coast in spring in immense schools, each group making for the very river and the self-same pond where it was hatched some previous year. Laws passed in town meetings of coastal villages to regulate the catching of these fish appear among the earliest town records of New England. The catch was often a communal affair, every male over sixteen being entitled to an equal share. Even in the twentieth century the maintenance of weirs when herring are "running" in some places still is a town right, proceeds going into the town treasury. Salted herring, dried herring, pickled herring were a staple food-store in old New England homes. Fresh herring was not much appreciated, because of the multiplicity of bones, but herring roe was a tidbit.

On the whole I enjoyed myself very much coming up the river. On board the boat I made the acquaintance of Dr. Gilman of Gardiner who told me all about the fossils and the Misses Allen's collection. I shall go down there tomorrow or next day. Some very beautiful residences on the river. Reached Hallowell at 2. Met little Carrie on the wharf and she carried me safe to her house. I say "safe," for it is a risk to walk through the streets of Hallowell, holes, gutters and all up hill. Funny-looking place. Received a most cordial welcome from Mrs. Agry and felt at once perfectly at home. Found this afternoon back of the house specimens of *Helix minuta*. Rather poor place for Shells, I think.

May 12, Wednesday: Rained hard. Could not stop for it to hold up, so I marched off for the Ferry. Hunted around in the rain for the Ferryman and at last found him. He paddled me across [the Kennebec River]. I hardly knew which way to go but seeing some hard-wood growth up the river, concluded to go up.

Captain Thomas Brown, in his *Conchologist's Textbook* advised his young readers that

land shells are found in woods, hedges and gardens, where they take up their residence either in the hollows of trees, crevices of rocks, holes in old walls, roots of hedges, under stones, amongst moss or adhere to the branches and stumps of trees and under the shade of the leaves, or amongst nettles or other weeds.

Ned Morse was thoroughly familiar with the possible habitats of terrestrial shells.

As the river was high, I went up the road and along the bank which was very high and steep and from which I slipt and went down faster than I intended. Did not hurt me, however, and again attempted to go on but was surprised at the size of the clay banks which towered up to the height of fifty feet or more. Piles and piles of clay; I should think it would slide sometime. Well, I wandered around in a drenching rain all the forenoon and got tired out but found only one specimen of *Bulimus labricus*. Got back sick of Hallowell as far as Shells are concerned.

After dinner it cleared up and I went up to Cy Lowell's house.

Stopped there a little while and then we went down to the Factory, an immense building four stories high and very long. It was a cotton factory where they were making cloth, a thing I never saw before. A most curious sight it was to see the girls, some of them very pretty, each one taking charge of a half dozen of machines, stopping them, adjusting something, then starting them again, the complexity of which I could not understand. Nor I don't suppose they could, either. I could not but pity them poor girls, to work in these heated rooms twelve hours a day. What slave ever worked like that? And after we had passed through the building and came out into the entry, to see the bonnets and shawls! No Sutlas or silk bonnets there. And compare their laborious lives with girls I am acquainted with in the city. Poor girls.

Borrowed "John Halifax, Gentleman" of Cyrus Lowell and read in it. This evening Miss Marshall and Curtis was here; played whist and had a nice time.

May 13, Thursday: Beautiful day. This forenoon I went up back of the house and found some very good Shells. *Pupa pentocton; Pupa simplex; Helix sayi, Monodon alternata, Milium minutissima annulata* [and] *labyrinthica; Chersina indentata* and others common. Found quite a number of *Helix milium.*

This after[noon] with Cy Lowell went up to Augusta. Very beautiful walk and I enjoyed myself hugely. The Capitol is the largest building I ever saw. We went all through it, into every room, up in the cupola where the view was grand beyond description.

The Cabinet has never been disturbed, or rather has been disturbed a good deal since Dr. Jackson left it in 1838. Dusty, dirty. Drawers all nailed up, specimens broken. Could find nothing that I wanted. The library looked far different. Everything looked nice, neat and clean. We examined the books and had a glorious time. In the library was a Letterbook with some of the most exquisite paintings of Maine scenery I ever saw. They were paintings in watercolor executed when the Geological Survey of the State was made. . . .

May 14, Friday: After Shells this forenoon and this afternoon. Never had such poor luck in my life; a miserable place is Hallowell for Shells. I find the *Helix milium* everywhere. . . . This evening [Cy and I] are invited to Miss Reed's. . . . Played Muggins, nice time. Drank pretty heady wine and acted rather lively. . . .

May 15, Saturday: This forenoon I went up back of the house for Shells; could not find many. Went down to the Boston boat with Cyrus. Capt. Agry came on it; *Eastern Queen*; runs on the outside route between Hallowell and Boston; a new boat and a fine one, too. . . .

May 16, Sunday: Finished reading "John Halifax." Played on piano, smoked, loafed and at 5½ P.M. went to Episcopal Church to oblige Ada [Agry].

In *John Halifax, Gentleman,* Dinah Mulock (Mrs. Craik), the English author, produced a book whose triple thread of manly friendship, unrequited love, and rags-to-riches career held appeal to young people through a full century. Ned Morse in reading it undoubtedly identified himself with the hero, Phineas, whose stern and godly Quaker father tried to dominate him. John Halifax, the hero's noble friend, would remind Ned of his own admired friend John Gould, and the lovely Ursula was a character to put fickle Lutie of Portland to shame.

May 17, Monday: After Shells with Pisidium net. Cyrus went with me. We tried in all the brooks between here and Augusta and got a few *Lymneas* and *Pisidium.* I tried in the Kennebec river; could not get anything. This afternoon went down to Gardiner in Steamboat; fine sail down. Called in to Dr. Gilman's. He has some fine fossils and minerals. Went over to clay banks and found some [fossil] ferns which go by the name of Pipe Stems. Gentleman I met there who gave me some fossils said that he thought it was the roots of old trees and this clay clung around it.

May 19, Wednesday: All packed up. Cyrus came down to the house; marshalled my things down to the boat. Bid goodbye to all. I came away from Mrs. Agry's with feelings akin to Sorrow. . . . Met John on the wharf. . . . Went over to Willie's this evening. . . . Saw his Aquarium which he has got started. Everything was dying and no wonder; he had it crowded with Bladderweed and the bottom buried with dead, rotten shells. . . . Saw Mr. Rowell of Panama, a fine-looking man. He showed me two immense bundles of Shells from Aspinwall, Panama and Acapulco which he is to give me in exchange for mine. Brought Aspinwall shells home and undone them. They were very nice and I sat up till quite late examining them. The other bundle is a good deal larger.

May 20, Thursday: Mr. Rowell came to the house tonight bringing with him a very large bundle of Shells from Panama and Acapulco. He was much pleased with my Cabinet. Is to have [buy of E. S. M.] 1,000 boxes. I sat up till one o'clock examining them. They were more interesting and in much better condition than the Aspinwall Shells.

May 21, Friday: Worked this evening part of the time on my Shells and done a lot. Went down to Fuller's. His collections getting along nicely. This Rowell exchange is the largest one I have ever received. Most of the shells are perfect and many of them are rare, some quite so.

May 22, Saturday: Mr. Julius A. Thompson got home from Cuba bringing for me a large bag full of land and fresh-water shells; over a hundred of nice Ampullarias, Helices, a large box; and Cylindella, 300.

May 23, Sunday: Arranged some of the Shells from Mr. Rowell. I can hardly find room in my Cabinet for them. Shall have to get more drawers. Busy all day.

June 8, Tuesday: John Sparrow's salery [*sic*] is raised from two thousands to five thousand dollars a year. Boiler makers struck for higher pay.

June 17. Had a spat with the King of Fools, John Sparrow, and started for Boston tonight on the boat.

The next day was his twentieth birthday, but he did not think of it, nor did he wonder what his family thought about his absence. Perhaps Deacon Johnathan walked the floor that night, dreading lest his boy had killed himself. There was no telegraph, no telephone, as yet for rapid communication or inquiry.

June 18. At Hinkley's Locomotive works this forenoon. O, by the way, I was completely worked up in the boat. . . . Called on Sprague, he was out of town. . . . Have no patience to write a thousand minor things. My stomach completely out of order all the time.

June 19. Frank Cummings told me I should have 400.00 a year if I staid which was my intention but on inquiring at quite a number of places I found board so intolerably high that . . . I should do better by staying at home, besides a man I revere and respect above all others, Mr. Charles J. Sprague advised me as a friend to stay where I was. Went into Bricher & Russell's wood engravers and staid there till train time, three P.M. . . . I think I shall try to learn to be a wood engraver.

So Ned went back to Portland, back to draughting under John Sparrow, back to a father who had promised Jane Beckett that he'd let the boy alone. Deacon Johnathan loved his children sincerely. It was his tragedy that he could not understand and joy in his boy's bent toward the new science of zoölogy, then achieving importance under the magic spell of Louis Agassiz, who had come to America in 1849 and was teaching at Harvard. It was sad that the steady procession of prominent men coming to Portland to see Ned Morse's collection of land shells gave the father no sense of pride. Jane Beckett to a maternal general sympathy with her son's interests added a genuine, if limited, interest in science. She made experiments in hybridizing flowers of the house plants in "the south window"; she assembled a collection of plants for which the Portland Natural History Society awarded her a certificate. She must have listened, glowing with pride, when strangers came to see Ned's shells in the room upstairs, which, despite the needs of the older brother and the three little sisters, she had set apart for Ned's cabinet.

"None so blind as those who will not see" was an old Yankee saying that might be applied to Deacon Johnathan. It is really surprising that the steady procession of prominent men calling at Spring Street to see Ned's cabinet did not give him an intimation that there might be something valid and important in this matter of shells.

July 15 [1858], Thursday: Beautiful day. When I got home to dinner this noon, folks told me that two gentlemen had called and said they would call again at 1 o'clock, which they did, punctual to the minute.

Surprised and pleased at finding them to be Judge Cooper and son of Hoboken, N. J. Judge Cooper is an excellent collector of shells and is well acquainted with Shells. Lea named his *Unio coopercanus* in honor of this gentleman. He was much pleased with my collection; said my specimens were nice and in fine order. Told me a lot of names and much information new to me. He stopped till four o'clock and I had a glorious time talking with him.

He is an oldish man, about 60, I should think, and still he has all the enthusiasm of a young collector. He was pioneer in the pursuit with Lea, Haldeman, Conrad and Gould.

He is now on his way to Eastport to dredge in deep water, to stay some time. Informed me that Stimpson was down this way; had left Conchology because the synonyms was so perplexing and had gone into Crustacea, Annelidae, etc. What a man Stimpson must be, to step in as he has and overload the already cumbersome load of Synonomy by publishing a new name to every New England shell, taking all the credit to himself, and then step out again . . . , declaring that conchology is perplexing!

Bland is a merchant in New York City. Mr. Binney lives on his income. Both together are preparing and collecting all of Say's writings which he is going to publish. Mr. Rowell told Mr. Cooper about me. Mr. Cooper is going to send me some Shells when he returns.

W. G. Binney was the son of Amos Binney, Jr., whose *Terrestrial . . . Mollusks of the United States,* edited by A. A. Gould, had been given Ned a year or so previously by John Thompson of New Bedford. Thomas Bland (1809–1885) was an English mining engineer whose residence in the Barbados inspired a study of mollusks. In 1852, on account of his health, he left the West Indies and settled in New York, making a comfortable fortune directing mining companies and pursuing his zoölogical studies. He published his *Land Shells of the West Indies* in 1861, wrote many monographs on Mollusca, and for years was associated with W. G. Binney in work on terrestrial mollusks of North America.

Thomas Say (1787–1834) was one of the earliest American naturalists. He was a son of Dr. Benjamin Say, the richest man of his day in Philadelphia. Young Say studied pharmacy, started a business, endorsed the notes of a friend, and thereby became bankrupt. He took refuge in science. Tradition said that he slept on the floor in the rooms of the Philadelphia Academy of Natural Sciences, founded in 1812, one of the first museums of natural history in the country. He cooked his own meals and lived on seventy-five cents a week while he studied entomology. In 1819 he accompanied Major Stephen H. Long's expedition to the Rocky Mountains as zoölogist and in 1823 went with Long again to explore the sources of the Minnesota River. Then, in 1825, he became a convert to the practical socialism of Robert Dale Owen, English reformer who had just purchased thirty thousand acres in Indiana for his New Harmony colony. The experiment lasted three years, and Owen sank $200,000 in it. When Owen returned to London, Say remained at New Harmony, and between 1830 and 1834 published six numbers of his work, *American Conchology*. *The Complete Writings of Thomas Say,* edited by J. L. LeConte, was published in 1859. The previous year W. G. Binney had edited *The Complete Writings of Thomas Say on the Conchology of the United States,* illustrating the volume with seventy-five plates of shells, drawn by Mrs. Say. Ned Morse's journal reveals that Bland was associated with Binney in editing the Say papers.

July 16, Friday: This forenoon was shrouded in dense mist which came pouring in from the ocean charged with saline odors. Things looked dubious. Fuller, John and I had intended going to Prout's Neck this afternoon . . . and I watched the weather somewhat anxiously. . . . I was there [on the beach] with a pail full of pie, cake, bottles, ropes, etc. We got under way about two o'clock in a passable four-wheeled chaise and a small horse, and rode on, Fuller enlivening us all the time with jokes, stories. . . . We came to a place where rotten logs and rocks were strewn, got out and overturned a lot, finding the big *Pupa* we were in quest of. Shouted to the boys, telling them my good luck. . . . We immediately set to work and overturned all the rocks in that locality. I found 12 specimens besides some young ones. John got six or seven and Fuller six. We found them clinging to the underside of rocks in comparatively dry situations and still not exposed to the sun. Singular shell. Soon after halted up to a brook where we got large specimens of *Lymnea deidiosa*. Kept on to Prout's Neck House, where Fuller led us to the original place of discovery where we also found *Vertigo milum* but could not find any big Pupas.

July 20, Tuesday: While in the Office this forenoon Mr. Lowell came in with a young man, rather short and slim. Short hair, sandy; dark complexion, clear gray bright eyes covered with a pair of gold spectacles. Introduced him to me as Mr. Stimpson of the Smithsonian Institution. I was exceedingly glad to see him.

We went up town after Fuller but he was not home. We then went to my house and looked over my shells. Pronounced a long *Chemnitzia* new and also my *Leda* from the landslide new; Fuller's *Pecten* and also the *Yoldia* of Fuller's.

Told me a lot of things new to me and I derived much valuable information from him. He has a sloop chartered to him in Boston and is on his way to Mt. Desert to return in two weeks. He has two companions besides himself with two men to take care of the boat. Has invited Henry Willis to go with him who is to join him at Boothbay. Wanted me to go with him; urged me to. Did not, however. Sorry I did not.

It is probable that, in declining Stimpson's invitation, Ned Morse triumphed over one of the greatest temptations of his life; but after the fruitless rebellion and vain trip to Boston he had honestly resolved to settle down and make a go of the work at the Portland Company in order to become self-supporting. Sparrow, he knew, would never give him two weeks off to go cruising up and down the coast with one of those bug-hunters; and Johnathan Morse would consider his son a chicken-livered quitter if he threw up his job for a conchological spree.

Yet William Stimpson (1832–1872) was one of the romantic

figures in New England natural history. He was the son of a Roxbury man, a dealer in stoves and ranges, with a store at the corner of Congress and Water streets, Boston. His father intended the boy to become a surveyor, but young Stimpson was engrossed in shells. He even plucked up courage, when he was fifteen, to call on Dr. Augustus Addison Gould at his home in West Street. Dr. Gould took a fancy to the earnest lad and asked him to carry a note to the state librarian. The note proved to be an order for a copy of Gould's *Report on the Invertebrata of Massachusetts.*

Then the fat was in the fire. Stimpson marched down the long granite steps from the Bulfinch front of the old red brick Statehouse under the golden dome on Beacon Hill, hugging the precious volume to his breast, and flatly refused to go any farther in surveying. Dr. Gould introduced him to Agassiz, and finally the parents were brought to countenance more schooling for the boy. He entered the Boston Public Latin School in 1848, made a scientific voyage to Grand Manan the next summer aboard a fishing vessel, and in 1850 joined Agassiz as a laboratory worker in Cambridge. Two years later he went on the North Pacific Expedition with Captain John Rodgers, U.S.N., journeying to Bering Strait and Japan, which had just been opened to foreigners through the negotiations of Commodore Matthew Calbraith Perry. At the time he called on Edward Morse, Stimpson was attached to the Smithsonian Institution in Washington, D. C., under Baird and Henry.

A tragic fate awaited this brilliant, attractive, persistent young scientist. In 1865 he was offered the post of director of the Chicago Academy of Sciences. He had just been married, and he went to the new post with the highest anticipations. He assembled in a new building not only all the property of the

Chicago Academy of Sciences but all his own materials for projected writings, rare volumes, and specimens borrowed from other institutions and from scientists in many capitals, who were glad to coöperate with him. In October 1871, when Mrs. O'Leary's cow kicked over the lantern, the city of Chicago was swept by flames, and Stimpson saw his all destroyed. Friends hastened to secure him an appointment with the Coast Survey for an expedition to study the Gulf Stream, but Stimpson died within eight months, heartbroken and swiftly consumed by tuberculosis.

It was the opportunity of a lifetime for Ned Morse to be invited to go exploring with Stimpson, and only the narrowness of his father's provincial mind prevented it. Yet it is significant to see how young Morse had matured since the wild explosion that sent him packing off to Boston. He did not ask his father or consult his mother; he simply declined the invitation on his own resolution and wrote in his diary, "Sorry, however."

Stimpson did not forget the noteworthy collection of shells he had seen in Portland. He told William Greene Binney about young Morse and his *Helices* later that summer. Ned Morse, thrilled at receiving an invitation to correspond on matters of science with one of the foremost conchologists of America, preserved carefully all his life the letter from Binney:

Burlington, New Jersey
August 25, 1858

E. S. Morse, Esq.

Dear Sir,

From what Stimpson told me the other day, of your researches in conchology, I am still more anxious to correspond. I fear my letter did not reach you, so I will try again. If you did not receive it, please hunt it up at the Post Office and also the pamphlets I sent. . . . Stimpson pointed out to me the difference between your *astericus* and his *exigua*. I am anxious to see it. I send you a circular of my Say reprint. . . . I am anxious to have it succeed in order to persuade Bailliers to publish the other American writings on Conchology. . . . Hoping to hear soon from you, I remain,

Yours truly,
W. G. Binney

Binney, as we have seen, was a noted name in conchology. Amos Binney, a graduate of Brown (1821) and of the Harvard Medical School (1826), was one of the founders of the Boston Society of Natural History in 1830. (The society on May 17, 1849, moved from Mason Street to a new building in Phillips Place, off Tremont Street, where Ned Morse attended meetings in 1858. Agassiz delivered the dedicatory address in 1849, speaking on the future of the natural sciences.) Independently wealthy, Binney did not practice medicine. He was elected to the Massachusetts legislature and in 1836 was a leader in securing a state appropriation to make and publish reports of geological, zoölogical, and botanical surveys. Attacked by a lingering illness, he died while on a trip abroad. His best-known work, with the elaborate title of *The Terrestrial Air-Breathing Mollusks of the United States and the Adjacent Territories of North America,* was edited by Augustus Addison Gould and published, with a chapter on anatomy by Joseph Leidy of Philadelphia, in 1851. His son, William Greene Binney, followed his father's course and in his turn became an authority on American land mollusks.

Augustus Addison Gould (1805–1866), whose *Invertebrata of Massachusetts* had meant much to Ned Morse, was to become a kindly friend to the Portland boy. His home, where Morse later became a visitor, was at 15 West Street, Boston. Gould was graduated from Harvard in 1825, earned money for a medical education by tutoring in Maryland, and received his medical degree from Harvard in 1830. His inexhaustible energy permitted him not only to carry on his medical practice but also to do much research in natural history, particularly conchology, and to write some of the standard works of the day, as well as to contribute frequently to scientific periodicals. He met Louis Agassiz soon after the latter's arrival in Boston, struck up a warm friendship with him, and collaborated with him on his *Principles of Zoölogy,* published in 1848. Amos Binney was another close friend; and after Binney secured appropriations from the state legislature, Gould wrote a *Report on the Invertebrata of Massachusetts* (1841), with beautiful illustrations of the land, fresh-water, and marine mollusks of the state, which he drew himself. He also made a study of the mollusks obtained by the United States exploring expedition under the command of Charles Wilkes in 1838–1842.

Gould was an early member of the Boston Society of Natural History and its president for several years, and a charter member of the National Academy of Sciences and the American Academy of Arts and Sciences. He died in 1866 of Asiatic cholera, a disease which occurred in Boston with some frequency for a period of about forty years. It first appeared in Boston in 1832 and was not under control until around 1870, when the leveling of Fort Hill necessitated tearing down many of the wretched, overcrowded tenements in which the infection throve. In 1849 an epidemic of cholera centering around Fort Hill caused 611 deaths. The Committee on Internal Health for

Boston reported 586 cellars in which from five to fifteen persons dwelt. "In one the tide had risen so high that it was necessary [for the physician] to approach the bedside of a patient by means of a plank which was laid from one stool to another; while the dead body of an infant was actually sailing about the room in its coffin." It is not surprising that Dr. Gould came to die, at the age of sixty-one, of cholera in Boston.

It was easier, Ned found at his draughtsman's table through the hot summer days, to make a gesture of renunciation than to live with the memory of the thing foregone. He had refused Stimpson's invitation because he knew his father would not approve of his leaving his job to go shell-hunting, but day by day the renunciation rankled, and he grew reckless.

September 3 [1858]: Ira Berry invited me to Stearns' this evening, his 21st birthday. He has also joined the Free Masons and invited the whole lodge. We had about fifty at table. Oysters very good. After the supper a few of us boys sung, much to the edification of the Masons. We kept up our singing and in the meantime imbibed immense quantities of Ale which made us feel pretty light and good. And when Ned Patten invited us up to his house, though after 11, how could we refuse? So up we went some eight or ten of us . . . and were rather noisy. We drank pretty freely of his liquor and then went upstairs for a song. As nobody could play the piano, your humble servant volunteered but to save his poor miserable gizzard, he could not see a single note or key and my fingers were as weak as any baby. The last thing that I remember is that some was standing round in a wavering ring, hands on each other's shoulders singing some tune. The next thing I remember was tumbling over and onto somebody's granite steps. Then I brought up against Park Street Church. Sam Berry carried me to his house and there I vomited off the liquor at my leisure. Devil of a time.

Indeed it was, and Deacon Johnathan, unable to pierce the veil of the future and know that never again in eighty-seven years would his son be helplessly drunk, did his parental duty in admonishing him. The result was that Ned wrote secretly to Boston to apply for the job as wood-engraver he had investigated in June.

In another month he was in Boston, writing home enthusiastically to John Gould, describing the first plumbing he had ever seen.

October 10, 1858

Dear Juan: I arrived in good order, not sick on the boat, was driven by the Hackman to 199 Harrison Ave. who charged me 35¢. I have a bed by myself in a large room with Steve Russell and Frank. Steve is the one I work for. Our room is furnished splendidly, marble mantel, very tall room, nice carpet and the walls hung with the nicest steel engravings. . . . In a little room adjoining is pipes bringing hot and cold water, a large bathing tub 6 feet long and two feet deep where you can fill it up in two minutes with hot or cold water or both by simply turning a cock. Also an arrangement which saves the travelling down two or three flights of stairs and the inconvenience of being half froze in the attempt. I intend taking a bath every week. . . .

November 10, 1858

. . . The first thing that would strike you about Boston, yes, strike an awful blow to your thinking apparatus, is the horrid din and incessant rattlety bang of heavy wagons, omnibuses and carts over the paved streets. Every court, lane and alley is paved, . . . cobblestones mostly used and causing the most infernal racket. . . . Another peculiarity of Boston is that they will never let a pavement rest. They are continually disturbing it, ripping it up and laying down something new. If I did not have to pay my four dollars a week (board and room) I should be flush. . . .

November 16

Have just returned from a meeting of the Natural History Society. I wish you could see Agassiz, John. Such a good-natured genial face and with all the scientific men there, Dr. Gould, Jackson and a lot of others, he appears like a Father. Have been downright sick with a cold, . . . in bed two days. . . . The carpet bag came. Ah, boy, it was a luxury to lie back on the sofa and read them letters in the carpet bag from Mother. . . . I ran my nose way into the carpet bag and snuffed in big breaths of home. I could smell the work of Mother's flatiron and . . . familiar odors of the kitchen. . . . What sweet recollections around the name kitchen; not only the heavenly odor of fried doughnuts and the more refined smell of Roast Turkey and the right jolly smell of baked beans which is the only force that operated on matter Sunday mornings and moved lazybones from their cozy couches after water applied cold to the face, kicks, thumps on the door have failed. . . . Oh well, what I am getting at is this, that the more civilized a place gets, the less attractive grows the kitchen. The kitchen in this house, for instance, the reigning nymphs are two bouncing ugly Paddies who kick and swear amid steaming washtubs and dirty clothes.

The financial depression that began in 1857 was severely felt in New England. Edward Morse had chosen unfavorable times for his sally towards independence. Like a Peri at the gate of Paradise, he had one glimpse into the heavenly world of science after which his soul longed; he described it vividly to John Gould.

Boston, December 5, 1858

Juan, Attended the meeting of the Natural History Society last Wednesday evening. Meeting was called to order by the president Dr. [Charles S.] Jackson who is certainly a homely-looking man; looks like a fuddy duddy. Dr. Winslow, a rather portly, pompous "slow win-ded" cuss, got up and read a long article on the "Tails of the Comet." [The Pons-Winnecke comet, discovered in 1819, had become visible again in 1858.] He thought it was similar to the luminous vapor round the sun. . . . The tail is two billion miles long and he noticed the curvature of the tail increased as it neared the Sun . . . he illustrated it on the blackboard, . . . the rays thrown from the comet go off in a straight line and the comet as it increases its speed leaves the rays further behind, making more of a curve. . . . Imagine a car stationary on a track and a stream of water starting from the side; it would play out straight. Now let the car be set in motion, and the faster it went the more curved would be the stream.

After a long and tedious harangue he got through and sat down. Prof. Agassiz instantly got up and said, "Dis is not der place for such papers and if such papers are read before the Society it will lead to unfortunate results. None are able to discuss the question and such papers should be sent to Astronomical Societies. I for one object to der paper. Dis is a Society of Natural History as der books in de shelves show." This came down on the meeting like a bombshell. Dr. Gould immediately got up, and he concluded with Prof. Agassiz that it was not right to read such papers and that the meeting came together to discuss Natural History.

Poor Dr. Winslow in the meantime wore an expression which it would be impossible to describe. Somebody, I can't tell his name, got up and said, "I think the paper was quite interesting and that such subjects are as proper as Geology and I will say that the conduct of Messrs Agassiz and Gould was improper and, if I may be allowed to speak, incourteous to speak their minds in such a manner right before Dr. Winslow."

Agassiz by this time was boiling with rage. He got up and said, "Science ought to be respected before individuals."

Dr. Winslow got up and set his endless tongue going. "Will Professor Aggasiz [*sic*] name his objections here?" Aggasiz, "I vill not speak of it here; dis is not der place for it." Winslow said he could see no harm, no reason why such articles should not be read here and says he, "Will Prof. Agassiz name his objections to the piece?"

Prof. Agassiz hopped up, leaned over the seat and spoke out quick,

"I vill not speak of it here, dis is no blace for it." [He] was red in the face, bit his nails and I thought at one time he would use them with the rest.

Dr. Winslow said he thought there was some private reason for this attack. Prof. Agassiz then spoke in a calm tone, calm from excessive madness, and said, "I have been to many meetings and I have seen dis gentleman (Dr. Winslow) violate de rules of der meeting. It vas at a meeting in Albany of the Scientific Association for de discussion of natural or unnatural phenomena dat ven Prof. Howe proposed der question of Spiritualism and Dr. Winslow said to dat venerable man dat dose who wished to discuss der question should repair to some Insane Asylum."

I tell you, John, if he did not bring it out well! The voice, the accent, was so queer and expressive.

Dr. Winslow caved in on this. He said he knew he spoke hastily and that if the good old man (Prof. Howe) was alive he would sit down and write begging his pardon. It was getting rather hot and personal. Prof. Agassiz was red in the face, Dr. Gould hitching about in his seat, Dr. Winslow saying he should not only withdraw the unlucky Tail but should withdraw himself from the Society altogether; members talking hotly together.

During his brief stay in Boston, Ned had accompanied some of the young men he met to the Parker House. On October 23, 1858, "we went down to Parker's and had a smoke, with Claret, ale, oysters, charlotte russe. Magnificent price; a long purse is necessary to dine at the Parker House once a day." Again, on Christmas Day, "Ira Berry and I drank wine at the Parker House."

By mid-December, Morse was penniless and with no prospects. He had received a gratifying mark of recognition from the foremost author of books on shells, W. G. Binney, but no cash with it. By the end of December he was ready to go home.

December 24 [1858]: Binney sends me a circular of a work he is about to publish, descriptions of land shells of the United States. He has drawings in plate of my *Helix astericus* and wishes me to draw *Helix milium* for insertion also. Good thing.

December 31: When the Hack came this morning for us we were in bed. We dressed ourselves in haste and came away without breakfast. Hungry and cross I felt. Billy Harwood and his sister came with me. I came by the lower route. Every station we came to we would get out and try to find something to eat. Two very pretty girls got in at Lynn and got out at Salem. While we were out of our seat, they tossed two large apples in our seats with pieces of paper. Marked on it was "Elijah in the Wilderness." The apples were devoured with many thanks to our pretty ravens. . . . Got home at 2. Glad to get home.

January 1, 1859: Another year gone, with many acts to be regretted but this is not the page to mourn last year's misdeeds. . . . I will have this page the starting point and I will strive to see how far I can get from the position I now hold. . . . I will also endeavor to improve my handwriting which is certainly open to a great chance of improvement.

A lifetime did not suffice to carry this resolution through to success. Thomas Bailey Aldrich, author of *The Story of a Bad Boy,* wrote to Morse thirty years later from the editor's office of the *Atlantic Monthly* in Boston:

March 1, 1888

. . . very pleasant to get a letter from you the other day. Perhaps I should have found it pleasanter if I had been able to decipher it. I don't think that I mastered anything beyond the date (which I knew) and the signature (which I guessed at). There's a singular and perpetual charm in a letter of yours; it never grows old; it never loses its novelty. One can say to one's self in the morning "There's that letter of Morse's; I haven't read it yet. I think I'll take another shy at it today and maybe I'll be able in the course of a few years to make out what he means by those t's that look like w's and those i's that haven't any eye-brows!" Other letters are read and thrown away, and forgotten but yours are kept forever, — unread. One of them will last a reasonable man a lifetime.

Francis Wilson, actor, who corresponded with Morse during his stage tours through the country, took delight in twitting Morse on his illegible chirography. Responding to an invitation Morse had written to him, Wilson declined it by pretending he could not decipher it:

March 27, 1899

My dear Morse: I have read your letter very carefully and conclude that you accuse me of stealing a shirt. As near as I can make out, it was flannel. I scorn to reply to the accusation. Go to hell! . . .

Years, as they passed, with thousands of pages of diaries, scientific notes, book manuscripts, and correspondence with hundreds of friends in America, Europe, and Japan, made Morse's handwriting worse and worse. Peter H. Goldsmith, editor of *Inter-America,* described it vividly:

March 25, 1922

I have just received your very charming and interesting letter of March 23, and after a labor of two or three hours and consulting all the office force about this or that detail, I have been able to read most of it. Your handwriting is still as charmingly "intriguing" . . . as it used to be. The point is that while at first the word looks absolutely and hopelessly unintelligible, when one studies it at various angles, uses a magnifying glass, thinks it over, backs and fills and then finds a bit of a hill on which to get a running start and goes at it with a bound and a leap, he almost always succeeds in getting something out of it. To my mind, this is fine writing. I was particularly delighted with "mollusca." I thought of all the nouns, — common and proper and improper, — from Morehouse to moth balls, and finally, remembering your dredging proclivities, I got mollusca out of it. I consider this one of the intellectual triumphs of my life.

Among Ned's correspondents in science was Mr. Whittemore of Cambridge, an employee at the Boston customhouse, who gave him, from government supplies, a handful of steel pens, after which the writing in the journal becomes clearer. In January 1859 Ned had a commission from Bricher & Russell to go to Taunton to make drawings of machinery at the factory of H. S. Fairbanks. Passing through Boston, he writes:

January 22: Saw Whittemore this morning, gave him 3.00 [for dues at the Natural History Society] and written description of *Helix milium*. . . . Had snow in Boston and as I neared Portland it increased till the snow was piled up five feet along the track.

February 10, Thursday: Worked on [drawing of] Lathe today. This evening went with Mother to Deering Hall. Mt. Vernon Association have given four entertainments for the Mt. Vernon fund. This evening I witnessed a splendid tableau of Washington at Valley Forge. Good thing. Also a farce acted. Good.

The Potomac estate of George Washington, descending through nephews to John A. Washington in 1847, was offered to the United States government for $100,000 but was not accepted. Edward Everett, pastor of the Brattle Street Unitarian Church in Boston at nineteen, Congressman, governor of Massachusetts, United States minister to Great Britain, and president of Harvard College, succeeded Daniel Webster as Secretary of State under Fillmore. In 1854 he retired to private life and plunged into a campaign to preserve Mount Vernon as a national shrine. A remarkable orator, he raised $70,000 by touring the country lecturing on Washington, and enlisted the support of patriotic women who, in 1856, incorporated the

Mount Vernon Ladies' Association of the Union. Such a cause undoubtedly recruited the efforts of Jane Morse, at once a romantic spirit and an indefatigable worker. The Mount Vernon Ladies raised the remainder of the $200,000 with which in 1860 the mansion house and its two hundred acres of river frontage were purchased and restored.

February 12, Saturday: Wrote to Binney; sent *Helix astericus,* few *vadia,* and many *harpa.* . . . Letter from Mr. Thomson with *Chemnitzia visinturalis, Helix Sayi* and *Cyclas occidentalis.* He says he has found specimens of *Helix astericus* among his shells from Fairfield, Me.

February 16, Wednesday: Invited out to Mr. Fernald's this evening. Dressed up and at 6 o'clock a whole load of us started from Mr. Fernald's store in a large open four-seated Omnibus drawn by four horses. The sleighing was superb, road excellent, moon bright and everything conducive to a rousing good time. After a while we got out to Mr. Fernald's. . . . Lutie, Hattie, Abbie Lowell, Georgie Little, and a lot of girls I knew was there. Soon supper was announced and we boys went in to help the rest. Baked beans, meat and everything. After they got through the waiters sat down and pitched into the good things. Everything nice and the Coffee splendid; excellent; good; superb. I wish I could always have coffee taste as good. Sang. Smoked. We danced cotillions, contras and about 1 o'clock had a splendid ride in.

A full moon, five feet of white snow over the countryside, shadows of dark pine trees, four horses trotting with tossing heads and shaggy fetlocked feet spurning crisp gouts of snow back towards the great boat-sleigh filled with buffalo robes and pretty girls in shawls and roguish bonnets! It is a veritable Currier & Ives print that Ned Morse evokes, sketch of an innocent country amusement which the automobile, asphalt

Johnathan and Jane Beckett Morse

and concrete highways, and motorized snowplows have extinguished.

John Sparrow had filled Ned's place at the Portland Company; he could find work neither in Boston nor in Portland. He was restless and dissatisfied with Portland after his city experience:

February 28 [1859]: This evening with John down to Mrs. B's. It recalled old times vividly. Played whist, Lutie and I as formerly against the rest. After that we sat down to Baked Beans. I enjoyed myself very much. Lutie, I think has changed. . . . Sweet little confiding Lutie of times gone by is not the same now. Once she was loving, graceful and tender. Her voice had a peculiar tone . . . that made one happy to hear it. Now she is bold and not retiring and timid. Then she laughed at wit and fun; now only at those who suffer infirmities. Then she was constant; now chasing after others she knows are not pure and virtuous. . . . My John of former days has changed; changed a great deal. . . . A rare shell is found and, though I visit his house many times while the shell lays in his drawer, still I have no invitation to look at it. . . . I am even twitted and laughed at for curiosity I take in wishing to see a Maine shell new to me.

March 10, Thursday: Went to Boston this morning in company with Adeltha Twitchell. Had a nice time on the way up. Got in Boston the usual time. This evening Charles Parley, George Simmons, Ive Brazier and myself went to the Museum. Saw "Our American Cousin," Warren, as Asa Trenchard. Most humorous and affecting piece. This afternoon . . . called into Everett's and saw some splendid paintings. One by Bierstadt $800.00. Magnificent thing. Frost has a very correct picture of Bethel from Sunset Hill and many other fine things.

Albert Bierstadt (1830–1902) had just returned to America after three years (1853–56) studying painting at Düsseldorf,

Germany. He specialized in panoramic views with almost photographically correct topography. He later gained much fame for his paintings of the West and the Rocky Mountains, and his historical pictures, "Discovery of the Hudson" and "Settlement of California," now hang in the Capitol in Washington.

The play, *Our American Cousin,* by Tom Taylor, the British dramatist who was editor of *Punch,* was new when Ned Morse saw it, having been opened in New York in 1858 by Laura Keene. William Warren, of whom Morse speaks, was to delight Boston audiences for many a year at the old Museum. And the title, *Our American Cousin,* was to be whispered about the globe on telegraph wires, mentioned in newspapers, magazines, and books for decades after April 14, 1865, when, as John Hay expresses it, "No one could ever remember the last words of the piece that were uttered that night — the last Abraham Lincoln heard on earth" in Ford's Theatre as he was assassinated.

April 1, Fools Day: I have been "sold" once or twice today. Received letter from Mr. Rowell in California, containing order for more boxes. . . . At Fuller's this evening had a bowl of *Venus mercenaria* chowder which was nice.

April 2, Saturday: Drew and played chess. Two weeks ago tonight my Aquarium was started and it looks healthy and good.

April 4, Monday: Pleasant today, windy. Very full meeting of the Natural History Society. Quite interesting. I proposed A. E. Verrill of Norway, Me., as corresponding member.

April 5, Tuesday: Election Day. Jewett elected against Holden, Dem. Legislature has granted half a township to the Portland Society of Natural History. Good. . . . Father had an operation performed on his heel this afternoon; a malignant wart cut out. He looks more relieved in mind than he has for a year.

April 9, Saturday: Three weeks since my Aquarium was started; all in good order. Saw Way Kimball of Boston this afternoon on the wharf to see reefing of topsails by the new process; Steamer Nova Scotian.

This was probably the introduction of the steam donkey-engine on the decks of sailing vessels that was to cut down crews, obviate the initiative and agility of the old-time able-bodied sailors, and thus change the economic picture in the coast towns of New England, where for two centuries all the likely lads had gone to sea in their teens.

Jane Morse's stratagems of oyster suppers at home (music on the piano by Carrie) and tableaux by the Mount Vernon Ladies were powerless to relieve the tension between father and son, the conflict between hard-shell Baptist theology and youthful liberalism. The unfortunate thing — to read between the lines — was that medicine in Portland before the Civil War could not diagnose the gnawing internal disease that subtly affected Deacon Johnathan's disposition. Even his young son could perceive that the excision of the growth on his father's leg made J. K. "look more relieved in mind than for a year"; but if Father was "cranky," the family attributed it to too many hot biscuits.

Deacon Johnathan had only a year more to live. He was straining every nerve and all his energies to make a living for his wife, three little girls, and a grown son through the hard times of the panic of '57. Grit and his courage to carry on came to him through the traditional channel of his generation — orthodox religion; but to his son Ned the well-worn pious phrases were infuriating. Dependent on his father for a home, bound by his affection for his mother to avoid open argument and conflict, Ned betrays in his journal the haunting sense of hatred for his father's religion.

April 10 [1859], Sunday: Today I shall always remember as a bright spot in my earth's existence. This morning at 8½ went up to Fred Robinson's . . . , met Charles Robinson who boards there. With him I rode over to Spring Point bridge and collected materials for his Aquarium. Had a very pleasant time. Got back at twelve, cleaned the things and arranged them in the Aquarium. Then we sat down to a sound, healthy dinner. Fred Robinson, his wife, Charles Robinson, George Pearson, Miss Card, Eliza Libbey and myself. All Free Thinkers and all Deists. It was a treat to set down with such ones and hear intellectual conversation untrammelled by stiff-necked orthodoxy. Those who could converse on religion with minds unprejudiced.

It was five days since Deacon Johnathan had had the malignant growth excised from his leg. He probably could not yet get his Congress boots on and was sitting in the parlor, foot propped up on a pillow, reading his Bible on the Sabbath. Ned had left the house early to avoid an argument about going to church; he stayed to dinner at Robinson's in order not to sit opposite his father's "stiff-necked orthodoxy." It was just as well for Johnathan's peace of mind that he could not know his scamp Ned was reading that work of the Devil, Tom Paine's *Age of Reason*.

After [dinner] we sang together. A mental as well as a physical feast. Our music was not confined to the Church line nor was our conversation constrained to Church Theology. I spent the rest of the afternoon there and returned home to tea. With Ira [Berry] this evening; talked long and earnest in Religious Theology. I have been reading Thomas Paine's "Age of Reason" and instead of finding it the gross writings of a low mind I found a writer who was a Christian in precept one who had a high and holy opinion of God, whose death was mild and serene, contrary to the Pious Frauds circulated from the Pulpit.

April 11, Monday: Carried drawings of [Taunton] buildings to the Depot. . . . Got some lobster eggs this afternoon [for the Aquarium]. This evening at a meeting of the Natural History Society. Elected Ad Verrill corresponding member. Fussed and talked about a building committee of three. I left early in disgust.

April 15, Tuesday: Down to Portland Company's wharf found a lot of Eulis. and other things. My Aquarium is dying, I think. Hermit crab left his shell. At Turnverein and then Lager [beer] with Fred, Ned and Ira. I drank a glass of Hungarian wine.

April 17, Sunday: Aquarium clouded up. *Eulis.* died. Water smelt bad and I knew the thing was done.

April 18, Monday: Received letter from Mr. A. S. Packard, Jr., Bowdoin College, asking if a collection of Insects would be acceptable to the Society; asking also information in regard to fossil Shells. . . . Wheeled in a barrow two bushels of fish entrails from the fish market this afternoon.

To read this boy's journal with foreknowledge of his career is like sitting in a concert hall, the stage filled with musicians all fiddling, blowing into reeds, tooting through trumpets, beating on drums, while you hold on your lap the open score of the symphony they are playing. The unschooled listener is aware only of a mighty confusion of sounds, a multiplicity of instruments, and the grotesque motions of the players; but he who reads the score recognizes here the introduction of a theme, there a new motif, the mighty surge and fall of symphonic movements.

What blind instinct, what unseen baton, led a Portland boy to pursue the motif of shells, through conflict and opposition, against every pressure of parental urging, of provincial tradition? He was out of work, he was scorned and nagged, he had no hope of economic or social recognition, and still on an April afternoon his destiny drove him to the unsavory task of wheeling home a barrowful of fish entrails to search through them for tiny deep-sea shells that he had no other method of procuring. Already, in the score, it was written that in twenty more days the shell motif would be revealed as the main theme of his life's symphony, but he could not turn the page.

The letter from A. S. Packard, Jr., was another motif in Ned Morse's symphonic score, although it seemed only a hesitant communication from a college student to the unknown secretary of the natural history society nearest to Bowdoin College. Young Packard was the son of Bowdoin's professor of Latin, Greek, and oratory. At fourteen he began to collect minerals and shells, at eighteen to study comparative anatomy. He took his A.B. degree from Bowdoin in 1861, was assistant on the Maine Geological Survey, 1861–62, and then became a student-assistant under Agassiz at the Museum of Comparative Zoölogy, Cambridge. He was a brilliant and earnest student, taking his A.M. degree at Bowdoin 1862 and an M.D. degree from the Maine Medical School in 1864. Immediately thereafter he was commissioned as assistant surgeon with the First Maine Veteran Volunteers, serving at the front until the end of the Civil War.

Returning to Boston, he was briefly connected with the Boston Society of Natural History and was shortly called to Salem as curator of the Essex Institute. He was to become the most accomplished entomologist in the country, author of the immensely successful *Guide to the Study of Insects* (1869) and *Text-Book of Entomology* (1898). He taught at the Maine College of Agriculture, at Bowdoin, and at the Massachusetts Agricultural College and at Agassiz's Penikese school before, in 1878, he settled permanently at Brown University. From 1862, at Cambridge, he was to exert much influence on the career of Ned Morse.

April 19 [1859], Tuesday: Examined [fish] entrails; Shells quite scarce; found some very good ones, however.

April 27, Wednesday: Shan't smoke today. . . . Worked hard today; made design of wheel press for Taunton and a design of iron bridge for Portland Co. which will amount to $8.00. Good day's work.

May 1, Sunday: A most lovely day, bright and warm. This morning I went up to Charles Robinson's; we walked over to the Cape [Elizabeth] after *Helix milium*. Passed ever so many going maying or schiving with their hands full of flowers. Many of these were new-made converts. Charles talked and I eagerly listened to every word he said. His mind is strong and clear and his reasoning above dispute. Stopped at his house to dinner and then over to Spring Pond ledge and staid till six o'clock.

May 2, Monday: Another beautiful day. . . . Met Ad Verrill this evening and with John went into Turnverein a little while and then to a business meeting of the Portland Society of Natural History. Plans of a suitable building were presented by Mr. Harding which were very nice; just the thing we want.

May 3, Tuesday: Ad Verrill and Syd Smith of Norway and I started for the Cape this morning. Syd is a young fellow Ad has interested in Shells; he appears to be a first-rate fellow. We found a lot of *Helix harpa*. This afternoon we went over to the ferry to Spring Point ledge; got a lot of things.

May 7, Saturday: Telegraphed to Boston for block this morning; got it this evening. At 7 P.M. started for Charles Robinson's store; he invited me to go to the Islands with him in the club boat. We walked down there. The boathouse is fixed up in fine shape. Soon all the members were present, in their uniforms of short caps and blue shirts. Eight of the crew manned the oars and we sat in the stern. It was grand to move along at such a speed. There was in all thirteen of us. Went down to Banges Island where Mrs. Storer got us up a rousing good supper of chicken pie and Baked Beans.[2] We played

[2] Baked beans could always draw a superlative from Ned Morse, and he lived in the brick-oven era of New England when beans were beans. The housewife put her beans to soak Friday night in "cold water to cover." They were small, hard white beans kept in a cool place in the back shed against any tendency to sprout. In the morning the water was changed and the beans brought to a boil in the pan, and rinsed off with cold water to wash away the frothy scum and crack the swollen skins. A small onion was peeled, and placed inside the two-quart earthenware bean pot, shaped like a squat jug, round-bellied, with tight-fitting cover and handle on one shoulder. Repeated firings in the oven, laced with bubbling bean juice, imparted a lovely Vandyke brown glaze to bean pots.

Into the pot, on top of the onion, went the wrinkled boiled beans, with a pinch of soda, a teaspoonful of salt, a sprinkling of pepper, and a saltspoonful of dry mustard. Over all was poured a cupful of dark New Orleans molasses, and two cups of boiling water. Then the cover was set down tightly in the wide mouth of the pot, and the bean pot, balanced on the broad blade of a long-handled fire shovel, pushed clear to the back of the cavernous oven built into the chimneyside of the kitchen hearth. Clang! went the black wrought-iron door, and for twelve hours the beans simmered and swelled, soaked up the sweetness of molasses, the tang of mustard and onion, the savor of salt, while a delicious odor filtered through the cracks of the door to fill the house. In the iron kettle on the hearth bubbled and jumped and knocked with cheerful tintinnabulation the bright, tight-covered tin pail holding brown-bread batter of Indian meal, rye flour, raisin plums, molasses, sour cream. By five o'clock the brown bread was steamed to solidity, removed from the pail, and set in the oven to bake a crust.

New England housewives, innocent of vitamins, nevertheless had a healthful habit of serving with the beans and brown bread piccalilli of chopped green tomatoes and spices; chili sauce with red peppers added for piquancy; or coleslaw, of chopped and pickled cabbage. For sustenance, savor, and social cheer it would be hard to beat the old New England baked bean suppers.

on the piano, smoked in the parlor, and carried on until 11 o'clock. . . . It was beautiful coming up the harbor.

May 8 [1859], Sunday: Drew a seal this afternoon. Mr. Carpenter of England at the house this forenoon to see my Cabinet. Singular man. John and I with him this evening back of the hill. He got his feet all over mud.

Thus knocks Fate at the door! And all that Ned Morse noted, on the afternoon that brought about the realization of his heart's desire for a scientific career, was that the angel who visited him unawares got his feet all over mud!

CHAPTER THREE

A CHANCE IS OPEN
1859–1860

THE slight, bearded English clergyman who knocked at the door of 23 Spring Street, Portland, on the afternoon of May 8, 1859, held the key to Ned Morse's future in his hand.

May 8, Sunday: Mr. [Philip Pearsall] Carpenter of England at the house this forenoon to see my cabinet. Singular man. John and I with him this evening back of the hill. He got his feet all over mud.

May 9, Monday: Mr. Carpenter at the house this forenoon. He labelled and re-labelled many of my foreign shells. This afternoon I went up to Mr. Carpenter's, who is stopping at Mr. Ned Dow's. His room was filled with papers. He showed me an immense number of drawings of the Mazatlan shells on which he published a long article. Showed me also a drawing of Coecums which he is preparing for a monograph.[1] He told me he collected them from sponges. This evening went down to Dr. Lunt's and got the sand from some sponge drawers. Examined it and found any quantity of Coecums.

May 26, Thursday: Today I received a letter from Mr. P. P. Carpenter, who has seen Agassiz and told him of my taste in Science and talent for drawing. Mr. Agassiz wishes to see me. Shall go to Boston tonight.

May 27, Friday: [By Steamer *Montreal* to Boston, breakfast on new South Market Street, then walked down Franklin Street towards Washington Street from the docks.] On both sides of the street are

[1] Carpenter, an amateur malacologist, was working that winter at the Smithsonian Institution in Washington.

huge massive elegant granite buildings, wholesale stores. In the centre of the street a beautiful row of Elm trees. It looks beautiful.

Of course it did. The sun was never so golden, the trees so green, the air so exhilarating, food so delicious, scenery so charming, as on the May dawn when one went to meet Agassiz.

I walked on the Common quite a while, then to Steve Russell's office [Bricher & Russell]; staid there till nine o'clock and then took the horse cars for Cambridge.

The ride through Cambridge is charming. Beautiful residences almost hidden by trees, shrubbery, etc. Soon found my way to Prof. Agazziz [*sic*] house. Knocked with heart anywhere but in the right place. Was shown to his sanctum. He greeted me cordially. Was "glad to see me; would be at leisure in a few moments." I walked around examining the library of Science, Fishes, . . . in fact a little of everything with pictures of distinguished savans [*sic*].

The "Fishes"—and how unerringly Ned Morse went to look for them on the shelves—was the early work which gained Agassiz instant fame, *Selecta Genera et Species Piscium* (1829–1832); the description of the fishes of Brazil collected by von Spix and von Martius on the Amazon River and turned over by von Martius, when von Spix died, to the nineteen-year-old Agassiz. Or perhaps Morse took down Agassiz's *Histoire naturelle des poissons d'eau douce de l'Europe centrale* (printed in parts, 1839–1842) or the *Recherches sur les poissons fossiles,* published 1833–1844 at his own press in Neuchâtel.

Soon he was at leisure and we walked over to the Museum which is confined in a wooden building and packed close with specimens. Prof. Agazziz smoked continually. Cigars. His draughtsmen all smoked. Two of the best draughtsmen in the country. I saw Ad

Verrill [from Norway, Maine] there and he showed me round. We had a long talk together. Shall see Prof. again in the Fall.

Abbott Lawrence, Boston merchant, financier and ex-minister to Great Britain, in 1848 gave Harvard an endowment for the Lawrence Scientific School, and Edward Everett, then president of Harvard, offered Agassiz the chair of zoölogy and geology. The college at that time could provide only an old wooden building on the muddy banks of the Charles River for the laboratory indispensable to Agassiz. During twelve years it had grown more dilapidated, more crowded with specimens, collections, draughtsmen, students. Always Agassiz had worked to realize his dream of a museum of comparative zoölogy.

Ned Morse met Agassiz at one of the most triumphant and crowded months of his life. A fortnight later, on June 14, 1859, he was to lay the cornerstone of his museum. Francis C. Gray, his friend, dying, had bequeathed fifty thousand dollars for a museum. The public had subscribed over eighty thousand dollars to supplement it, and the legislature of Massachusetts had appropriated another hundred thousand dollars. Immediately after the laying of the cornerstone, Agassiz was to sail for Europe to purchase, as the nucleus of the new museum, the collection of his old teacher, Professor Bronn of the University of Heidelberg.

Agassiz, at the moment when Ned Morse met him, was beset by many problems. He was a Harvard professor, with seminars and lectures to prepare for; Commencement time, with its attendant examinations and gradings was in the offing. His salary was only fifteen hundred dollars a year, plus the use of a house the college had built for him on Quincy Street; yet he had a wife, three children, and a small army of draughtsmen, assistants in research, lithographers, and other dependents to maintain. He was deeply in debt but fanatically committed to purchases and expenses for increasing his collections. He had become so much interested in the experiment of higher education for women, after his wife, Elizabeth Cary Agassiz, started a girls' school on the upper floor of their home to add to the family income, that he lectured almost daily to the young women. He was writing his *Contributions to the Natural History of the United States,* huge scientific tomes. He was consulting with architects and engineers over plans and estimates for the museum.

Small wonder that, after walking with Ned Morse to the old museum, he turned the youth over to Ad Verrill with the injunction, "See me in the Fall," and disappeared into a maelstrom of pressing matters.

Morse, returning to Portland, impatient for the three months that must elapse before he could go to Cambridge to work with Agassiz, spent three weeks in June on the long-planned expedition to Eagle Lakes in northern Maine with John Gould, seeking fresh-water shells. A collecting trip in July, among the islands of Portland Harbor marked the beginning of a lifelong friendship.

July 5 [1859], Tuesday: Hurried around this morning and got a pile of provisions ready with shell boxes, greatcoat, blue shirt and everything necessary for an Island trip. Ned Waterhouse came after me in a team. At 10 o'clock we were on board the sloop *Eagle,* Fred Kimball, Ned Waterhouse, Ira Berry and myself. A good stiff breeze soon carried us away from the city with its dust and noise;

we were off for a week. Our pipes were lit and going down through the harbor we sung and smoked. Ned sailed the boat and showed good skill. Long about 2 or 3 o'clock we hove in sight of Goose Island. As we passed between it and the Gosling, we could see the tent and as we hauled up on the beach, Hudson came down to greet us, looking like Robinson Crusoe with his hat over his eyes. We had a great shaking of hands and then proceeded to land our stuff. We tossed things from one to another. Corned beef dropped on the bank and rolled down, collecting dirt, sand; plates were broken but we did not care for anything in our enthusiasm. We busied ourselves in fixing the tent for the night, getting supper. To bed about 10 o'clock.

The pages for July 6 and 7 are blank in the journal. Perhaps Ned was working on the discoveries which were reported and printed in the *Proceedings of the Portland Society of Natural History,* volume I: "Excavations at Goose Island, Me., and Occurrence of Rare *Helices* beneath Indian Deposits."

July 8, Friday: A. S. Packard, Jr., came to the island today. He appears to be a first-rate fellow. He was accompanied by his Scotch terrier and a net to procure insects with which he collected a great many. He has a good deal of scientific knowledge and we talked a long while on those things. In the afternoon, Fred [Kimball], Packard and myself walked over to the little island at the other end. The tide was coming in and we forced to wade back across which was tough enough, but when I got across, I discovered that I had left my sketchbook behind. I had to wade back and found the tide within an inch of my book. On my return back I was up to my middle in water, the tide had risen so rapidly. I left my pants on the trees nearby and walked through the woods to Todd's home where I procured a pair of ragged overhauls. We hunted shells a little but the mosquitoes were so thick we had to give up. Got home [to the camp] about 6. Hudson went to Bath today and will not get back till Monday.

July 9, Saturday: We were fearfully exercised by the flies last night. Poor Packard was kicking all night. His friends came after him today. Fred and I dug a pile of clams for a clambake. A pile of rocks were heated, then a layer of [wet] seaweed in which were placed clams, and then covered up with seaweed to cook 20 minutes. Of all the delicious messes, baked clams beat all. We fairly stuffed ourselves with them. Ned and I fixed up the boat and had a sound delicious sleep.

By September it was time to think of going to Cambridge.

September 22 [1859], Thursday: Got a letter from Ad Verrill today inquiring why I did not come up. Rains very hard today.

September 23, Friday: Rained very hard today. Went up on the engine to Paris and walked over in drenching rain to Norway. Been with Addison Verrill all day. Have had a long talk with him. Everything looks smooth and happy ahead. I hope in God and the Powers in Him that I may go to Cambridge. What a chance is now open to me. If Agassiz will only take me, and there is every chance of it now, won't I work and study! Ad gave me a lot of Shells from Grand Manan. Stopped with him tonight.

September 24, Saturday: Rains yet. Came home this afternoon tired.

September 27, Tuesday: Dissected animal of *Bulima undatum*. This evening Fred [Morse] and I started for Boston on the boat. Loafed around on deck, read paper and then went to bed. The motion of the boat though very slight gave me the usual sensation of dizziness and distaste for everything.

September 28, Wednesday: Fred and I got breakfast in a cellar near the markets and then I walked with him up through Franklin Street and at last to Bricher & Russell's where, after wasting some time, we saw them. From there we went to Williams and Everett and saw Rosa Bonheur's pictures, "The Muleteers," "Oxen on the Highlands," and a portrait of her. It was a treat to look at them; would be impossible to describe them so I won't attempt it. We looked at them for more than an hour.

Ned Morse does not give the address of Williams. An old Boston *Business Directory* lists "Williams, E. & Co., 23 Long Wharf, commission merchants," and "Williams, Dudley, Looking glass, Picture Frames," at 234 Washington Street; it was probably the latter that had Rosa Bonheur on display. The French animal painter had begun exhibiting at the Salon in 1840, but her fame rose from the Paris exhibition of 1855, when she showed "Hay Harvest in the Auvergne." Her popularly known "Horse Fair" at the Metropolitan Museum dates from 1853.

Ned continues: "I then went over to Cambridge. Saw Prof. Agassiz and had a long talk with him. Made all the arrangements and got everything settled. Shall be off with Ad Verrill for Cambridge in a few weeks. Saw the new building intended for a museum in process of erection."

Ned Morse does not specify, but internal evidence in later pages of the journal indicates, that Agassiz engaged him as a student-assistant, with the privilege of attending lectures, certain duties in connection with the museum arrangement and scientific research, and a salary of twenty-five dollars a month, with board and lodging to be furnished when the building was ready.

September 29, Thursday: Got in [to Portland] at 2 o'clock this morning. Fred and I came straight home and had a good sleep.

October 1, Saturday: Rather cool today. Drew [tobacco] pipes for Lowell & Senter [in return for his new meerschaum pipe] in wood. This afternoon Bicknell and I went out on train to Penobscot to see them put down Iron Bridge. Staid all afternoon. Interesting and busy sight. . . .

October 3, Monday: Saw Steam Fire Engine trial this afternoon; new engine for the city; done very well. Tonight I have been at the house arranging my Maine Shells and singing with the Piano.

October 4, Tuesday: Entered paintings of Shells in Fair and got contributor's tickets.

October 16, Sunday: Endeavoured to clear up my room but the confusion fairly discouraged me. Henry Dennett in a little while, played two games of chess.

October 17, Monday: Worked on [draught of] Seed Sower today; nearly finished it. Got a letter from Ned Waterhouse today, Mobile, Alabama.

October 18, Tuesday: News of an insurrection at Harper's Ferry. Finished draught and got $6.00 for it.

John Brown's little band of eighteen negro emancipators seized the United States arsenal and armory at Harper's Ferry, at the confluence of the Shenandoah and Potomac rivers in Virginia, on the night of October 16, 1859, and surrendered Tuesday, October 18, to a company of marines from the Washington Navy Yard, commanded by Colonel Robert E. Lee. Ten of Brown's men were dead, two of them his own sons. Brown, severely wounded, was captured with six others. To Ned Morse and thousands of other Americans it was an incident; to historians it marks the dawning of a bloody era of civil war.

On the first day of November 1859, Edward Sylvester Morse, aged twenty-one, left his father's home in Portland to begin formally a career in zoölogy. With the sympathy of his mother,

Jane Beckett Morse, he had finally won a five-year struggle against his father's religious and economic objections to such a career. His father's acquiescence was granted as a result of his engagement, at three hundred dollars a year "and found," as a student-assistant to Professor Louis Agassiz at the new Harvard Museum of Comparative Zoölogy. Agassiz's notice of the youth crowned seven years of solitary study of the fresh-water and land shells of Maine, during which young Morse had identified two new species of land snails, and received recognition from the leading conchologists of Boston, New York, Washington, and England.

November 1, Tuesday: Everything ready and, bidding the folks goodbye, started for Boston on the morning train. . . . Got in Boston at the usual time and had my trunk carried to the Express office and after seeing Bricher & Russell, started for [Cambridge].

Went to the laboratory and saw Hyatt, who went round with me to find a room but could not find one. I then tried to find Prof. Agazziz [*sic*]; went down to the new building and waited, but he did not come. Was astonished to see the progress they had made on the building on four weeks. Returned to Boston. Went to the Boston Theatre and saw the "Revel"; quite laughable. Stopped at the Adams House [Washington Street] overnight, making quite a big hole in my funds.

Alpheus Hyatt (1838–1902) was the son of a leading merchant of Baltimore. His interest in fossils dated from boyhood, and he became a distinguished zoölogist and paleontologist. He entered Yale in 1856, but his mother's desire that he should enter the priesthood resulted in his spending the following year at Rome. His devotion to science was dominant, however, and in 1858 he returned to enter the Lawrence Scientific School of Harvard and, under Louis Agassiz, dedicated himself to natural history. He interrupted his studies to serve with the 47th Massachusetts Regiment in the Civil War, being discharged at the close of the war after attaining the rank of captain, and returning to Cambridge as curator of fossil cephalopods at the Museum of Comparative Zoölogy.

From 1867 to 1870 he formed one of the Salem group of naturalists that included Putnam, Verrill, and Morse; in 1870 he went to the Boston Museum of Natural History, where he remained until his death. He also was in charge of the cephalopods at the Museum of Comparative Zoölogy, and taught zoölogy and paleontology at the Massachusetts Institute of Technology (1870–1888) and at Boston University (1877–1902). He was prominent in the founding of the Marine Biological Laboratory at Woods Hole, Massachusetts, serving as first president of its board of trustees.

November 2, Wednesday: Over to Cambridge again this morning and Hyatt at last found a nice place for me and Ad Verrill when he comes, Mrs. Cleaveland's, Winthrop Square; appears to be a nice woman. Fixed my room and got everything ready to go to work in the morning.

This afternoon Prof Agassiz went with me to Boston and introduced me to Cutting at the Aquarial Gardens. Prof A. wishes me to make drawings of the Gasteropods and Acephala in their live state.

The Aquarial Gardens in 1859 were in Bromfield Street, operated by James A. Cutting. Towards the end of 1860, as reported later in the journal, a new building was opened on Central Court, off Washington Street (240 Washington Street to Avon Place). On June 16, 1862, the management passed to P. T. Barnum, and it became Barnum's Aquarial Gardens, with dog shows, baby shows, etc., as added attractions. It passed through various hands, becoming the Theatre Comique, a variety house, and finally the Adelphi. It was burned on February 4, 1871, and rebuilt for business purposes.

Under Cutting the Aquarial Gardens had a semi-scientific atmosphere, with the blessing and patronage of the great Agassiz, which made the place one of the few in the Boston of the fifties to which a young woman might be escorted without a chaperon. It seems to have been used for respectful and respectable flirtations or courtings, as well as for the study of live gastropods.

November 3, Thursday: At Boston this morning; made drawings of *Bulimus undatus* animal while clinging to the side of the glass. This afternoon Prof set me at work drawing the *Mya arenaria* [clam] with the syphon extended. We all dine at Mrs. Magee's, a nice woman. There are about 30 Students dine there so we have a room full. At the laboratory this evening.

November 4, Friday: Drew more on *Mya* and the rest of the day we have all been upstairs tacking on labels securely, preparatory to removal.

November 7, Monday: Feel almost relieved of my cold; the sweat last night did me good. Worked hard today moving things from the old building to the new. Thousands and thousands of jars will have to be removed.

The cornerstone of the Museum of Comparative Zoölogy had been laid on June 14, 1859, after private individuals and the Massachusetts legislature had subscribed over two hundred thousand dollars to construct a fireproof museum under Agassiz's directorship. The plan called for a magnificent rectangle open on the eastern side, the main part to be 364 feet by 64 on the western side, with wings 205 feet by 64. Under the urgent necessity of removing his valuable collections from the old wooden building on the river bank, Agassiz had expedited the construction of the first half of the north wing. In less than six months from the laying of the cornerstone it was capable of occupation, and the moving of the collections was begun. Later generations of Harvard students and their parents, sisters, and sweethearts were to identify the group of accretions around Agassiz's original building as "the place where the glass flowers are."

To call the roll of Agassiz's student-assistants for the year 1859–60 is to recite the names foremost in American natural history forty years later, authors of standard textbooks, curators of splendid collections, directors of great museums in every large city of the Eastern seaboard. If youths of genius gravitated naturally to the feet of Agassiz, equal weight must be given to Agassiz's wise selection. Certainly it was not chance which produced the constellation made up of Verrill, Hyatt, Ordway, Shaler, Morse, and Putnam.

Frederic Ward Putnam (1839–1915) was born in Salem and entered Harvard in 1856. From 1857 to 1864 he was closely associated with Agassiz, leaving to become curator of vertebrata at the museum of the Essex Institute in Salem; since 1859 he had been curator of ichthyology at the Boston Society of Natural History. He was to be president of that society from 1887 to 1891. When George Peabody endowed the Peabody Academy of Science in Salem, Putnam became its director, serving for four years, 1869–1873. In 1874 he succeeded Jeffries Wyman as curator of the Peabody Museum of American Archaeology and Ethnology, for which the south wing of Agassiz's museum was built. He pioneered there in sending out expeditions for the study of southwestern and Central American archaeology. After 1894 he gave half of his time to the duties of curator of anthropology at the American Museum of Natural History in New York.

November 11, Friday: Busy packing fishes by the million.

November 15, Tuesday: Ad Verrill came this morning, bringing with him [from Norway, Me.] two large Owls, alive. He brought me a letter from home from Dr. Somerville, Philadelphia, stating that he had proposed me as corresponding member.[2] Attended this noon Prof. Wyman's lecture. . . .

Addison Emery Verrill (1839–1926), of Norway, Maine, boyhood friend of Morse, entered the Lawrence Scientific School of Harvard in 1859. In 1864 he was called to Yale as professor of zoölogy, where he remained until his retirement as professor emeritus in 1907. He was curator of the zoölogical collection in the Peabody Museum of Yale during these forty-three years. From 1871 to 1887 he was also in charge of the scientific work of the United States Commission of Fish and Fisheries in southern New England. His most noted work was in the field of corals and coelenterates.

The detailed account which Morse gives of his first lecture at Harvard, under Professor Jeffries Wyman, illustrates at once the paucity of preparation with which a student could reach a college classroom in 1859 and the material which in that day was new to physiology.

He [Wyman] stated that we commonly respire and inspire about 20 cubit [*sic*] inches of air and that after we have expelled that

[2] Of the Academy of Natural Sciences. Somerville had inspected Morse's shell cabinet at Portland, July 20.

quantity by natural effort, we can after that expel by force 100 cubit inches more. Furthermore after we have inspired the natural amount of 20 cubit inches, we can inspire by force 100 cubit inches more, making 220 cubit inches. Now then, after the lungs are cut out from the body and are perfectly collapsed, 100 cubit inches more can be forced or squeezed from them. Such a fact, he says, is true and yet how few know it.

November 16, Wednesday: Received letter from Fred [his brother] stating that Father's store was broken into last night and $2,000.00 worth of furs were stolen. I feel blue enough about it for such a loss is pretty heavy.

November 17, Thursday: Attended meeting of Boston Society Natural History with Putnam, Shailer [*sic*] and Verrill. Went in to Aquarial Gardens. From there we went to Parker's and Shailer gave us a game supper.

Nathaniel Southgate Shaler (1841–1906) was born in Kentucky, son of a doctor who had been graduated from the Harvard Medical School. A frail child, he was educated at home, soundly grounded in the Greek and Latin classics, German philosophy, and the southern tradition of skill with sword or pistol. Only nineteen in 1859, he was several years younger than the others, but with his breeding and sophisticated education must have presented a striking contrast to such New England country lads as Ad Verrill or Ned Morse. He was lively in conversation, generous and gracious in manner. Inviting his new associates to "a game supper at Parker's" would be a natural impulse with him. His father, incidentally, gave him an allowance of fifteen hundred dollars a year at Harvard.

Shaler went to Harvard at seventeen but was speedily enticed from classical studies to Agassiz's laboratory. He was graduated from the Lawrence Scientific School in 1862 with a B.S. *summa cum laude.* He spent two years in the Union army as captain of the 5th Kentucky Battery, returning to Harvard as Agassiz's assistant in paleontology. From 1869 he was professor of geology at Harvard and from 1891 to his death dean of the Lawrence Scientific School.

The influence of Agassiz was a contagion in those days in Cambridge; epidemics of "collectors' rash" were breaking out in all walks of life. Even the son of a boardinghouse landlady succumbed and brought home a pailful of shells, as Morse's diary shows:

November 18, Friday: Into Boston this P.M., bought pair of ready-made pants for $4.00; nice ones too. Worked and sweat all day. . . . Raining hard and warm. We have got the rooms all cleared. In Will Cleaveland's room; he has taken an interest in Shells and he had a whole pailful of stuff, Lymnea and all kinds of fresh-water shells.

November 19, Saturday: Received letter from Mr. W. G. Binney asking for the *Helix astericus* and *Helix harpa.* Tore down the cases in upper room this forenoon. This afternoon Put, Ordway, Hyatt and myself cleared out the attic. In clearing up the things we bundled them round rather roughly. Prof came in and saw apparently great disorder. It vexed him somewhat and he spoke, not harsh, but quick. The hardest expression he used was "deuce take it" and told us we handled things as if they were of no value. He then went home for the day. After we worked sometime longer, we went down in Putnam's room and talked about different things. About half an hour after he had left, Prof came back and sat down with us and talked more than an hour socially and even more pleasant, if such a thing was possible. Conversed with each one of us about what a nice time we should have.

In fact, I think, after he had left it occurred to him that he had spoke to us a little cross and he had come back from his house to wipe away any unpleasant feeling that might have arisen from the vexed state he was in.

Agassiz, at this time fifty years of age, was carrying a tremendous load, delivering college lectures, conducting original research, writing scientific works, giving public lectures, overseeing the construction of a museum and negotiating for its support from many quarters. For several years he had been displaying symptoms of overwork, perhaps the result of high blood pressure; Marcou writes that "after 1853 he quite often got in a passion, even losing control of his words." Ned Morse gives us a very charming picture of the great teacher striding back through the early November dusk to atone to a group of youngsters for a natural and justifiable "explosion."

Ned Morse went home for Thanksgiving, but it was but a dull feast to him, with his father gloomy over the loss of two thousand dollars worth of furs in the hard times after the panic of 1857.

November 25 [1859], Friday: Not much today. Blue and fairly sick of Portland. Packed up Shells.

November 28, Monday: Took first train this morning for Boston. Time passed very quick on board the train. Found they had just commenced to haul the old building.

The old wooden laboratory near the chemical laboratory of the Lawrence Scientific School was hauled across Cambridge to the ground where eventually the south wing of the museum was built to house the Peabody Museum of Ethnology. For several years it was used by Agassiz as a barracks for his assistants, among them Ned Morse. The students dubbed it Zoölogical Hall.

November 29, Tuesday: Prof set me at work on Florida Shells.

December 1, Thursday: Prof Aggazzi [*sic*] told me this morning to make a complete series of Shells, representing their varieties, form, growth, etc. Commenced on it at once. This week we have worked in the cellar of the new building.

December 2, Friday: All day we have been busy unpacking an invoice of two barrels and a box from California. Some of the Crabs, Starfish, etc., were most beautiful. This evening naming some species of Florida Shells.

December 3, Saturday: Stove in the cellar smoked so today that I took some Shells to my room. Yesterday it was so hot I was sweaty all day, could not wear my overcoat. Today it is so cold I can hardly keep warm. . . . We were all vaccinated tonight.

December 4, Sunday: Woke up and found the ground covered with snow. Ad [Verrill] and I have kept the room today, he arranging the birds, of which he has about 300 specimens; and I busy on Shells belonging to the Museum.

December 5, Monday: At home today fixing Florida Shells. Drew Bryozoa under microscope.

December 7, Wednesday: Attended meeting of the Natural History Society of Boston. Putnam proposed me for a resident member.

December 9, Friday: At the Museum. Prof gave me a lot of minute fossil shells from Paris Basin to look over and seperate [*sic*].

December 10, Saturday: Worked away on Pliocene fossils; find more work on them than I anticipated.

A letter to his mother, dated December 10, 1859, gives many details:

I am getting along first rate, making good progress in my studies and feel happy as a *Mya arenaria* at high water. Prof has given me still another branch in Natural History. He wishes me to pursue it because he wishes me to learn the use of the microscope. The branch is called Bryozoa; all the species are very minute and microscopic. I have made several drawings of some of the species under the microscope which please Prof much and he was not slow to show his pleasure.

For the last week I have been at home working on a large collection of Florida Shells. For the next two weeks I shall be engaged in my room on a collection of fossil shells from the Paris Basin, seperating [*sic*] the species. It is nothing but fun and pleasure to me.

Every week a large barrel or can filled with specimens of Natural History from some distant part of the Globe is received at the Museum. Among the material I pick out some beautiful and singular Shells which I select out and put in bottles. Each student takes hold and takes care of his department and we have jolly times over such lots when they come.

Verrill and I take our supper in our rooms, we have rich new milk and crackers. We eat like sin, or in other words consume immense quantities. Tell Father the money arrived all right. Send up the [Portland] *Transcript* as I and others wish to finish the story "Woman's Sacrifice."

I have made several first rate acquaintances here and feel perfectly at home. Before the first of January we shall be all settled in the new building [Zoölogical Hall].

There is no better man in this world than Prof Agassiz. We all feel towards him like a son to a Father. All call him Prof.

The recreations of a pupil of Agassiz were innocuous enough.

December 14, Wednesday: Snowing hard today; in the house all the time. About 4 this afternoon who should come in the house but Ira Berry [his Portland chum]. He is going to Philadelphia to work. We sat down together; had a good smoke; then . . . took the Cars for Boston; supper at an eating house. Stopped all night [at his aunt's].

December 15, Thursday: After a good breakfast of beefsteak, Ira and I started for Somerville in search of Sophia Whitman. After a long while riding [on the horse cars], we walked up through snow and wind to a ridge on which was the "Boarding School for Ladies." Saw Sophia and a Miss Sweetsir. Social time for about 20 minutes, when we took the cars back to Boston. . . . Took a parting drink. Bid Ira Goodbye and returned to Cambridge. Found Ad [Verrill] sick in bed with all the symptoms of Smallpox.

Smallpox apparently caused no alarm to the Harvard authorities or to Agassiz in those days. Ad Verrill continued to lie abed in the boardinghouse and Ned Morse to go back and forth to the Museum. It was not until the severe epidemic of smallpox in 1872, when 738 persons died in Boston, that the community took official cognizance of the situation and in 1873 set up its first Board of Health.

Ned did not even mention the smallpox to his mother, writing on December 18. He was anxious to know how his father was getting on with packing up his cabinet of shells, which were later sent to Cambridge and amalgamated with the museum collection.

I have been busy all this week in the same work as before. My black suit came safe; two pockets in the vest were minus, or rather were not sewed up at the bottom. You must excuse this scrawl. It's a pity you can't read my writing; you will have to get used to it though for I shall endeavour to write every week and perhaps oftener.

How does Father get along with my Cabinet? I am getting along gloriously. I can't help contrasting my position now with that of my Portland Company style of doing things; having now a director [Agassiz] advisor and friend, one who is pure, good, noble in every attribute.

The journal for the next few days is concerned chiefly with Ned's diversions:

December 19, Monday: Attended Mr. Vanderhoff's dramatic reading tonight, very interesting. He read "As You Like It" and I was highly delighted with it. Today I have worked on fossils. Saw the Boston Lancers, full uniform, go to Mt. Auburn escorting a funeral.

December 21, Wednesday: Still busy about fossils. Prof gave me more work on fossils to read. Attended Vanderhoff's reading . . . ; still more interesting and humorous than the first subject: Shylock in Merchant of Venice and Bardell vs. Pickwick in Dickens' writings.

December 22, Thursday: Received letter from Boston Society of Natural History informing me of my election as a resident member.

The *Proceedings* of the society for December 21, 1859, record that "Messrs A. E. Verrill, E. S. Morse and J. L. Foley of Cambridge chosen resident members." All three were students under Agassiz. Fred Putnam, Shaler, and Albert Ordway were already members, and Agassiz himself, of course, a regular attendant and the chief commentator at all meetings. On December 21, while Ned Morse was listening to Shylock's "my ducats, my daughter," Theodore Lyman, a favorite pupil of Agassiz's, was reading a paper before the Boston Society of Natural History, entitled "Descriptions of New Ophiurida Belonging to the Smithsonian Institution and to the Museum of Comparative Zoölogy at Cambridge." Agassiz, listening with interest and pride, made a little speech afterwards suggesting that his pupil's name should be appended to a species he had ably described — *Ophiura longicauda lymani;* "in reference to this paper," says the report, "Prof. Agassiz made some remarks on the principle he thought ought to govern Zoological nomenclature — viz: that each species should be indicated, not by the name of the first describer or by the one who established the true genus, but by his who combines originally or afterwards the true generic and specific names."

In addition to attending Shakespearean and Dickensian readings, Ned Morse on the evenings of December 21 and 22 wrote a long letter to his mother:

In three weeks more the building in which we are going to live will be ready for occupation and in the course of a month we shall move. The bedrooms and studies are very high and of good size. Verrill and I room together; we have a nice large bedroom opening directly into a larger room for us to study in evenings. It will be a sitting and reception room. We shall have a gas chandelier and droplight. All these rooms open into a large hall-like room where Prof says he is going to have a large stove put "so that Students may sit around and enjoy their cigar and pipe before they go to bed." I give you a plan of the second story showing my rooms, etc.

We [Verrill and Morse] have not yet decided whether we shall call on Prof for money to furnish our rooms or not. Your anxiety in regard to my undershirts amused me. Change my undershirts every week, and our bed-clothes are well aired every morning. We keep our coalfire going all the time, so in the morning I bring my clothes out of the bedroom to the fire and dress myself comfortably. I will take account of stock of my linen and send it to you so you can see whether I have lost anything. I know I have not lost anything yet and don't believe I shall. I shall need one or two more flannel shirts as my other ones are getting rather ragged. As for my greatcoat it is going to seed rapidly; shall have to call on Father pretty soon for money to purchase it. If Father can't let me have it, I shall borrow it of somebody else. When I see boys here in college whose Fathers are poor and yet actually sending them through college; tuition, board, clothes and all for four years; and think that Father can't let me have enough to pay for my clothes and a little change for about 18 months, it looks queer enough to me. However, if Father has not got it, all right; I shall be a salaried assistant by that time and then I shall have money of my own. Your speaking of doughnuts made me long for a bite of one; when you get a chance send me up an enormous pile of them.

Today I have been at work in the Museum. I could never begin to write you what a truly good man Prof Agassiz is. He shows it in a thousand ways. He has such a jolly "Good Morning" and his "goodnight" breathes of kind wishes. He always says it; I can hear him now "Good night, Mr. Morse, good night." His voice is tender. Ah, he is one of the best men that ever lived.

Putnam, the assistant, told me yesterday that Prof told him I was to be the assistant in the Museum, that it was his intention from the first. Isn't that jolly? I was sure of it from the manner and earnestness in which he has given advice and direction to me. Prof told me the other night he wished me to read Shakespeare thoroughly in order, he said, to be able to write classically. Tonight he gave me earnest work to do. It is this: there are thousands of species of shells found fossil in England and France which authors have considered the same as recent living species. Prof don't believe they are the same and he wishes me to examine closely and make comparisons of those species in order to prove their difference.

It is seldom in human affairs that the exact moment of the planting of a seed can be traced, but here Ned Morse, writing to his mother, records the day, December 21, 1859, and the place, the Museum of Comparative Zoölogy, when Louis Agassiz made to his young pupil a remark that was the genesis of such permanent objects as a granite monument in Japan, a marine biological laboratory at Enoshima, a gallery of rare old pottery at the Museum of Fine Arts in Boston.

Brachiopods, an isolated class of marine animals, with shells resembling to the casual glance those of the common scallop, were the group in which Ned Morse undertook to trace the development from fossil to recent and living species. Fossils were numerous, but in twenty years of search on the Atlantic coast from Maine to North Carolina he found few living species, and in 1877, in quest of Pacific species, he went on what was to be a brief collecting trip to Japan. How he remained to organize the teaching of zoölogy in Japan, to write a number of classic books on Japan, to make the first and greatest collection of Japanese pottery, belongs to the story of his later years. Here is the beginning. Henceforth in his journal or his letters when Ned Morse speaks of *Terebratulina,* he is talking of brachiopods and engaged in the inquiry Agassiz suggested to him.

December 24, Saturday: Worked like the Devil at Museum. Prof talked with us about our Iceland expedition. He is bound we shall go next summer to that region. . . . Up late tonight till 12 o'clock, wished all a Merry Christmas and gave my presents. How happy I have been this evening and yet I look back a year ago this night and find myself blue and desponding in Boston and wondering how I shall be a year hence. Time surely changes everything.

December 25, Christmas: I have had a regular Merry Christmas today. We laid abed late and Mrs. Cleaveland sent us up a pitcher of hot coffee, loaf of hot brown bread and butter. We made the best breakfast since we have been here. . . . This afternoon skated on Fresh Pond; have not had skates on before for seven years. Got along first rate.

December 26, Monday: Very busy at the Museum carrying upstairs from the cellar Tertiary Fossils.

December 28, Wednesday: All the students were at work when the Committee came to examine the Museum. This afternoon we all congregated in Put's room and talked over next summer's trips. Prof came in and it was definitely settled that we should go in two parties. Hyatt, Putnam and anyone else who cares to shall go to one

Edward S. Morse in 1863

Nellie Owen Morse in 1869

of the Great Lakes and work through to Hudson's Bay; while Wheatland, Ordway, Verrill and myself shall go to the coast of Labrador and work up to Hudson's Straits and perhaps into Hudson's Bay exploring on our way back to the rivers, etc.

There are several possible reasons for Agassiz's interest in the Hudson Bay region at this time. In 1856 Dr. Elisha Kent Kane had published his *Arctic Explorations,* an account of the search for Sir John Franklin, which was eagerly read — read aloud — in thousands of homes in America. In 1859 M'Clintock had just learned the tragic end of Sir John Franklin and the 129 men of the *Erebus* and the *Terror* who had sailed in 1845 to search for the Northwest Passage. But for twenty years longer the world was to ponder, advance theories, send out exploring parties on the chance that, somewhere, some of Franklin's party survived and were trying to make their way to civilization. It may be too much to infer that Agassiz and his keen youngsters had in the back of their minds the scheme of penetrating far enough north to question Indians and Eskimos for news of Franklin's men, but King William Island, where the boat of the *Terror* and the skeletons of the men were found, is only some hundred and twenty miles north of Hudson Bay, and it would be reasonable to think that survivors, if any, would be toiling southward across Arctic barren and forest wilderness.

A more likely explanation is that Agassiz, with his brand-new museum, was envious of the collection of Arctic flora painfully brought back through dire vicissitudes by Kane for the Academy of Natural Sciences in Philadelphia which Agassiz's friend, Asa Gray, the noted botanist, had helped to classify. Kane's collection of fauna, including, of course, shells, destined for the Smithsonian Institution, which had helped finance the Kane expedition in search of Sir John Franklin, had necessarily been abandoned when the brig *Advance* sank in the ice. Thus, in the amiable rivalry of science, it would indeed have been desirable to acquire for the Cambridge museum the first valid collection of Arctic fauna and shells.

January 1, 1860: Prof has given me the most unlimited control over Shells in the Museum. My arrangement of fossil shells last week appeared to have pleased him very much. The stoves have been put into our new rooms and will be kept burning for three or four weeks before we move in, in order to dry the plastering.

January 3, Tuesday: At work in Lyell's collection. For the past week we have had tremendous cold weather.

January 4, Wednesday: S. H. Beckett [librarian of the Portland Athenaeum] here today. I showed him round the Museum and astonished him somewhat at the extent of the collections. This evening we all went in to Tremont Temple to hear Prof Agassiz lecture. The hall is very large and elegant. Prof stood on the platform without a desk or anything in front of him and spoke for two hours. He had a large audience who seemed much interested in the subject which was the value of a Museum.

January 5, Thursday: Worked on Duval's collection of fossils.

January 6, Friday: This forenoon we received a live cow from Salem. We endeavoured to kill her with chloroform. Put held a sponge to her nose and after exhaling it [*sic*] three minutes she dropped down on her side. We kept the sponge to her nose a long time which put her in a deep sleep but had no other effect. We then put Prussic acid on her tongue and injected some in a vein but that would not kill her. At last we cut her throat and she quietly bled to death.

January 7, Saturday: At work on Duval's collection. They have got the tables partly finished and some of them painted. Discovered a pretty property in *Pecten concentricus* tonight; the width of the ears would go round the circumference of the shell just exactly five times.

Pecten was a scallop shell, and "ears" was not literal but referred to the flanges at the top, where the two shells of the bivalve are joined.

January 9, Monday: At work at Duval's fossils. All the students collected and Prof gave his first instructions to us.

The construction of the new museum and the work of moving collections from the old building had naturally interfered with Agassiz's regular routine of lectures.

January 10, Tuesday: At work labelling fossils. Into Boston today. At Aquarial Gardens was much interested in watching the monkeys act. One monkey had a rat in its arms and was taking the same care of it as it would a baby of its own.

January 11, Wednesday: News today in the papers of a frightful accident which occurred in Lawrence, Mass., last night. Pemberton Mills fell, crushing with it about 600 operatives. Two or three hundred killed and hurt. I went down in the afternoon. The Depot in Boston was filled with a homogeneous mass of rowdies, anxious faces and drunkards. The train consisted of 13 cars filled dense. When we got in Lawrence found the place filled with persons, some crying, others yelling and singing. Visited the ruins. There were three streams yet playing on it. I never saw such a wreck in my life. Just imagine two hundred human bodys mixed up in a mass of iron and brick. Horrible.

Ned spared his mother the details of the Lawrence disaster. His letter of January 15, 1860, is chiefly on museum life:

The building is approaching completion. Last week a nice large blackboard was put up in the students' room, made of two enormous slates, perfectly smooth; presents a splendid surface for drawing with chalk.

You have heard perhaps through the papers how prevalent Small Pox has been and is now in Cambridge and Boston. Dr. Plimpton, a medical student who I see at the house quite often, vaccinated me some time ago with the very best matter in two places but it did not take. Since then, he vaccinated me again with the purest matter which had produced, as he said, on others "Magnificent pustules." It had no effect on me and he says I might sleep in the bed with the worst case of smallpox that was ever known and not be harmed by it. Shan't try it though.

Last Monday all the students collected in the Student's room for the first time. There was Mr. Barnard, Mr. Lyman, Scudder, Hyatt, Putnam, Wheatland, Ordway, Verrill and myself.

Prof wished to make a few remarks to us so we all took chairs and sat down around him. He commenced his remarks by saying that we were together to pursue the study of Natural History in its true sense, earnestly and scientifically. That we had an opportunity never offered to anyone before. Students in Natural History hitherto have had lectures now and then and a library to go to, but no specimens to handle.

He said that in Europe no one can get access to specimens unless he collects them himself. In the British Museum no one is allowed to take a specimen out of the case unless he is really a naturalist and even Cuvier, when he wished to handle a specimen, was obliged to do it in the presence of one of the Officers of the Museum. He said one is obliged to study [specimens] through the glass of the closed case. He said that we should have the most unlimited license in the use of specimens; to handle when we pleased but that we should bear in mind that a specimen which is small or common should not be viewed in the light of being worthless.

He said that a specimen worth one cent played as important a part in the museum as that worth a thousand dollars and we must treat it in the same manner. He wished to impress upon us the fact that we had a duty of profound interest to perform. That this museum was unlike any other museum ever founded. That it should be carried on in such a liberal manner as to set an example to others and do away with that illiberal spirit which is seen here and there. And in order to make it great, he depended on us to work with a will and an interest.

He spoke of the arrangement of the Museum and what he was going to have. One room, as we enter the building, was to be devoted to the epitome of the Animal Kingdom. In it were to be placed types of the Vertebrates, Articulates, Radiates and Mollusks. He was not going to have it crowded, but it should be arranged in such a manner that the prominent features of the Animal Kingdom could be seen at a glance. Then he was going to have a suite of rooms for Vertebrates; another for Mollusks, and so on. One large room was to be devoted to Corals. But he would tell us about this another time.

Just then some of the joiners in another room burst out in roaring laughter at some of their jokes. Prof said, "You perceive the difference in the laugh of a cultivated man and the laugh from one that is not."

The rest of his remarks were scientific and so I won't bore you with them. Tomorrow he speaks to us again in the same manner. Here I am eating crackers and milk with one hand and writing with the other. You can't imagine how much I enjoy my supper in my room of crackers and milk.

I think I shall have to get a working suit soon. I suppose my shells are all packed. I will send Father the cards to tack on them and he can send them with the bill of the boxes as convenient. Let him charge the Express man to take great care of them.

The journal continues:

January 17, Tuesday: Busy moving fossils from Lecture room upstairs. It is a luxury to have two stout men at your bidding to do as you choose and direct. . . . Putnam has got a queer animal from South America which looks like a miniature bear except that he has a long tail; a very long one. He is as playful and mischievous as a monkey; we had him in our room tonight.

January 20, Friday: Prof spoke to me about the manner I had of jumping from one thing to another and that I must guard against it. I shall strive to do so. Received pants, vest and drawers from Mother, also letter from her.

January 21, Saturday: I went in [to Boston] to see Church's "Heart of the Andes." It would be impossible for me to describe the picture. It is certainly a masterpiece. Everything drawn with that fidelity which bespeaks a close student of nature and the distance, middle and foreground are so admirably shown that one can guess the distance of any object. It is certainly a wonderful work and shows, as Prof says, what intricacy, patience and industry can accomplish. . . . At Aquarial Gardens. Met Ad [Verrill] there and he introduced me to two young ladies from Somerville School, Miss Hatch and Howe. At Museum this afternoon. Received telegraphic despatch from John [Gould] to meet him at Haymarket Square at 8 P.M. and we met and came together again . . . as of old. Went round Boston with him, and to Aquarial Gardens and he came out with me and engaged a room for three or four days.

A feud which Ned Morse had conjured up a year earlier with his best chum over a parcel of shells from which Gould did not offer Ned "next best" choice was thus extinguished. John had never been aware of it, since it existed only in Ned's mind, and he naturally found no difficulty in meeting in Haymarket Square "as of old." With Morse himself it had been a scapegoat for friction with his father, despondency at being out of work, and misery at being ordered by his father to give up "wasting time" on shells. Now that Agassiz had given him "unlimited license" with the vast collection of shells in the museum, the feud dissipated into thin air and the friendship was renewed, to last as long as both lived.

When Edward Morse was eighty-four years old, he began an article in the *Nautilus* magazine with the words: "Recently I received the right valve of the common clam *Mya arenaria*, from my friend Major John M. Gould," and proceeded with a dissertation of an abnormality of the shell, which the two octogenarians had studied as eagerly together as when they were barefoot lads digging clams on a Maine beach.

January 23 [1860], Monday: Arranged recent mollusks. Prof gave us another lecture this forenoon. Ad [Verrill] was called up to the blackboard for report of remarks. John, Ad and I went in to the Opera tonight. The opera was *Martha* and it was truly splendid. I was in a perfect trance after we came away until we went to bed, continually singing "The Last Rose of Summer," which was rendered grandly at the close of the Opera. John never saw anything like it before and was highly delighted with it.

Flotow's opera *Martha* was first performed at Vienna in 1847. Like his eight other operas it has failed to hold the interest of producers or the affection of the public except for a couple of arias. In 1938 the tuneful lilt "Heart so-O-o pure," with the accompaniment rearranged in "swing time," made an astonishing success on the radio and records in America; and "The Last Rose of Summer" was kept alive in thousands of hearts well into the twentieth century by the noted contralto, Mme. Schumann-Heink.

January 24, Tuesday: Very busy at Museum. John I introduced to Prof. He [John] went in to Opera tonight.

January 25, Wednesday: John went home. I went in with him this evening to see him off on the boat. While there saw O. Hayes get some money out of a man under false pretenses. Very busy at Museum. I weigh 146 pounds with greatcoat off.

So John went home, to his employment as a clerk in a Portland bank, his recreation of collecting and studying shells. He had had a wonderful visit, hearing opera for the first time, meeting the great man Agassiz in person, watching Ned's work in the museum. They had had hours in which to talk confidentially, and John had learned of Deacon Johnathan Morse's inability to supply funds to Ned, had seen the ragged greatcoat, the frugal meals of crackers and milk in his bedroom. The journal records nothing of these talks, but thereafter on the first of every month it mentions "Letter from John. $8.00 enclosed." We know that John Gould was getting a salary of five hundred dollars a year at the bank, roughly eight dollars a week, and the inference is obvious. He sent his old chum Ned a quarter of his earnings regularly, as a loan to be repaid "when convenient," to see him through the years of study under Agassiz. John Gould was the type of the loyal, understanding friend who believes in one and stretches out a hand for a lift over a hard spot in the road.

January 27, Friday: Received today at the Museum many large boxes of fossils from Europe, embracing five or six Ichthyosaurus, very valuable; also box of trout, rabbit, etc., from Norway, Me. [home of Ad Verrill], and a barrel containing specimens in alcohol from Acapulco [Mexico]; two of them containing specimens in alcohol, the other contained dry specimens. We were all engaged in unpacking them all day.

January 28, Saturday: At the Museum the forenoon; busy. Last night I did not get to bed till three o'clock. Dinner today of trout, rabbit and Guava, very nice. [The guava perhaps came in the "barrel containing dry specimens" from Mexico.]

January 29, Sunday: With Lina and Willie [Cleaveland] and Charlie to a universalist lecture in Cambridgeport. I was astonished at the liberality he expressed; his lecture was on Swedenborg and he spoke in the manner of one who is liberal to the fullest extent.

About this time Ned wrote a long and informing letter to his mother.

Dear Mother: Yesterday[3] we ate a moose. Prof is going to have it stuffed and a skeleton made of it. It was skinned today and we sliced off a lot of venison steaks. I carried Mrs. Cleaveland home a good lot of it.

Yesterday the tables were moved in the Students' room and we shall all get to work there shortly. The whole place looks deserted, more like a meeting house on weekdays. Of the thousand [Harvard] students that were here Tuesday not more than thirty are left. The term has closed for a vacation of six weeks. And, I tell you, if one wouldn't meet happy faces Tuesday morning, why then, I never saw one. The students are not allowed to smoke in the streets while the college is in session, but as soon as the term closes, they have a right to do as they please. An hour after the term closed, the streets were crowded with students, all with cigars or meerschaums in their mouths so it was a funny sight. The students, arm in arm, would walk right under the nose of the tutors or Professors, puffing smoke in their faces.

Prof is down on the college students who go through the classical department. He says most of them come there to get a varnish or gloss, so that they can mingle with great men. That they care nothing about the studies and the whole thing is irksome to them and hence their intense satisfaction when vacation comes. He says that he knows *his* students have a vacation all the time; and so we do, for if we go away for a while it is only to dip harder in our persuits [*sic*]. He lets us go away when we choose because he knows we shall only continue our persuits with the same interest.

John was here and we had some good times together. Spent most of Sunday in the Museum with John showing him the collection.

[3] The letter is dated January 29, 1860, but it was evidently begun several days before.

He got a room next to mine for three or four days. The $10.00 bill [received in a letter] was very acceptable and I am much obliged for it. It came just in time, as my boots have all gone to pieces.

Prof says he intends that each student shall have a key for every lock in the building, cases and all, so that we can go in anytime. I sent a pile of ragged clothes down in carpet bag by John; when you send it back be sure to fasten it secure as it gets unlocked very easy.

Verrill has a habit of laying down on the sofa and getting to sleep when it comes late and I have some fun with him. Our room is always in great disorder as you may imagine when I tell you that Verrill is a good deal like myself in those things. Well, as I was saying, Verrill gets to sleep, and when he wakes up he is awful sleepy. While he is asleep I put chairs, table, boots, spitbox, coalhod and shovel in such a way that he has to go over them in order to get to the bedroom. I then take the lamp in the bedroom, undress, open the door about an inch and yell out for him to come to bed. I then jump into bed, keeping ears open for his execrations.

He tumbles over one thing and says "Gosh!" Then he strikes his toes against something else and says "Devil on it!" When he brings up against the coalhod he sometimes says "Damn!" but his other expressions are ludicrous enough. His phiz when he gets into the bedroom is laughable. He looks sleepy and the light striking on his eyes causes them to close and his features are distorted in every shape by the pain or perplexed by his clumsiness. I generally laugh myself into a fit of hiccups and Verrill wants to know what there is to laugh at. . . .

I guess I shall come down the last of March and bring up with me my Cabinet.

Here it is Saturday [January 28] and I have finished another week. We have all been busy as usual at work in the Museum. Things righted now and in the course of putting the specimens where they belong. Speaking of cases, the expence [*sic*] for that item alone will be $9,000.00. Just think of it! Nine thousand dollars! There are four different concerns endeavouring to get the job, each one is making a sample of what he can do and the samples so far are excellent.

Yesterday we received at the Museum eighteen boxes of fossils from Europe forming part of an extensive collection that Prof bought [the Bronn collection from Heidelberg]. All the boxes were as large as any goods box and four of them were ten feet long, containing fossil Ichthyasaurus [*sic*], a rare fossil. We received five specimens. Did you go to Dr. Boynton's lectures? If you did, you know what they look like. They are mounted in heavy wood frames like a picture. The specimens are valued at $300.00 apiece and we have five of them, from six to ten feet long.

One of the students asked Prof what he gave for them and he said "ten cents apiece." We were naturally astonished when he said this but he put on a very sober face. Guggenheim, the Dutchman, commenced to count the bones in the specimen before him, when Prof told us he gave $5,000 for the collection, comprising fifty thousand specimens, which would make ten cents apiece.

Besides receiving the above, we have a large box from the North of Maine containing a number of new species of trout, rabbits and birds; a barrel containing specimens in alcohol from San Francisco; three barrels from Prof's son [Alexander Agassiz] collected in Acapulco. Two of the barrels were filled with dry specimens and the other jam full of alcoholic specimens. You can't imagine what a pile of specimens there were. All of us were engaged yesterday in sorting them over as they belong—Fishes, Shells, Crabs, Starfish and everything.

Prof was jolly enough over them. As we would take them out of the barrels, it was fun to hear his ejaculations.

"Ah, what a beautiful Pleuronectes!

"Look at that, gentlemen, isn't it splendid?"

And he would hold it up for us to see more fully. I got hold of something in one of the barrels which looked like the animal of some shell. I examined it quite a while and at last threw it down, exclaiming that it was a leaf. Some of the students laughed at me, but Prof exclaimed, "You need not laugh, gentlemen; for I, not long ago, puzzled over a piece of a pair of old breeches for two hours!" and then we all laughed with him.

Today we had a nice dinner from specimens received in the museum. Roast goose and onions; Rabbit pie; and nine large undescribed species of Baked Trout. I eat so much I was scarcely able to move from the table.

Last Monday John Gould, Ad Verrill and myself went to the Opera and I tell you it was grand. The opera was *Martha* in which is sung the Porter's Song "I want to ask you, can you not tell me," etc. It was grand. "The Last Rose of Summer" is in the same Opera and it was very plaintive.

Last Wednesday I was in Boston and saw Sprague.[4] He said he was real glad I was out here [with Agassiz] and told me he had paid much credit to me [? MS torn] in Fungi lately. I was engaged in literary work. Governor Banks was there. When he was shown through the Museum last Friday, I had an introduction to the Governor, who took off his hat very politely to me [at the second meeting, in the Natural History rooms]. Further appropriations will be made to build the rest of the Museum Wing in the Spring. I saw by the proceedings of the Philadelphia Academy of Natural Sciences that I was elected corresponding member; has a certificate been sent me yet? I think I shall come home the end of February to bring up with me my Cabinet and other things.

When you write, give me all the news about my friends. Have you seen any accomplished lady who has diligently studied Greek, Latin, German, French and Spanish? Who has studied Conchology and Botany? Who is at some *Future* time going to eclipse Rosa Bonheur? Who is going to travel over the world a la Ida Pfeiffer or Bayard Taylor? Who can skate, dance, ride a horse, a locomotive and forty other hobbys? Who was going to be a missionary and learnt her trade round Gorham's Corner distributing tracts? If so, tell me about her, as I wish to keep posted up about that distin-

[4] Charles J. Sprague, for whom Ned had collected fungi at Bethel several years earlier.

guished satellite who attracts around her all the Englishmen, Bluenoses and Theatre Actors whenever they visit Portland. Has she any new hobbys to ride?

The "distinguished satellite" and "accomplished lady" was, of course, Jane Beckett Morse herself and the paragraph an epitome both of her life and of her influence on her youngest son. Gorham's Corner was the little village of Gorham, Maine, to which she retired permanently after the death of her husband. Jane Morse lived to be an old, old woman, but to the end of her days she carefully preserved this letter of January 1860 because of the affectionate, lightly teasing, deeply loving tribute her son paid her for twenty years of motherhood.

Next Summer [the letter continues] I am going to explore the coast of Labrador and way up through Hudson's Straits with three other students, Verrill, Ordway, and Wheatland. We hire a schooner and start about June 10 to be gone all Summer. We shall go far enough North to see the sun shine all night. Won't that be fine? We shall go to the head of Ungava Bay and explore the river that enters the head of it. Look on the map and you will see what a nice tramp we shall have. Prof is bound we shall go and wants very much collections from that region. When I come home, I will tell you more about it. Now, Mother, write, if you don't write but a line, as I wish very much to hear from you.

CHAPTER FOUR

DARWINIAN DEBATES
1860

SCRAWLED STAVES, bars, and notes from a leaf of Beethoven's sketchbooks are faint indications indeed of the intangible thought, the ultimate torrent of sound, that is a sonata. So, likewise, are the impersonal records of a meeting, the jottings of a student's journal, the merest dried leaves — souvenirs of a controversy that, beginning dynamic as a seed, put out roots, branches, blossoms invisible, and died down again to fertilize the thought of a new generation.

The coincidence of Morse's arrival in Cambridge in the autumn of 1859, when Charles Darwin's book, *The Origin of Species*, was published, made him a spectator at one of the battles of the human intellect, an acquaintance of both proponents and opponents of the Darwinian theory. In his immature mind was waged, in miniature, the intellectual strife of the world of the nineteenth century. Not that a country boy of twenty-one years, from the provincial background of Maine, could so appraise his situation. Morse realized only that two men he admired, Louis Agassiz and Asa Gray, held opposite concepts. Some rugged streak of intellectual integrity preserved him from settling the issue by personal loyalty. He was to spend thirteen years in individual research to find his personal answer to the problem; but for the moment he was still a boy, his interests divided among family, girls, and allotted tasks.

February 1 [1860], Wednesday: One-twelfth of this year gone already and gone quickly, too. Tenth of January this year I signed a pledge against spirituous liquors and have kept it strictly. I have been busy today in Genera. Have seperated [*sic*] the so-called genus *Ceritheum* into half a dozen distinct genera. Here is what Prof [Agassiz] says in regard to Family, Genera, Species. Family is a form determined by structure. Genus depends on the ultimate details of structure. Species . . . ornamentation, distribution, form, etc. Prof says he is going to keep me a year on it. So cold I thought I would not go in to meeting of [Natural History] Society, so I staid at home. . . .

February 2, Thursday: Very much interested in the genus *Ceritheum* which I have cut into several groups of genera. It should by good rights be a family by itself.

Ceritheum is a form of fossil mollusk of the Eocene geological era, a gastropod genus whose series of mutations were important in the study of the stratigraphy of the Paris Basin.

February 3, Friday: At Museum, busy. This evening I got an alarming letter from Fred [Morse] stating that Mother was taken sick. I think I shall go home next week.

Judging by the next entry, his chestnut-burr prickly nature had got him into trouble again, threatening a new friendship, but, as usual, Ned Morse's friends liked him warmly enough to refuse a quarrel.

February 5, Sunday: In the house all day. Blue, sick, desponding. With my happy position here and everything in science to make me comfortable, still I am unhappy because it seems to me as if I had lost a dear brother and sister. . . . Last night before I went to bed, I wrote a note to Lina and in it I mentioned the reasons for the course I was obliged to persue. . . . [Tonight] Willie came up and asked me downstairs; I stayed but a little while; it was perfect torture. I went back to my room. At 10 o'clock Willie, the dear good fellow, came up after me again. I went down and found Lina weeping on the sofa. What could I do? . . . In ten minutes I was perfectly happy. We were all happy, happy, happy. Harmony was restored. Willie and I went home with Lina and afterward we walked till one o'clock on the bridge.

February 6, Monday: Perfectly happy. At the museum today worked like a trooper.

February 7, Tuesday: Willie, like a true brother, helped me pack my carpetbag. From Prof $10.00. Started for home this noon; ride very tedious. Expected to find Mother confined to her bed and to my astonishment found her up, dressed and in good health. I was completely happy.

February 8, Wednesday: At Mrs. Brazier's. Found Lutie changed a good deal. I am afraid she will be an old maid.

February 9, Thursday: Rec'd letters from Philadelphia Academy [of Natural Sciences] announcing my election as corresponding member and demanding five dollars for a diploma. Find Portland extremely dull and stupid. Portland Natural History Society have a fine building fitted up for a collection.

February 14, Tuesday: This evening I went to a dance at Union Hall [Portland]; had a real good time. Got acquainted with two nice girls, Miss Newhall and Miss Owen. Enjoyed their company very much. Joe Starr fell while dancing, poor fellow; I pitied him. I have kept my promise firmly in regard to drinking. Last evening at Abbie Lowell's, Mrs. [Ned] Emery, I should say, with John [Gould]; had a real pleasant time.

A letter he had sent to his mother on the date of learning of her sickness contains more details of his life in Cambridge, which by contrast made Portland seem so dull:

This snowstorm has come plunging us into Winter again. I had even gone after Shells before this snow came, and found quite a number of [helices] and pupa under leaves. It is colder than Greenland here now and Ad and I have moved the table up close to the stove. For the last two days the water has frozen in our pitchers, though we sleep warm as chickens, not big grown-up ones but little fellows three days old. We have as many clothes on our bed and in the morning dress in front of a rousing coal fire.

We have had another member added to our Museum. He is to be Secretary, write up Synonyms and catalogue Collections. He is thoroughly conversant with all the principle languages spoken. He is to have a room in our clubhouse. He is very jolly and pleasant, and I had not been acquainted with him ten minutes before I laughed at his fat belly. For he is certainly the fattest man I ever saw; he is twice as fat as this sketch represents him. He is going to teach us German for nothing. His native place is Copenhagen.[1]

I have been engaged for the past two weeks in Genera, have taken, for example, the Genus *Ceritheum* and detected seven or eight genera in it. It is a great work and Prof is highly pleased with manner and zeal with which I do it. He hinted to me tonight that he was going to keep me a whole year on it. I like it very much and by the way he talks I should judge he intended that I should write something extensive on it. The Museum will soon commence publishing articles and papers in Natural History and it is Prof's intention to have all the Students prepare papers for it.

Ned Morse never published an article on *Ceritheum*, nor did he ever make an expedition to Hudson Bay by schooner. This letter illustrates a situation frequently occurring in the career of Agassiz, in Europe as well as America, which arose from two factors, the adoration given him by his students and the pressure on his time and attention from the multiplicity of his scientific projects. Agassiz was wont to think aloud, expressing as they occurred to him random, fugitive ideas, inspired by the

[1] Morse evidently refers here to Hansen, a Danish artist who made scientific drawings.

circumstances of the moment. His rapt pupils accepted every word that fell from the lips of the great master as Delphic utterances, almost ordained to be realized. After months or years, when the projects had not materialized, the pupils would grow bitter and tax Agassiz with disappointing or misleading them. And Agassiz, the incident long vanished from his memory under an avalanche of new details, completely bewildered and upset, would charge his pupils with ingratitude.

February 16, Thursday: At the Museum. All glad to see me. At work, arranging as usual. Putnam has got his room fixed up in fine style.

February 17, Friday: Severe snowstorm. Have got a bad cold. My boxes [his cabinet and collection of shells] came tonight.

February 20, Monday: Interesting lecture from Prof this morning on Geological Periods. I liked it very much.

February 21, Tuesday: Busy at the Museum on arranging the Strombida [*sic*]. Unpacked part of my collection which has come. This evening at the Zoölogical Hall. Put and Ordway have got their rooms fixed and tonight they gave a supper. I went there this evening with my dressing gown and slippers. We had there Hyatt, Emerson, Scudder, Mr. Clark, Dr. Wheatland, Putnam, Ordway and myself. We played Dominoes, Cards, etc., and then went into the Hall where Oysters and other things were devoured with a relish. Afterward came the Liquors; I drank nothing but ale of which I consumed three or four bottles. Sung and told stories till one. Jolly time.

Samuel Scudder (1837–1911) was the son of a wealthy Boston merchant. He entered Williams College at the age of sixteen and became interested in natural history through a case of butterflies collected by another student. He was graduated from

Williams in 1857 and went to the Lawrence Scientific School to study under Agassiz, taking his B.S. degree in 1862 and remaining with Agassiz until 1864, when he became custodian to the Boston Society of Natural History. He was assistant librarian at Harvard, 1879–1882, and in 1886 was appointed paleontologist to the United States Geological Survey. He was a noted authority on fossil insects and a distinguished entomologist, whose writings reflected the charm of his highly cultured personality.

Theodore Lyman (1833–1897) was the son of the Boston mayor and philanthropist. He took his B.S. degree at the Lawrence Scientific School in 1858 and 1859 was elected one of the original trustees of the Museum of Comparative Zoölogy. His relations with the Agassiz family were of intimate friendship. In 1857 he was one of the Florida expedition, bringing back a fine collection of echinoderms and corals which he presented to the museum. He was abroad on scientific work from 1861 to 1863, but returned in the autumn of 1863 to serve as aide-de-camp to General Meade for the remainder of the Civil War. In 1866 he served as chairman of the Massachusetts Fisheries Commission, and in 1882 was elected to Congress on the civil service reform issue. His war letters were edited and published by George Russell Agassiz, grandson of "Prof. Agassiz," under whom Lyman and Morse were fellow-students.

H. James Clark was originally a favorite pupil of Agassiz's, a graduate of the Lawrence Scientific School and a skillful and original research worker in histology and morphology in both plants and animals. Between 1856 and 1862 he worked with Agassiz on his *Contributions to the Natural History of the United States*. Most of the histological and embryological portions of the work seem to have been his; and more than half the plates illustrating the embryology and histology of the turtles and acalephs have his signature. In 1855, when the Quincy Street house was building, Agassiz had a special stone pillar mounted in a north light for Clark's microscope, and again, in 1860, a special pillar was put into the Museum of Comparative Zoölogy. In the latter year Clark was named adjunct-professor of zoölogy at Harvard, undoubtedly through Agassiz's influence. There was little money, however, in all this appreciation and honor, and Clark, who had married, finally became embittered through financial stringency. He complained that his years of research and work entitled him at least to have his name mentioned as collaborator with Agassiz in the *Contributions to the Natural History of the United States*. Whatever the original verbal agreement had been, Agassiz felt that Clark was presumptuous in this claim. Clark, in despair and bitterness, in July 1863 published a pamphlet, *A Claim for Scientific Property*, maintaining his claim to recognition for his researches. Finally, upset by the constant altercations, Agassiz told the trustees of the college that either he or Clark must go. This required little deliberation on the part of the trustees; in 1865 the younger man was dismissed.[2]

Albert Ordway presented a dashing contrast to the figure of Clark, stooped over a microscope. Ordway was a born explorer whose war service led him away from a life of science, in which he showed real promise. Ordway had charge of crustacea at the Museum of Comparative Zoölogy and was doing notable work on trilobites when the Civil War broke out. He promptly en-

[2] On Clark, see Marcou, II, 52–54, and the *Dictionary of American Biography*.

listed and rose to the rank of brigadier general in the Union army.

The scientific societies of Boston were an arena secondary only to London for the battle of beliefs precipitated by Darwin's publication of his *Origin of Species.* As early as 1857, Darwin in a letter to his correspondent Asa Gray, Harvard botanist, had postulated his theory of the origin of species. Gray is frequently quoted by name in *Origin of Species,* which appeared in book form on November 24, 1859, and which had been summarized for the Linnean Society in July 1858. Asa Gray was a pillar of the Boston Society of Natural History, as was another early supporter of the Darwinian theory, William Barton Rogers, the geologist. The two had many an argument on Darwinism with their associates in Boston.

Agassiz, following up preliminary brushes and skirmishes over the new idea, read the *Origin of Species* critically and, on February 15, 1860, publicly delivered his verdict before the Boston society:

> Prof. Agassiz [say the *Proceedings*] made a verbal communication in opposition to the theory of Mr. Darwin, recently put forth in his work on the origin of species. Mr. Darwin he acknowledged to be one of the best naturalists of England . . . ; but this great knowledge and experience had in the present instance been brought to the support, in his [Agassiz's] opinion, of an ingenious but fanciful theory. . . . Animal representatives were as numerous and diversified in early geological periods as now; he instanced the brachiopods.[3] In the lowest beds of the Potsdam sandstone we find *Lingula prima,* and allied species are found in the Silurian, Devonian, Carboniferous, Permian, and Triassic, . . . up to the living species. . . .

[3] This remark by Agassiz influenced the entire subsequent career of his pupil, Edward Sylvester Morse.

About forty species of fossil *lingulae* are found in these beds; only seven species of living *lingulae* were known to exist until he recently added an eighth (*L. Ravenelli* Ag.) from South Carolina. . . . He thought the persistence of this form through so extensive a period, the last no more perfect than the first, was a fatal objection to the theory of gradual development.

Prof. Rogers admitted that the persistency of lingula, and other similar cases . . . , were formidable objections to this theory; but he thought that Darwin would meet such objections by the fact that the vital characters of some animals fit them for resisting change and extinction better than more plastic natures . . . ,[4]

and so on, at some length.

Jules Marcou, friend and scientific associate of Agassiz, wrote in his *Life, Letters, and Works of Louis Agassiz*:

Under the leadership of the botanist, Asa Gray, and the geologist, William B. Rogers, almost all the naturalists [in Boston] became at once strong Darwinians. . . . Agassiz was too much a naturalist to accept a number of mere suggestions until they were scientifically proved by exact observations. . . . At the meetings of the American Academy of Sciences and of the Natural History Society of Boston in 1860, objections against the acceptation of Darwin's theory led to several debates between him and Asa Gray and William B. Rogers.[5]

Some of the public criticized Agassiz for retaining the Harvard chair of geology instead of asking Rogers to take it. At any rate, Marcou continues:

Darwin's "Origin of Species" became a thorn in his side. His pupils in a body turned against him. . . . To the disgust of Agassiz, they turned from their master to applaud all the articles on evolution and origin of species, published in American periodicals by Asa Gray, Chauncey Wright, and John Fiske, the last two not even naturalists.[6]

[4] *Proceedings*, VII (1861), 231–232.

[5] II (1896), 106–109.

[6] *Ibid.*, II, 109.

Marcou is incorrect in at least one detail of his memories. Ned Morse did not turn against Agassiz to follow Darwin. He faithfully and literally worked for years testing for himself the theory of evolution in the very class which Agassiz had proposed as an example — brachiopods. It was not until 1873, at a field meeting of the Essex Institute, that Morse finally and publicly announced himself a believer in the theory of Darwin.

It is almost impossible for men and women of the twentieth century to conceive how stern a mold of Calvinistic orthodoxy surrounded New Englanders in the old days. In Boston the rise of a wealthy, leisured class had already cracked the mold. The accumulation of large private libraries, the pursuit of scholarship, and travel abroad had created a nucleus of liberal-minded spirits, some like Gray and Rogers pacemaking with Darwin's mind, others ready to accept at sight. In the rural communities, however, untouched by travel or European culture, Calvinism made a last-ditch stand. Its followers believed dogmatically in the revelation of the Bible; they embraced passionately stern theological strictures on "free thought." When Deacon Johnathan opposed, with parental injunction and audible prayer, his son's collecting shells, dissecting, and studying under the microscope, it was not in a jealous spirit. He sincerely believed, and all the power and formalism of his church endorsed and tutored his belief, that to fly in the face of God and the Bible was to be condemned to the eternal tortures of hell-fire. He opposed natural history to save the soul of his child.

It was written in the Bible: "In the beginning God created the heaven and the earth. . . . So God created man in his own image. . . . And out of the ground the Lord God formed every beast of the field and every fowl of the air; and brought them unto Adam. . . . And the Lord God caused a deep sleep to fall upon Adam . . . and he took one of his ribs and closed up the flesh instead thereof. And the rib, which the Lord God had taken from man, made he a woman. . . ."

That was the revelation in which Ned Morse had been nurtured for the first twenty years of his life. To understand him as he was in 1860, it is necessary to list the things he had never known. Ned Morse, educated in a middle-class home and in the rural academies of Maine, had never read a book on history, or philosophy, or art. Architecture, music, and sculpture were terra incognita to him. Living in Cambridge, in the year 1859, he never knew, by sight or by reputation, the liberals of the day who passed him on the street. He had not read the writings of Hawthorne, Emerson, James Russell Lowell, Thoreau, Bronson Alcott, Prescott, Motley, Longfellow, Whittier, Oliver Wendell Holmes. Ole Bull, Horatio Greenough, Samuel G. Howe and Julia Ward Howe, Charles Eliot Norton, might have been living on another planet, instead of within ten miles of the Museum of Comparative Zoölogy, so far as Ned Morse was concerned.

It was no affectation, no priggishness, that filled the pages of his journal with references to religion. His life, up to his twenty-first year, had been like one of the brooks of his native Maine in springtime; a foaming, flooding torrent coursing beneath a crystal sheet of ice that shut it from the universe without. Every ray of light and warmth, every idea and thought, reached him only through the perverting refraction of that cold, crystalline barrier.

This Maine boy, striving solitary, thwarted, poverty-stricken, to burst the icy bonds of theology, had an extra barrier set before him in the one channel he had been able to find. Agassiz, the teacher he adored, the great master to whom he had given allegiance, was himself fettered by the glacial bonds of Calvinistic theology. Agassiz was the son of a Swiss pastor, and Agassiz could never bring himself to accept Darwinism, to reject the first chapter of Genesis.

Men of Calvinistic stock, whether in the mountains of Switzerland or on the stern and rock-bound coast of New England, had all the faith-justified cruelty of a Torquemada, the rigidity of an inverted Golden Rule, forcing others to do as they did. Perhaps only born New Englanders can understand the staunch, stark, stern rule of life according to which the Pilgrim Fathers founded a colony and their descendants lived in it. Its terrible strength, maintained through nearly three centuries, lay in its self-righteousness, its consciousness of rectitude. It was a Herculean task to oppose a man who believed, completely and sincerely, that Jehovah personally carved his code in letters on stone tablets.

Ned Morse cannot be understood until one forgets the Boston, Cambridge, and Concord groups of the eighteen-fifties, with their famous names, and recognizes Morse as typical of the great majority of New England youth of his day.

Meanwhile, all unconscious of such implications, he was for the moment merely a homesick youngster with a bad cold. For more than a year previous, the journal from time to time has recorded a clear picture of infected tonsils, although the streptococcus was as yet an "undescribed species" to medical science, like the trout which Agassiz's students had so blithely eaten. Ned wrote to his mother on February 19, 1860: "Here I am in Cambridge all stuffed up with a cold, sore nose, eyes all bunged up, throat sore and everything uncomfortable." A week later, as another letter shows, he was still not well.

February 26, 1860

I have been in the house confined with a bad cold. Mrs. Cleaveland has been a mother to me, sending me nice tea and the nicest of home-made bread which is a luxury after eating baker's chips. Our rooms are papered and tomorrow Ad and I go in to buy carpets. Now I want you to send by bed and bedding by next Thursday. Putnam and Ordway have their rooms filled up and are in them. Mr. and Mrs. Glenn are all settled downstairs and in a week we shall have our breakfasts and suppers furnished us and it will not be long before we shall have our dinners also in the same way.

Prof was in the building where our rooms are yesterday and said, "How I wish I could have a room here with you, too."

February 29, Wednesday: This evening I went with Lina into Aquarial Gardens, Boston. Had a splendid time. I admired to look at her when she was watching some motion of the fish. She is a splendid girl and one which has a rare mind. Very busy at the Museum today.

March 1, Thursday: Our carpets are being put down and Verrill and I have been busy buying things, furniture, etc., for our rooms. Arranging my collection in the showcases. Left Mrs. Magee.[7]

March 2, Friday: My collection came out from the boxes in capital order. I have them most all arranged, and Prof likes the looks of them first rate.

During the week-end he wrote to his mother about his new quarters.

March 4, 1860

Dear Mother: Send immediately pillows, pillowcases, sheets, blankets, comforter and towels. I shall buy me a bedstead and mattress

[7] The boarding place where he had eaten since November. Meals were now to be provided for the assistants at Zoölogical Hall.

in Boston tomorrow afternoon. Our rooms are all fixed up in grand style. Prof gave us sixty dollars apiece to furnish our rooms and we fixed them up in style. On our study we have a splendid dark green carpet, $1.00 a yard. Our paper is very handsome. We have a nice large new black walnut bookcase, a lounge, two cane chairs and black walnut tables. In our bedroom we have a large wardrobe bureau with six drawers; wash stand and so forth. I tell you, we are comfortably situated. My Cabinet is entirely unpacked and Prof expressed his admiration at their coming out so safe, not a single shell was broken or misplaced. I want you to send those things immediately, send them by the Tuesday night boat. O, those doughnuts were delicious. I . . . nibbled them one at a time, and that maple sugar is so jolly!

March 5, Monday: Prof exhibited a large and elaborate geological map of France and explained it . . . as clear as anything. . . . Slept in the room for the first time tonight.

March 6, Tuesday: Received from Prof $88.00. [Paid] Mrs. Cleaveland [for lodging] $2.50; Magee [for board] $30.00; tables, $15.00; Association [Philadelphia Academy of Natural Sciences for diploma of membership] $5.00; Crockery [for washstand] $1.13. Had our breakfast and supper in our big room [the common lounge, Zoölogical Hall].

Ned was so busy getting settled in Zoölogical Hall and working out the genera of *Ceritheum* that his journal for the next fortnight is blank. He wrote to his mother, on March 11:

We have at last got settled, and I tell you, we are perfectly happy. Our rooms look as nice as any in Cambridge. We have a splendid black walnut bookcase, two nice black walnut tables, a black walnut mantel and a lounge, two nice large armchairs in our study. Our carpet is a dark green and black check and looks tiptop. We have nice chintz curtains. I wish you could peek in and see us. We have our breakfast and supper in the Hall and shortly we shall have our dinners here also.

I would have written you a longer letter but I have been writing all day on a paper to read before the Club [informal organization of Agassiz's student-assistants] and my hand is so cramped I can hardly hold my pen.

Writer's cramp is a disease unknown to the twentieth century, with its fountain pens and typewriters. In the fifties men were just giving over sharpening quills to use the new steel pens. When Whittemore at the customhouse slipped a couple of Uncle Sam's steel pens into Ned Morse's pocket, it was enough out of the ordinary to be entered in his journal.

March 19, Monday: Procrastination and this is the effect; two weeks that I remember nothing of. Have commenced and filled twenty cases of Mollusca, all catalogued and arranged generically with the species seperated [*sic*]. . . . Yesterday found *Helix milium* in woods back of Divinity Hall; *Astericus* also, Cooke has from Salem. Saturday Prof came in and spent the evening. He took off his hat and coat and passed his nice cigars around and chatted away there in nice style. It is very pleasant to be associated with one in such a way. He is a great, good man.

March 20, Tuesday: Received paper [Portland *Transcript*] from Father today. . . . Attended lectures of Prof Wyman and Prof. Agassiz. Prof gave his first lecture in the Museum, over a hundred students present. Drunk. Not me but the rest.

Jeffries Wyman (1814–1874) was graduated from Harvard in 1833 and from the Harvard Medical School in 1837. In 1840 John Amory Lowell, trustee of the recently established Lowell Institute, made Wyman curator and one of the first lecturers, and from 1847 he was Hersey Professor of Anatomy at Harvard. Somewhat later he was also associated with his brother, Morrill Wyman, in conducting a private medical school in Cambridge.

In 1858, with Captain John M. Forbes and George Augustus Peabody, he made an exploring trip to South America, up the La Plata River and across South America to Valparaiso, returning by way of Peru and Panama and bringing magnificent collections to the Harvard museums.

He was a great friend of Asa Gray and of Agassiz, and, beyond his medicine and anatomy, cultivated an interest in shells which, through shell heaps in Maine, Massachusetts, and Florida, merged into an interest in archaeology. On the foundation of the Peabody Museum of Archaeology and Ethnology at Harvard in 1866, he became its first curator. He died of tuberculosis in 1874. He had been president of the Boston Society of Natural History from 1856 until his disease necessitated his retirement in 1870.

March 21, Wednesday: Worked at Museum. . . . This evening at meeting of Boston Society of Natural History. Prof Agassiz and Rogers discussed Geology all evening.

The *Proceedings* of the Society show that on March 7, 1860, Agassiz had elaborated on his opposition to Darwinism: "Professor Agassiz made a communication on consecutive faunae and their corresponding number of geological formations, as furnishing arguments against the theory of Darwin." [8] Professor Rogers rebutted vigorously. On March 21 the Society's secretary noted: "Prof. W. B. Rogers explained his views of the stratigraphical relations of deposits formed in an ocean, under each of the three conditions of a *stationary*, a *subsiding*, and a *rising* position of the sea-bottom. . . ." [9] Professor Agassiz, however, maintained that the facts did not justify the conclusions.

[8] VII (1861), 241.

[9] VII (1861), 246.

In a letter dated March 20, 1860, to his older brother Fred, Ned Morse enumerates the activities of a student-assistant:

I am attending Prof Wyman's Lectures on Embryology which occur every Tuesday and Thursday forenoon; also Prof Agassiz' Lectures which occur on the same day in the afternoon, beside the lecture which he gives to us [his student-assistants] each Monday forenoon. We all take notes of these lectures and write them out and it keeps me pretty busy. Beside our regular work and study at the Museum, we go in to the meeting of the Boston Society of Natural History twice a month on Wednesday nights. And then we have formed among ourselves a society which we call the Agassiz Zoölogical Club. We meet every Thursday night and take turns in preparing a paper. I read one at the last meeting on Physical Causes. Enclosed is a photograph of Agassiz. Price 25 cents if the rest of the boys want one.

The journal continues:

March 22, Thursday: Attended Prof Wyman's lecture today; very interesting. Subject, the brain, and he is giving us fine things in this line. Prof Agassiz lectures, too. . . . I note [these] down in a separate book. . . . Meeting of the Club tonight; Hyatt read an interesting paper on Ammonites and considerable discussion arose from the remarks made by Agassiz and Rogers last evening [on Darwinism].

March 23, Friday: John Richardson [of Portland] here to see me this morning; he is now boarding at Father's at home; he has my room. I showed him round the Museum and then bid him goodbye. Going to dinner who should I meet but [Ned] Bicknell [of Portland]. I showed him to my room and after dinner showed him over the Museum. He was exceedingly delighted with the collection and appeared to be well acquainted with almost all the large fossils. After awhile he left and I resumed my work, that of lugging shells from the students' gallery into the room above. Horace Mann, the son, is here as a student and Prof has put him on shells under my charge.

Horace Mann, senior, the educator, had died on August 2, 1859, while president of Antioch College, in Ohio. A native of Franklin, Massachusetts, graduate of Brown, and leading lawyer, in 1827 at the age of thirty-one years he began a remarkable public career. As member of the Massachusetts House of Representatives from Dedham, and later president of the Senate, he was responsible for four measures between 1827 and 1833: a temperance bill; the abolition of lotteries; the establishment of state hospitals for the insane; and the establishment of a state board of education. Mann became executive secretary of the latter board and revolutionized public education, founding the first normal schools for the training of teachers and advancing the standards of common-school curricula. In 1852 Mann, a Whig, was the unsuccessful Free Soil candidate for governor of Massachusetts, and in 1853 he accepted the presidency of Antioch. His son bore one of the best-known names in Massachusetts in that day.

March 24, Saturday: Cold and uncomfortable. Received $25.00 of Prof [his monthly salary]. Had a great walk today. Into Boston this afternoon at Aquarial Gardens a little while; saw live kangaroos for the first time. Met Lina and Mary and Clem there. . . . Shaler home today. Shaler and Foley called with Mr. Bolton, a friend of Shaler's. The two latter gentlemen, with Cooke, imbibed enormous quantities of Bourbon whiskey and after a while Bolton felt sick and he and Foley went home. Soon Cooke began to act decidedly groggy and after a while commenced to vomit on his clothes, our carpet, spittoon. Soon he merged into a beastly state of intoxication and in this condition we led him into his room, where he managed to get into bed. . . . Ah, it was a beastly sight. Bah!

Fine goings-on in the temple of science, with a pupil of Agassiz's as the victim, but Nat Shaler, fresh from Kentucky with a cargo of bourbon whiskey, was dynamite to these New England lads who had not been weaned on mint juleps nor taught to carry their liquor like gentlemen.

March 25, Sunday: Cooke today has kept his bed. . . . Played . . . chess with Shaler and beat him. To bed early tonight.

March 26, Monday: Prof spoke to us today on our duties, that it was up to us whether the museum should be a success or not. . . . Shaler's carpets are here and he will be here shortly.

"Prof's" little lecture went home to his pupils' consciences, and they devoted themselves soberly and industriously to their work the next few days.

March 27, Tuesday: Attended Prof Wyman's and Prof Agassiz' lectures today. Worked away on the Fossil Shells; they are mostly all arranged. At home this evening.

March 28, Wednesday: Still working on Shells in Students' Gallery; have arranged most of them. It is surprising how poorly and artificially Shells are grouped together in families and genera. Many Genera such as *Buccinum* and *Cytherea* contain at least half a dozen genera apiece. . . . Went to bed with a roaring headache.

March 29, Thursday: Attended Prof Wyman's and Prof Agassiz' lectures today, very interesting both of them. Got a hedgehog and a woodchuck in cages as commencement for a Zoölogical Garden. Ran down to Somerville tonight to a fire.

March 30, Friday: Splendid day, warm and bright, in fact Glorious. . . . At work hard at the Museum. I am getting deeply interested in grouping together different forms constituting genera. I find in the so-called Genus *Buccinum,* of which there are so many species, only three or four true *Buccinum.* This evening Ad and I went over to Somerville [to the Boarding School for Ladies]. I saw Sophia Whitman, Miss Howe and Miss Dennison. Had a first rate time. Sophia sang to me one of the vocal beauties of *Martha,* "How so fair Stood she there." Saw large fire in Medford on our return. Had a nice time and Shall Go There Again.

March 31, Saturday: [About 12 o'clock] Putnam came in and told me to look out of the window. . . . Saw the glare of a conflagration. After a while got up and went into the Hall with the rest of the boys; at last we concluded to go and so prepared ourselves; I put on my blue shirt and Put, Nat [Shaler], Cooke and myself started. Cooke and Put walked. Nat and I ran all the way. We crossed Brighton bridge, as the fire was in Brighton. The sight from here was magnificent. It looked like the miniature burning of Moscow [which he had seen in the Panorama]. It was two miles and a half from where we stood and yet the sparks reached us and we could feel the warm breath of the fire. On our way we met and passed firemen and others. I felt like running and after we got there I felt no bad effects at all from it. The sight was splendid, three or four buildings already burning and six others in ashes, besides a million feet of lumber in one grand sheet of flame. We came back about 3½ o'clock and sat up till almost five talking about it.

April 4, Wednesday: Democratic victory in Portland, Howard elected 48 majority; largest vote ever thrown in the State. How glad my Democratic friends must feel. . . . At the Boston Society of Natural History tonight. Prof. Agassiz and Rogers continued their discussion tonight and it was interesting.

The secretary of the Society wrote in his notes for April 4, "Professor Rogers reported that he had listened with extreme surprise to the statement just made by Professor Agassiz." Gray and Rogers did not spare their old friend in debate. Marcou has said: "Some of his [Agassiz's] antagonists were excellent debaters and they annoyed him exceedingly by shaking their heads, or even saying a few words aloud, disconcerting him," and Darwin wrote of Asa Gray, "a born reviewer, a capital reasoner, a poet, a hybrid, a complex cross of lawyer, poet, naturalist and theologist; was there ever such a monster seen before?" Gray himself, writing to a friend after one of these debates, said gleefully, "I knocked out the underpinning from Agassiz' theories about species and their origin, showing the high probability of single and local creation of species, turning some of Agassiz' own guns against him." [10] It was bear-baiting, in short, with Agassiz tortured far beyond his friend's understanding and the young men watching as gleefully as though it were a pit with a bear beleaguered by a band of bulldogs.

April 5, Thursday: FAST DAY. Most of us in the house this morning. Splendid day. . . . To walk.

This New England holiday was of religious nature, although by 1859 the religion had mostly evaporated. It originated as a Pilgrim custom in 1623 when, on July 16, a day was set apart for "prayer and devoute exercises" during a drought in Plymouth. The gratifying sequence of refreshing rains the next day, saving the summer's crops, influenced the Massachusetts Bay Colony later to designate a day in spring annually for prayer for the crops; Thanksgiving Day in the autumn was the corollary, to return thanks for the harvest. Fast Day endured in Massachusetts until 1894, when the legislature substituted for it Patriots' Day, April 19, anniversary of the Concord and Lexington battles that opened the American Revolution.

April 6, Friday: Mann and I have been at work on the Alcoholic mollusca [from Acapulco]. I find that I can get along much faster

[10] Marcou, II, 109.

with someone to help me. Tonight after I came home, I found Prof Agassiz in the Hall with Shaler and Verrill. I sat down and listened to his conversation which for a wonder was off of scientific subjects. He told us of the necessity of associating with intellectual and refined women and it was grand to hear him talk of Humboldt, Arago, Hugh Muller and a host of scientific men all of whom he has been intimate with.

Dominique-François Arago (1786–1853), a leading French physicist, through the friendship of Laplace, began his scientific career as an assistant at the Paris Observatory. His journey to Spain in 1806 to make meridional measurements culminated in his being taken prisoner off the Balearic Islands by a Spanish corsair and finally being rescued from Spanish prisons by the good offices of the Dey of Algiers. He subsequently became director of the Royal Observatory. During the Revolution of 1848, as minister of war and marine, his notable reforms included abolition of flogging in the French navy and of slavery in the French colonies. His chief scientific achievements were in connection with the polarization of light and experiments establishing the validity of the "wave theory" of transmission of light.

"Muller" was probably Johannes Müller (1801–1858), pioneer physiologist, professor at the German universities of Bonn and Berlin and author of the *Handbuch der Physiologie des Menschen.* His later years produced significant work on marine fauna and comparative anatomy which would have been for Agassiz "right up his alley," in students' idiom.

The journal gives no direct clue to the Agassiz lecture on the "necessity of associating with intellectual and refined women," but it is obvious that Agassiz took a genuinely fatherly interest in his students and was pretty well informed of their extra-curricular activities. Ned Morse evidently took literally any injunction of the master, for by the next day he had already equipped himself with an Aspasia.

April 7, Saturday: Started for Boston this morning to make drawings of live *Pecten Magellicanus* at the Aquarial Gardens. Lina went with me and I enjoyed myself very much in having her present.

Young Nat Shaler, also, had been deeply impressed by "Prof's" lecture.

April 8, Sunday: In the house all alone this forenoon. This evening in Shaler's room. He read aloud to Ad and me one of Theodore Parker's sermons.

It is unfortunate that the journal does not specify which sermon, but some of Parker's most powerful and noted sermons were built around his discovery of the fact that Boston in the fifties numbered two hundred houses of prostitution. "Prof," in commending to his students association with intellectual women, may have quoted Theodore Parker. It is difficult to imagine how else a gay young blade of old Kentucky would turn from game suppers at Parker's restaurant as a Sunday evening diversion to reading aloud from the sermons of another Parker, the Reverend Theodore.

In four weeks Agassiz had exhausted his enthusiasm for a Hudson Bay expedition; now he was interested in a collecting trip in the Caribbean Sea. Ned Morse wrote to his mother on April 1, 1860:

Have been rather negligent about writing for the reason that I have had just as much as I could do in the scientific line. I have five lectures to attend every week of which I write full reports, besides evening meetings of the Club. And now I have under my charge Horace Mann, a young student and son of the immortal Horace Mann. I am teaching him how to group together shells.

Our northern trip is indefinitely postponed. Prof has promised the Trustees of the Museum to have the collections arranged by the first of September and in order to do that we shall have to work like sin. A gentleman has offered to Prof that he will furnish a nice vessel and pay part of the expences [*sic*] to enable us to spend the winter in the West Indies for collecting; and Prof has accepted the offer. The Museum will pay the rest of the expences. It is the intention now to cruise all around the Island and in the Gulf of Mexico.

The journal returns to daily routine.

April 15, Sunday: Very cold and pleasant. This week I have worked on alcoholic Mollusca down cellar, assisted by Horace Mann. It will take a month yet to arrange all of them. Everything goes along nicely in the Club. Our dinners are nice and our apartments and departments comfortable.

April 16, Monday: Prof gave us a lecture preliminary to one he is going to give next Monday on Varieties. It was Truth and I lamed my fingers taking notes.

April 18, Wednesday: Pleasant day. Society's meeting tonight.

The deep absorption of Agassiz's students in the Darwinian debates is plainly shown by Ned's terse record of the meeting of the Boston Society of Natural History when, according to the secretary's minutes, "Mr. [Paul] Du Chaillu gave an account of the habits of the Gorilla and other anthropoid apes from personal observation on the west coast of Africa." No gorilla could outweigh for an Agassiz student the charms of brachiopods as evidence for or against Darwin's theory of the origin of species. A little professional jealousy may have entered into it, also. Young Paul Belloni Du Chaillu had just returned from an expedition sponsored by the Philadelphia Academy of Natural Sciences, in which he had explored the Ogowe River and the Gabun delta on the west coast of Africa. He had brought back with him the first gorillas ever seen in America, and his books on the gorillas and Pygmies of equatorial Africa's forests were among the best-sellers of their day.

In April of 1860 Agassiz must have been well content. After a hurried summer trip to Europe and seven months of tremendous labor in getting his new museum under way, he had everything organized and running smoothly. His invaluable collections had been removed from the hazards of the old laboratory to the new fireproof building; his enthusiastic, brilliant staff of young assistants was settled in their barracks next to the museum; his routine of lectures was in full swing. Ned Morse gives the picture in a letter to his mother, April 8:

We have had a space behind the house [on Divinity Avenue] plowed up for a garden where we are going to raise tomatoes and squashes. Our breakfasts, dinners and suppers are luxuries now; everything is cooked up in the best manner and everything is of the best. Several of the Professors have visited our building and looked at our rooms and say they are not only the best rooms in the University but the best in Cambridge.

Night before last Prof was here and spent the evening. He talked of some of the distinguished men of Europe, told us anecdotes of Arago, Humboldt, Muller and others. You can't imagine how interesting it was to hear one talk who has been intimate with those.

Yesterday I made a drawing of *Pecten Magellicanus* alive in the Boston Aquarium. You must come up this way and see the Aquarial Gardens and the Museum.

Prof was in this afternoon. He comes in, perches himself on the end of the table and chats away as familiar as any of us. He told me he was going to put me on dissecting shortly. He wants me to determine the families in Mollusca which are now made out very poorly. . . .

We had a clubmeeting tonight and have discussed various questions in Natural History. Afterwards we turn to and drink ale and have a social, pleasant time. We hold these meetings every Monday evening; all the members composing it belong to the Museum and the papers read here are of the right stamp, namely original investigation.

Everything goes along gloriously and I now fairly begin to realize that I am a fixed institution here from the manner in which Prof lays out my studies.

The wet weather we have had so long cleared up today [April 17] and is now hot and sultry. Grass is all up and leaves bursting out in high style. Last night we had our regular Club meeting. It was Putnam's birthday, 21 years old. So when the meeting adjourned, Putnam asked them to leave the hall and come into our room for ten minutes. Then Put asked us out again and we were all surprised to find the large club table laden down with as rich a supper as I ever saw in my life. The Piano was piled with bottles of wine, all kinds; we had oysters, a rich pudding and everything one could desire. There were only eleven of us Museum boys and we had a rousing time. Danced, sang, played leapfrog, fenced and made a rousing racket. Did not get in bed till two o'clock this morning. We live a big sight happier than kings. We have got our garden ploughed, exercise in it every evening an hour. I have not drank any tea or coffee for almost two weeks and I have kept my pledge on liquors and always intend to.

I shan't get a chance to send the medicine by a friend; will try to send it when I send my Cabinet to John [having removed his own shells to add to the collection of the Museum of Comparative Zoölogy]. I shall have my photograph taken soon; Whipple of Boston takes them for 25 cents apiece and don't charge for the negative. I shall come home, I think, the 29th of May as there is a recess of the college and I shall have no lectures to attend. No use . . . I can't bring my mind down to a letter to save my gizzard. Nothing but [three sketches of shells] are in my mind. So I must wait until I see you.

April 23, Monday: Lecture by Prof this morning on Varieties. Working like sin in the cellar. . . . Wrote Mother.

April 24, Tuesday: Labelled a lot of bottles today. It will take Mann and me two or three weeks more to catalogue all the alcoholic specimens of Mollusca. . . . This evening saw the eclipse of Venus by the Moon.

April 25, Wednesday: Cold; awful cold. Sent [Cabinet] drawers to John. For the past week we have worked in the garden.

"Prof," with Swiss frugality, being obligated to furnish board and keep, as well as salaries, out of his own pocket to his student-assistants, not only served up moose steaks and new species of trout from the museum specimens but set his young men to the healthful exercise of planting a vegetable garden on Harvard University grounds.

April 26, Thursday: Splendid lecture by Prof this morning on the absurdity of believing that Adam and Eve were the first created and the only ones. It was a masterly lecture and was listened to with great attention.

The barbed witticisms of Asa Gray and William Rogers had finally driven Agassiz to jettison the first chapter of Genesis, although he still refused to install Darwin as author of a new Scripture. The debates at the Society of Natural History had reached such a pitch that "Prof" felt it necessary to try to hold his pupils' belief by rebuttal in the classroom of shafts launched in the Natural History rooms.

May 1, Tuesday: Morning not very promising to the May parties. Cloudy and unpleasant. Verrill's Father and Grandfather came here last night. Got up this morning with mouth tasting . . . as if I had come up on the Boston boat from Portland. Eat one piece of bread and a glass of milk. After breakfast back and head began to ache like the Old Scratch. Felt shivery and miserable. Laid down on the lounge and about noon went to bed. Dr. [Jeffries] Wyman was called and said I had sore throat, which was a fact. He said I might drink as much as I wished. Very restless.

May 2, Wednesday: In bed all day. . . . Eat nothing but an orange.

May 13, Sunday: Intensely hot today. . . . At five o'clock I walked up to [Lina's] house. It made my legs ache terribly. Staid to tea and eat a lot of nice homemade bread. After tea she walked back with me. My back ached tremendously and I had to stop every five minutes; thought I should have dropped down. Lina was very anxious about me. When I got home, I flung myself on Nat's lounge, smoked two cigars and felt an intense relief.

May 14, Monday: Over to the Museum a little while. How strange it is that I have such a hanker for Spinnage [*sic*], Asparagrass [*sic*], and other vegetable greens.

Not particularly strange, since he had just passed a winter on a New England diet of pie and doughnuts for breakfast, crackers and milk for supper, meat "all cooked up" and baked beans for dinner. He was as near scurvy as any of the Arctic explorers of the era. Dr. Elisha Kane, setting off in 1852 with his men on the hermaphrodite brig *Advance* for three years in the Arctic, laid in "two thousand pounds of pemmican, salt beef and salt pork of the navy issue, hard biscuit and flour." He admits candidly, in retrospect four years later, that "our stock of fresh meat was too small and we had no preserved vegetables; but my personal means were limited." Vitamins did not enter into dietetics until McCollum and Davis in 1913 published the results of their dietary experiments on rats.

CHAPTER FIVE

BUTTER-PAT REBELLION
1860

CHAUCER, or the old Scot, or even King Solomon of the orthodox Scriptures could have warned Ned Morse of what was about to happen to him. "For lo, the winter is past, the rain is over and gone; the flowers appear on the earth; the time of the singing of birds is come, and the voice of the turtle is heard in the land. Arise my love, my fair one, and come away," sang Solomon. But Ned Morse's biology was concentrated on the embryology of *Lymnea* and the lingual ribbons of *Helices*. He went to call on Ellen Owen and only noted casually, "Shall call again if I get time." Something had happened to him, but he could not diagnose the strange pain.

May 22 [1860], Tuesday: Raining hard which is much needed in this vicinity [for the Zoölogical garden of vegetables]. . . . Got out sections of the Lingual of *Physa heterostrophe* and *Planorbis embus* or something like that. *Physa* teeth arranged so [sketch].

May 23, Wednesday: Splendid day. Trees out in grand style and Summer has fairly come. At the Museum a little while. . . . Abbott Cummings [of Portland, working in Boston] out this evening and we called on Miss Owen at Mr. Richardson's, Dana Hill. I made her acquaintance at a dance last February and have not seen her since. She is a fine girl and I like her appearance very much. Enjoyed myself tiptop. Shall call again if I get time. Invited her to go into the Aquarial Gardens Friday.

May 24, Thursday: Blue and sick. . . . Heard band play in the square. The music affected me strangely.

May 25, Friday: Went into the Aquarial Gardens with Miss Owen at nine o'clock this morning. Had a nice time in her company. Mr. Cutting [proprietor of the Gardens] has got home. He expects his collection in a day or two. We were there a long time; then we went to Williams & Everett's and saw the pictures; returned to Cambridge at noon. . . . Walked home with Prof. He said that it would be better if there was no Natural History Society in Boston because we go in there and see at once that we know more than they do and it makes us rambling in our studies.

May 26, Saturday: Received letter from Fred [Morse] saying that Father was taken quite ill. Shall go home Monday.

May 28, Monday: Prof Agassiz' Birthday today; he is 53 years old. Packed up things and started for Portland. Nothing important occurred on going. . . . Felt glad to get home. Mother told me that Father had had a fit from which she thought he would not recover. I went up and saw him. He laid in bed on his right side, showing the first symptoms of paralysis. He shook hands with me but did not say much. . . . I sat on the side of his bed some time. He smiled often but that was all.

Several pages of the journal have been left blank at this point; Ned Morse never could bring himself to fill them up afterwards. Between the blank pages lies a small paper packet with a dusty brown withered spray of myrtle leaves and immortelles from an old funeral wreath.

His father, Johnathan K. Morse of the firm of Byron Greenough & Company, died on June 20, 1860, aged fifty-eight years and was buried on June 21, with a funeral at three o'clock from his late residence, 29 Spring Street, Portland. "The hat stores in the city were all closed this afternoon as a mark of respect to Mr. J. K. Morse, one of the firm of Messrs. Byron Greenough & Co., whose funeral took place this afternoon, the dealers attending his remains to the grave," said the Portland *Transcript.*

"He did not speak much. . . . I sat on the side of his bed and he smiled often," wrote Ned Morse.

There was no use saying much, Deacon Johnathan must have thought, lying there a-dying, with his scamp Ned beside him — Ned whose scientific ardor had turned his young ambition from business to ill-smelling, profitless shellfish; Ned whose fiery sceptic mind drove him from his father's church to follow the snares of the Devil in spiritualism, Theodore Parkerism and Darwinism; Ned whose restless spirit had led him far from Portland and the fur business and the family. Charles, the beloved first-born, had died in his bright youth. Fred, steady and responsible, would take on his shoulders the burden of watching over widowed mother and young sisters. What was there for Johnathan to say to Ned? He only smiled often, and Ned understood at last what his father was trying to tell him; understood it with a pang that halted his pen when he tried to write about it in his journal.

So Johnathan Morse died and was buried. The Spring Street home was to be broken up. Fred would help the mother and the girls move to the little village of Gorham. Ned Morse went back to Cambridge, immediately after his father's funeral. His hopes of a "glorious" expedition to Hudson Bay had been dashed by the trustees' insistence that the arrangement of collections in the museum be completed for public exhibition by autumn.

June 23, Saturday: Arrived in Boston this morning from Portland by boat. Was not sick a bit. Had a delightful ride to Cambridge in an open car; shrubbery and all make it so perfectly delightful. Cambridge is perfectly charming . . . , trees, hedges and all. Boys all glad to see me and I to see them. . . . At the Museum; don't seem to be much work going on.

June 25, Monday: Tiptop lecture from Prof today on Darwin. Arranged a large collection of Shells received from the Smithsonian Institution. This evening chased by two robbers.

June 27, Wednesday: At work sorting shells . . . species . . . from Smithsonian.

June 28, Thursday: Prof Agassiz' son returned home this morning. Prof introduced me to him. Fine-looking fellow.

Alexander Agassiz (1835–1910) had taken his degree of B.S. at the Lawrence Scientific School in 1857 and for the past year had been with the United States coast survey, of which Alexander Dallas Bache was chief. He was to carry on his father's tradition notably as a marine biologist and, simultaneously, display great business acumen in the Calumet and Hecla copper mines in the Lake Superior region. Of the fortune thus amassed, he gave liberally for the continued development of the institution to which his father had devoted so many years and so much work, the Museum of Comparative Zoölogy.

June 29, Friday: Verrill, Hyatt and Shaler started for Trenton, Maine, this afternoon.

Verrill and Hyatt had perhaps the evenest natures and the most earnest temperaments among the student-assistants. In their quiet way they had contributed much to the smooth functioning of Zoölogical Hall. Agassiz, at this time, in addition to the excitement of his son's return with great accounts of the explorations on the Mexican coast, had just been through the trying period of the Harvard Commencement and was engaged in another campaign to secure funds for the museum. With Verrill and Hyatt gone, a domestic revolution broke out, involving Mr. and Mrs. Glenn, who managed Zoölogical Hall.

July 2, Monday: A day that will always be remembered by the students connected with Zoölogical Hall. After dinner Cooke and I were with Putnam and Ordway in their room when Katie [Mrs. Glenn], our cook, came up crying, saying that Glenn had insulted her about us. We of course went down immediately and demanded of him what we had done thus to provoke him so against Katie. He said that Put had spoken disrespectfully at the table because Put told Bertie that the butter should be brought on in smaller pieces as it would please Prof Agassiz when he dined with us. This was all. Putnam, in his love and respect for Prof Agassiz, saw that huge, unwieldy lumps of butter would not compare with the supposed nicety of Prof's former living.

Of course then Put asked Glenn why he should be so harsh on Katie for that. Glenn only replied by telling us that we were mean, snivelling scoundrels and a pack of rascals that had conspired against him. He got into such a way and with such foolish and absurd threats that we could only treat it with contempt. His wife, even, caught hold of him and told him to stop as we were all laughing at him. Glenn shook his fist at us, threatened Put with death. Says he, "A man never grappled with me but to the death!" Stamping his foot, as he spoke it, says he, "I have got you; you can't do a thing," implying that we were under his foot and he should keep us there, in language that could not be misunderstood.

We immediately came to the conclusion that, unless Prof removed him, we would vacate the house, never dreaming for a moment that Prof, after listening to our story, would not expel Glenn at once from the house, thus getting rid of a man who is heartily despised by Guggenheim, Owen, Hansen, Verrill, Foley and Katie, in fact, by all. His disagreeable qualities so emanate from him that persons intuitively form a dislike to him. Livermore, the expressman, told me that he always disliked the looks of the man.

Ordway was sent for by Prof and this evening he went. To Ordway's grief and astonishment, Prof after having listened to Glenn would not hear one single word from us, simply saying that if it did not suit us we could leave. That Glenn was a valuable man and was put in as housekeeper and that we should everyone of us leave if we did not like it. This from Prof, who [*sic*] we have loved and adored as a Father! Have done everything to oblige and please him! And now, when we run to him as a family would to its head for redress and advice, he tells us to hold our tongues, that he will not hear a word from us. Is this morally right? Is this consistent with what he has taught us in his lectures? No, it is not; and now Prof has fallen in our eyes from a God to a man, even to the lowest and most unreasonable of all; for men who have no brain are not expected to listen to reason.

"Prof" Agassiz had schooled himself to bear slings and arrows from a board of trustees, a state legislature, a society of Darwin-minded scientists, but rank insurrection and mutiny over a pat of butter was too much for him.

July 3, Tuesday: All of us highly indignant in regard to the Row. We all went over to the Museum this morning, Foley with us, who is on our side. We sat together talking about it when Prof came in and commenced talking with a good deal of feeling and a little anger. He said he was very, very sorry that this had occurred, etc. Putnam then began to speak but Prof shut him up as before, saying he would not hear a word. He went up [stairs] and we immediately came to the conclusion that if much more injustice was heaped on us, we would leave the Museum. Jim Foley talked with Prof two hours this afternoon and spoke very plainly with him. Previous to this, however, we had given all the Provisions, tablecloths, napkins, etc., to Katie; and Prof had come in perfectly beside himself with rage and taken the Museum keys from Putnam, Ordway and Cooke. Foley offered his keys, which led to the talk.

July 4, Wednesday: Into Boston this morning; followed soldiers round a little while and then up . . . to see Steve [Russell] and his wife. . . . Passed a very pleasant time; stopped to dinner. . . . Then we saw the Balloons go up from the Common. Three of them went up.

July 5, Thursday: Not much work done at the Museum. Prof is still unjust. Last Tuesday Prof dined with us for the first and last time, as it was the last dinner of the club in the house. He is very agreeable company. It seems now as if Prof was going to turn us all out of the house, so outrageous does he act.

Poor "Prof" had evidently invited himself to dinner at the Zoölogical Club the day after the Butter-pat Insurrection began, hoping by eating the butter, whatever the size of the pat, and ignoring the topic to bring his young rebels to their senses. He had about as much success as statesmen do in conferences with belligerent minorities.

July 6, Friday: Still unsettled. Put and Ordway have engaged rooms out of the building. Lina invited me up to her grandmother's so . . . I walked up to North Cambridge. They live in an old-fashioned house . . . surrounded by trees. The grandmother is 82 years old and very active for one so old. Aunts are nice ladies. I eat cherries, walked with Lina, and had a nice time; we had a grand good supper, strawberries, ice cream, brown bread, and everything cosy. . . . We returned about 10 o'clock. When I got back Put said he had received a letter from his folks in which they advised him not to leave the building for Prof's sake; which advice he took in the right way.

July 7, Saturday: Putnam told Prof of his intention of remaining in his room which pleased Prof immensely. He told Put he should be friends always. Lina and I went to the Aquarial Gardens this afternoon. Saw some living tropical shells for the first time. I could not help remarking the beauty of the eyes of the *Strombus* and spryness with which he handled his proboscis. Most of [Cutting's] Florida things are dying.

Strombus is the large tropical conch shell. Morse's note about "the spryness with which he handled his proboscis" indicates that he may have been studying specimens of *Strombus pugilis*, commonly called "fighting conch."

"Prof" had now entered on the Reconstruction era of the Butter-pat Insurrection. The students were to continue to lodge in Zoölogical Hall but had to find boarding places outside.

July 9, Monday: [Commenced eating] at Mrs. Woodbury's today. . . . Opened large collection from Mendocino, collection by Alexander Agassiz.

July 10, Tuesday: Went in swimming tonight for the first time this year. . . . The water was warm but I felt weak on leaving it.

July 11, Wednesday: Up at half past four this morning, which is bright and splendid. Foley and Levitt down here this evening; we had songs and a good time. . . . Killed Putnam's dog with chloroform. I tried it on a striped snake and it caused his neck to scrivel [*sic*] up in a surprising manner.

July 12, Thursday: Helped Foley skin the dog. Up at six this morning. . . . Letter from Horace Mann, he will be home tomorrow.

July 13, Friday: Splendid morning. Finished work on Smithsonian shells. Museum is going to have a lawsuit in regard to glass jars from Germany. Almost every one leaked badly and hence was not suitable for the purpose intended. Lost my meerschaum pipe.

July 14, Saturday: At breakfast this morning, Mrs. Woodbury informed me of an incident which occurred to the little girl who waits on table. The little girl was washing her hair in alcohol when it caught on fire. Mrs. Woodbury in extinguishing it burnt her arms. The little girl burnt her head very bad. I watched with them tonight.

July 15, Sunday: Wrote letter to Mother. Watched again tonight.

It was an old New England custom, "watching" with the sick in the days before nurses were trained or hospitals largely patronized. The underlying reason was the family's fear of being alone at night if death occurred, but it degenerated into a morbid social expression that had little comfort or aid for the dying person. One has only to read of the number of persons who crowded into the little Washington bedroom opposite Ford's Theatre, to stand for seven hours while Abraham Lincoln was breathing his last, to realize the intolerable invasion of privacy that had become a social rite.

July 16, Monday: At Museum this forenoon; felt drowsy all day. Tonight slept sound.

July 17, Wednesday: Commencement Day. Saw the Lancers in escort to Governor Banks who came out. Saw Douglass and he truly is a "little giant," a head perfectly enormous. Sumner I disliked the looks of very much. Banks is again truly named — "the Iron man," for if anybody exhibited firmness he does.

United States Senator Stephen A. Douglas only two years earlier had stumped Illinois in a series of public debates with his rival candidate, Abraham Lincoln, on the slavery question. Lincoln's theme was expressed in his convention speech: "He [Douglas] had done all in his power to reduce the whole question of slavery to one of a mere right of property." On that thesis, Douglas, by popular vote was acclaimed the winner, being elected senator; few could realize that the debates in which Lincoln was defeated would ultimately elect him to the presidency of the United States. Douglas in June 1860 was a recognized candidate for the Democratic nomination for president.

Harvard University displayed a scholarly impartiality towards politics in having Douglas, the "Little Giant," on the same platform with Governor Nathaniel P. Banks, the Free-Soiler who in eighteen months more would be Major General Banks of the Union Army. Banks had earned the "Iron Man" sobriquet by the strenuous efforts of his rise from penniless bobbin-boy in a Waltham cotton factory to governor of the Commonwealth of Massachusetts.

Ned Morse dismissed slavery and abolition, Republicans and Democrats, from his mind casually and went back to his shells.

July 19, Thursday: Unpacked a large collection of dry shells from Garrett, Sandwich Islands; some very fine things among them. Prof Bardwell and I sang together a lot of tunes tonight.

July 21, Saturday: At work on Sandwich Island shells. This afternoon went into Boston with Shaler to see the Zouave drill. Got an awful ducking; did not see the soldiers. Visited new Aquarial Gardens; one bank finished and its proportions are gigantic.

July 22, Sunday: Received $25.00 from Prof.

August 3, Friday: Prof Agassiz went to Newport to attend the scientific convention last Tuesday and has not yet returned.

August 6, Monday: Finished up down cellar. Prof introduced me to Professor Carr of South Carolina who wishes to study shells for about four weeks. I am afraid he will be a great source of annoyance to me. . . .

August 15, Wednesday: Work, work, work. It seems an interminable job to get all the Shells arranged. I have the end cases almost finished and it makes a fine show.

August 18, Saturday: Sore throat. Took Willie, Nellie, and Emma Barry in to Aquarial Gardens.

August 29, Wednesday: Most splendid weather. At Museum a little while today, and it gave me sore throat. . . . Home and talked politics with Shaler, Foley and Putnam. Bet Foley a hat that Pennsylvania would not go Lincoln and Hamlin. Bet Foley $1.00 that New York would not go Lincoln and Hamlin. Bet Foley $5.00 that Kentucky would give Bell and Everett 25,000 majority.

August 30, Thursday: Splendid morning. Republicans have nominated John A. Andrews [*sic*] for Governor of this State, which I think will be injurious to their issue.

August 31, Friday: Making drawing of *Strombus* from *Voyage de Astrolabe.*[1] There is still a great deal of work to be done on the shells in the museum yet.

Young Mr. Morse, whose twenty-first birthday had occurred two days before the death of his father in June, was not a very good political prophet. Jim Foley would collect his bets on Lincoln and Hannibal Hamlin; and John A. Andrew would be elected governor of Massachusetts on the Republican platform of "the Union Forever and Down with Slavery." Andrew, a native of Maine himself and a graduate of Bowdoin, had been so zealous in raising funds for the defense of John Brown that he was "investigated" by a Washington committee. He had been chairman of the Massachusetts delegation in the Republican national convention at Chicago that nominated Lincoln for the presidency.

Bell and Everett were the 1860 candidates of the Constitutional Union party, nominated at Baltimore May 10. Abraham Lincoln and Hannibal Hamlin were the Republican nominees; the Democrats, split over the union-or-disunion question, nominated two tickets, headed respectively by Stephen A. Douglas, the "Little Giant," and by John C. Breckinridge of Kentucky.

[1] Jules Dumont d'Urville's *Voyage de découvertes de l'Astrolabe exécutée par ordre du roi pendant les années 1826–1829* (Paris, 1832–35).

John Bell of Tennessee (1797–1869) had been secretary of war under Harrison and a United States senator, 1847–1859. He was one of the two Southern senators who voted against the Kansas-Nebraska Bill in 1854 and against the admission of Kansas under the Lecompton pro-slavery constitution in 1858. He was a strong "Union" man in principle, and even when Tennessee seceded after the outbreak of hostilities he remained neutral.

There was no real basis in Ned Morse's experience for his championing Bell for president, but with Edward Everett it was different. Had not Everett been president of Harvard? Was it not Everett who had invited Agassiz to accept the chair of zoölogy at Harvard? Had not Everett organized the Mount Vernon Ladies, in which Jane Morse had been active? Indeed, Edward Everett's name had been a household word in the Morse home for years. This Abraham Lincoln, who was he? Some backwoods lawyer from Illinois; never heard of him. Edward Everett, now, was a noted man. He had been governor of Massachusetts, minister to England, too.

Maine was so remote, geographically, and so provincial that Ned Morse had not been subjected to the Abolitionist fervor of Massachusetts; his journal shows no trace of interest pro- or anti-slavery, and he had not even read *Uncle Tom's Cabin.* Maine folks, only a generation old in statehood themselves, were strong Union advocates, though; and Bell and Everett was the ticket for Ned Morse.

September 1, Saturday: Into Boston this afternoon at the new Aquarial Gardens. They will be opened to the public in the course of a month. It is a grand place for entertainment and pleasure. Mr. Cutting has already got four young seal at the new place. He has had them only three weeks and yet they already understand what is wanted and obey most wonderfully. Got home and found . . . [letter] from Fred saying I had better come home next Monday.

September 3, Monday: Received $25.00 from Prof. . . . Interesting lecture this morning. This noon started for Portland; had a very pleasant ride. Mother looking first-rate.

September 5, Wednesday: Went to Gorham, Maine, with Mother this morning; arrived about 8½ o'clock, and started for the house. Our place is not five minutes' walk from the Depot, situated in a very pleasant street. The house is shaded in front with thrifty shade trees, has a nice fence in front and a good large yard. A great lot of apple trees and in fact we have a tiptop place. Large house, sheds, barn, etc. . . . Came home in afternoon and went swimming with John tonight in the canal.

September 6, Thursday: Democratic meeting tonight. There seems to [be], in fact is, intense interest felt in the election.

September 7, Friday: Folks think of moving about the last of the month. Started for Bradford, Massachusetts, with Fred this afternoon. Got there about 7 P.M. . . .

His great-grandfather, Deacon Thomas Morse (1749–1829), and his grandfather, Thomas Morse, who married Priscilla Kimball of Bradford, both had lived in the country town on the Merrimack River, near Haverhill.

September 8, Saturday: Fred and Carrie came into Cambridge with me. Introduced them to the Students and showed them all over the Museum. Rained all the forenoon. They rode down to the car office and took three o'clock P.M. train for Portland.

September 10, Monday: Glorious lecture from Prof this morning. I shall endeavor to write out the lectures in full hereafter.

September 11, Tuesday: Republicans have carried Maine by a majority above 17,000. Attended Prof Wyman's lecture this morning.

September 13, Thursday: Have got to the last end of the Gasteropods in Students' Gallery. The workmen commenced putting in cases on Students' Gallery.

Another invasion of students' liberties in Zoölogical Hall was attempted by the Glenns. This time Katie, the cook, was the aggressor.

September 15, Saturday: Met little girls in the garden today who asked me if they might have some flowers. I told them to help themselves, and came upstairs. From the window I heard Mrs. Glenn [Katie] tell them to go away, and I distinctly heard the little girls tell her that I had told them they might help themselves. The little girls ran off, but I went down and called them back and saw that they took what they wanted.

September 16, Sunday: Splendid weather. Sore throat. Made dissection of *Echinus granulatus* and made drawing for notes.

September 19, Wednesday: Splendid day. Have got almost all the shells arranged in the cases. Prof told me explicitly that as soon as I got the shells arranged, I was to take up the families of Strombida and Cypraedae. At Leavitt's room a little while. Attended Bell & Everett meeting tonight.

September 21, Friday: Getting along finely with collection; have got the cases almost filled up. . . . Prof invited us all to his house to supper next Friday night.

September 22, Saturday: Attended Bell & Everett meeting tonight.

September 24, Monday: Foley and I in Boston, went to New Aquarial Gardens. . . . Will be opened a week from today. Prof's lecture postponed till tomorrow. Gas put in [at Zoölogical Hall] Saturday and I got a chandelier and droplight. Quite an improvement over kerosene oil. Nat Shaler was taken sick with measles last week and is just getting well. Cooke will go to Africa in about three weeks. Scudder will be home by the 10th of October. . . . Prof gave me $10.00 for gas fixtures.

September 25, Tuesday: Prof gave us a very interesting lecture today. . . . Also lecture from Prof Wyman. Jim Foley and I walked into Boston with a Bell & Everett procession. We were attacked just as we got across the bridge [on Cambridge Street, near the Charles Street jail building] by a lot of Irishmen, and several stones were thrown. Three of our company was struck by them, but nothing more occurred. We occupied the third gallery in Music Hall. There were many clubs present. The scene from the gallery was splendid, a perfect mass of human beings filled the floor, all very enthusiastic and it was pleasing to Bell & Everett men and frightened the Lincoln men somewhat.

September 26, Wednesday: Have got the shells all arranged in my room [in the museum]. This evening Leavitt, Boone, Bradford and Scudder were here; we played whist, sang, danced and had a good time. Verrill made an ass of himself before the fellows by sticking his head out of the door in a rage and jabbering off something which amused and disgusted the boys present.

For one who in later life could not bear the noise of steam whistles and locomotive bells, Ned Morse could be extraordinarily insensitive to the effects of a noise he was helping to make. Ad Verrill, the earnest student, probably did not approve of the company who had called on his roommate.

September 27, Thursday: Lecture from Prof Wyman. Prof [Agassiz] introduced me to Edwin Booth, the tragedian, this afternoon. He has a splendid-looking eye. Prof gave me directions to crowd the bivalves. Verrill's father here tonight.

Edwin Booth (1833–1893) was barely five years older than Ned Morse when Agassiz introduced the two, but he was already famous, partly through association with his noted father, the actor Junius Brutus Booth, and partly as a result of his own great histrionic powers, manifested when he stepped into his father's role as Richard III during his father's last sickness in 1851. During the years 1852–1856, Booth had made a successful tour in California, Australia, and the Sandwich Islands, returning to create Shakespearean roles in New York at the Winter Garden Theatre. In the winter of 1864–65 he played an unprecedented run of one hundred consecutive performances of *Hamlet*, ending March 22, 1865. Three weeks later his ill-starred brother, John Wilkes Booth, assassinated Abraham Lincoln in Ford's Theatre, Washington. Edwin Booth's personality and genius survived even this terrible association. In 1926 his name was placed in the American Hall of Fame. His home at 16 Gramercy Park, New York City, was left in perpetuity to the Players' Club.

September 28, Friday: Prof introduced me to Professor Henry of the Smithsonian Institution. Eight hundred shells all arranged in Shell room.

September 29, Saturday: Prof told us today that Professor Henry had promised to give us everything we wanted from the Smithsonian Institute. He was so deeply impressed with the manner in which we are doing things here that he is going to give us what we wish from the Smithsonian.

Joseph Henry (1797–1878), born at Albany, New York, was apprenticed at thirteen to a watchmaker, but a book on natural history which came into his hands inspired him to complete his education, teaching in country schools betweenwhiles to earn his way. In the field of physics, he discovered in 1829 the properties of insulation in connection with electromagnets, thus pioneering in the development of dynamos; before 1831 he had set up a mile-long circuit of telegraphy, and in 1832 he published an account of some researches in self-induction, which seem to have antedated Faraday's. His laboratory work, while he was a professor at Princeton, lecturing on physics, mathematics, chemistry, mineralogy, geology, and astronomy, helped to establish the principles on the basis of which electrical transformers and radio transmission have been developed. When the Smithsonian Institution was founded, he was made its first secretary, in 1846. He established the United States Weather Bureau; later, as chairman of the lighthouse board, his researches brought about the system of fog signals and the substitution of kerosene for sperm oil in lighthouse beacons. He was the second president of the National Academy of Sciences and a founder of the American Association for the Advancement of Science.

Nowhere in the material by or about Agassiz appears any record that the great naturalist envisioned or predicted any notable career for his pupil Edward Sylvester Morse. Certainly, when the genial Agassiz escorting the greatest American scientist of the day through the new museum, said, "And this is Mr. Morse, my assistant in conchology, Professor Henry," the introducer had no idea that he was presenting to Henry a future president of the American Association for the Advancement of Science.

September 30, Sunday: [William Greene] Binney here this afternoon. He looked over Helicidae with me at the Museum. He made me an elegant present, being a copy of his father's work [*Terrestrial Air-Breathing Mollusks of the United States and the Adjacent Territories of North America*], a work I have been long desirous of obtaining. He told me that the operculum of *Paludina subcasinata* had an orbicular operculum. If such is the case, that shell shows an affinity to Valvata and the extreme young is strongly cornated like a Valvata. He drew Prof and myself maps of the United States with distribution of some of the species of Helices marked over on them.

The distribution of the so-called *Helix alternata* was remarkable. The common form, smooth, ranged over a wide district north. The strongly ribbed variety so-called formed a band along the Gulf of Mexico, while a cornated form occupied a space in Tennessee, including the Cumberland Mountains. Prof suggested that the *Alternata* might be generically distinct from *Alcobira* and that these varieties might be species of this genus occupying the range above-named. He wished me to make a microscopical examination of the epidermis, as there were strong generic marks in this part of the shell.

October 1, Monday: Worked hard at Museum; got all the Air-breathing Pulmonates arranged. This evening went into Boston to witness private exhibition of Savages at Aquarial Gardens. The Savages, five in number — a Zulu, Caffre, Hottentot, Bushman and Fingo — are wild and present themselves in their native style. They went through some of their performances before us, comprising dances mostly. They stamped with their bare feet on the floor incessantly. In the wedding-dance they articulated words like these, "Tom-a-long-Tommy." Another dance was accompanied by words sounding very like these "Ha-He-Ha-He-He."

In 1860 the era which had regarded the black men of Africa as chattels and draft-animals, to be captured, transported, sold and set to labor, was very near its bloody conclusion in the Civil War; and naturalists were just beginning to observe them scientifically as possible species of the genus Homo.

October 2, Tuesday: Finished collections upstairs and commenced unpacking and arranging Kingsmill Group Shells from Garrett.

The Kingsmill group of coral atolls, just north of the Equator on the 180th meridian in the south Pacific Ocean, is today called the Gilbert Islands. They were discovered in 1765 by John Byron, taken under a British protectorate in 1892, and formally annexed by Great Britain in 1915. Inhabited by gentle Polynesian people, in the old sailing-ship days they were a port of call for water and fresh breadfruit, fish and tropical shells, to whaling ships or vessels voyaging from Honolulu southwest to Brisbane, Australia.

October 4, Thursday: Worked hard. At Professor Wyman's lecture we had a fine subject. Prof W. had made a dissection showing the muscles on the breast, arm and side. Rec'd circular from Prof [Agassiz] giving free admittance to Inauguration of Aquarial Gardens. I . . . took in Miss Isabella Barry. . . . Had a nice time. Prof addressed the audience for a short time. We all came back about 10 o'clock and played whist till 12.

October 5, Friday: Rainy and uncomfortable. Arranging Garrett's collection in new boxes, . . . white. Grand dressing up at Zoölogical Hall. We all prepared for Prof's supper tonight and about nine o'clock all of us started in a bunch for Prof's house. He welcomed us in a genial style. We were shown into the parlor and talked and chatted a while, when we were summoned to supper. The table looked very fine; Roast Turkey, Roast Duck, scalloped oysters, Tongue and Ham. Four large decanters filled with wine stood at the corners of the table. We pitched in with appetites such as none but Zoölogists have, sharpened by our omission of our usual supper. After, the ice cream and cake were seperately [*sic*] brought on with Italian wine. The wines were splendid and I drank a great deal. . . . We sat down at nine and did not leave the table till quarter of one. Prof told us many things which I should wish the time to note. . . . Returned home and Foley, Cooke, Bowditch, Fred Putnam and myself went into Cooke's room and endeavoured to smoke Bickmore out, a student who makes himself particularly disagreeable to us all. He knew the trick though, and stood it like a martyr. As we could not make smoke enough with our pipes, Jim Foley raked out a large coal from the fire and sprinkled fine-cut tobacco on it and I kept fanning it. The smoke now was infernal. I had to get down on the floor to save my eyes. About three we left him.

October 6, Saturday: Into Aquarial Gardens this afternoon; had a good deal of fun with the Savages. . . . Met a lady, one of Prof's students, who [*sic*] I have met before at our snake cage. Conversed with her some time.

October 8, Monday: Interesting lecture from Prof; still in Radiata. Scudder got home today from his trip to Lake Winnipeg and the Sascatchewan [*sic*]. He has raised quite a crop of whiskers, mustaches, since he has been absent.

October 9, Tuesday: Lecture from Professor Wyman today [on anatomy]. Body at lecture smelt rather strong. Received from Prof $25.00.

October 10, Wednesday: Great victory of Republicans in Pennsylvania. Examined land shells brought home by Scudder. They are the same as I found in Aroostook.

October 11, Thursday: Called on Miss Nellie Owen at Mrs. Richardson's this afternoon. At Mrs. Barry's this evening, . . . passed a very pleasant evening. Fruit very nice. Folks all good. They are well acquainted with Miss Owen.

October 14, Sunday: Have been busy all this week making ready for the Prince [of Wales] who will be here next Friday.

October 16, Tuesday: Forgot to attend lecture today, so busy was I at work. This evening walked in with Bradford [a student] to see Republican torchlight procession [in Boston]. Got jammed in the crowd; met Leavitt and Boone [students], and we three went around together. They had a rousing big procession. Portland Wideawakes and Portland drum corps were in the procession; saw many Portland friends that I knew. Got home about two o'clock.

October 17, Wednesday: Prof's girl students were in the Museum today; some very pretty girls among them. Glenn has been setting up the skeleton of the Irish elk and it looks finely. Attended meeting of the Boston Society of Natural History this evening.

The *Proceedings* for October 17, 1860, record: "Mr. S. H. Scudder gave an account of a recent visit to Lake Winnipeg and the Saskatchewan River, having ascended the latter about 150 miles."

Uncle Charles [Beckett] out to see me this morning. He says he enjoys drumming [with the Portland Drum Corps] first rate. Bell-Everett men had a great torchlight procession this evening. It was not so large as the Republicans but far more attractive and it was not augmented by three hundred and eighty Negroes nor represented by five other States [like that of the Republicans the previous night]. The illuminations were far more numerous and the enthusiasm greater. The continual jingling of little Bells with the clashing of the big Bell formed a singular feature.

October 18, Thursday: Splendid day. Holiday in College and all around on account of the Prince of Wales who is to review the troops on Boston Common. All went in this afternoon to see the soldiers parade through the streets. . . . Parade very good. Everybody was out. Streets were crowded and all to see a boy nineteen years old who, if he lives, will be King of the most stupid conceited nation of this earth. I pity him. Met Edith Barry and we walked home together, not being able to find a car. Stayed at her house to tea and . . . passed the evening playing whist.

Albert Edward (1841–1910), eldest son of Queen Victoria and her consort Albert, Prince of Saxe-Coburg-Gotha, had finished his studies of chemistry at Edinburgh, Oxford, and Cambridge; had traveled through Italy and Spain in 1859, and in 1860 visited the United States and Canada, avoiding royal honors by passing under the name of Lord Renfrew. After the death of the Prince-Consort in December 1862, he was to assume governmental duties, marry the Princess Alexandra of Denmark, and in 1901 mount the throne of Great Britain and India as Edward VII.

To judge by the next entry in Ned's diary, something must have gone wrong for Agassiz. Perhaps he had been out late at the ball given for the Prince of Wales in the Boston Theatre, with a dancing floor built for the occasion level with the stage, above the orchestra chairs. He had probably been carrying through a great press of last-minute jobs in all departments of the museum to have it looking well for the Prince and the dignitaries of state, city, and college who were to visit it. At any rate something aroused his quick temper, and it seems to have been in connection with the shell galleries, in part at least:

October 19, Friday: Prof showed his weakness and foreign temper today by blowing up a few of us. Though he is a great man, he is only great on one point and that is science. His benevolence is large and hence he is good-natured and in the scientific life, where everything suits him, he is the most genial man alive; but take him out of his sphere and he is not only ungrateful but actually impolite, rude and unjust. Not considerate and a mere boy in business matters.

The Prince went through the Museum today with a whole troop of city officials, Governor Banks, Edward Everett, the Duke of Newcastle, and the whole pack. Had a fine chance to see him. Looks quite like his pictures and very much as he has been described in the papers.

October 22, Monday: This morning Prof gave us a rather strong lecture on behaviour, etc. Told us we must not emulate the rich; we must give up all our social ties and in fact not care for dress or anything; be real diggers. At the same time, seek refined and cultivated society. With ragged breeches, for instance! Told us, poor as he was, he would not change his position for any man living. I thought of the income he is receiving from his book and school; his $1,500.00 a year as Professor; his wealthy wife; the house in which he lives for nothing; his children already engaged to the wealthiest. O, consistency, thou art a jewel!

The hardships and rigid economy of Ned Morse's early life could not be better illustrated than in this naïve estimate of fifteen hundred dollars a year as wealth. After the "blowing up" from "Prof," Ned dutifully endeavored to seek some "refined and cultivated society" and bethought him again of the girl who had attracted him in the spring. Ellen Owen lived at Cape Elizabeth, close to Portland, but she was an expert seamstress and evidently in the spring and fall turned her skill to account by a series of visits in Cambridge homes.

October 23, Tuesday: Lecture from Professor Wyman, very interesting. At Mrs. Richardson's to call on Miss Owen. Mr. Richardson was well acquainted with my Mother, Father and uncles; has sung with them and knew them all very well, so he told me. Nellie and I played whist against Mr. Richardson's son and his wife; had a very pleasant time.

October 24, Wednesday: At Mrs. Barry's this evening; played whist and had a jolly time. Lecture this afternoon from Prof on Geology, coral reefs, etc.

October 26, Friday: Thunder shower this afternoon. At Mrs. Richardson's for Nellie Owen, and we went into Boston to Aquarial Gardens. Enjoyed ourselves very much. Coming out, we got out of the cars down to the Cambridge port and walked up. Splendid moonlight.

October 30, Tuesday: Splendid lecture from Professor Wyman. . . . I am busy at present labelling Duval's collection from his catalogues. Jim Foley invited me to the Howard tonight. Went with him and enjoyed myself hugely. On my way called in to Nellie Owen's a moment.

The old Howard Athenaeum, opened October 13, 1845, on Howard Street off Bowdoin Square in the west end of Boston, is still in operation, although the latter part of its century of operation has seen it decline to the dubious status of a burlesque

Two Drawings by Morse

house. In Ned Morse's day the noted actor E. A. Sothern had just concluded a short term as manager, and E. L. Davenport had brought to the Howard the stock company from the Boston Theatre, which was in financial difficulties.

October 31, Wednesday: Singular things have occurred to me tonight and yet not singular, but the most natural. February fourteenth of this year, while at a dance in Portland, a face there attracted my attention, why I knew not. I was not in love or smitten; but the sight of that face, so good, so sincere and honest, caused me to seek an introduction to her which I effected after a while through Joe Starr. Unfortunately her card was full and I could not dance with her. So I thought our acquaintance was ended and with rather unpleasant ideas that I should never see her again, I left her as I had nothing to talk about and considering myself particularly stupid before strangers.

By fortune, and a misfortune, I was requested to dance out a gentleman's card who was unable to do so owing to pressing engagements. I can remember very well how pleased I was to find Miss Owen's name on it. I danced with her and bid her goodnight, not dreaming of ever meeting her again. [In the latter part of this month] she came and stopped at Mr. Richardson's on Dana Hill, and the second time I saw her I called on her with Abbott Cummings and invited her at the time to go into the Aquarial Gardens, which we did. By that time we were thoroughly acquainted and I longed to call her my sister. I wished for an honest female friend, one whom I could look up to with respect. I did not care for more than a friendly, sisterly love. In fact I think I shall never love a girl, never; I shall surely be an old bachelor. But I wished to feel a deep interest in one who would care the same for me, and I have found it.

Nellie thinks well of me. She says she has no sisters and has always wished for an older brother, and now I am the one to fill that place. . . . I feel perfectly happy. She is one to whom I will look for advice, one who will lead me to a higher scale of thought, and one I shall love to please by good behaviour.

CHAPTER SIX

VOTING AGAINST LINCOLN
1860–1861

BEYOND each peak, Ned Morse discovered in his twenty-first year, there is only a descent and painful climb towards the summit of another range of heights in the topography of Life.

Since leaving Bethel Academy in 1856, his every hope, ambition, effort had been directed towards achieving a career in science. The year 1860 had seen him standing on the summit to which he had so long looked up. He was in Cambridge, attending university lectures in anatomy, biology, geology; he was an assistant at the Museum of Comparative Zoölogy, responsible for the arrangement of shells and mollusks; he was associated with Agassiz, outstanding naturalist of the day.

As 1860 neared its end, examining his position, he found himself but poorly situated. His ambition was to work as a naturalist; yet most of his hours at the museum had been utilized in the most prosaic of tasks — lugging jars of specimens in alcohol from the cellar up to the gallery, unpacking and sorting the contents of barrels, copying in longhand catalogues. The death of his father, leaving a widow and three young daughters, had made more imperative than ever his need for economic independence, and his engagement to Nellie Owen added a deep desire to achieve marriage and a home; yet his financial position was worse than when he lived in Portland. There he had had a salary of two hundred dollars a year, with an opportunity for advancement as a draughtsman in the Portland Company. Now, his engagement as assistant to Agassiz, although nominally at a rate of three hundred a year plus living, actually during the first twelve months had yielded him two hundred and eighty-eight dollars, out of which he was paying board at the rate of four dollars a week, leaving him eighty dollars for clothes, medical expenses, travel to and from Portland, and cash in pocket.

In addition to his financial problem, the year 1861 brought two issues, one local and one national, which affected him distractingly. The first was the relation between Agassiz and his student-assistants. After a lapse of a hundred years, the reader of the journals can clearly see the elements of this situation; can judge that the young men owed a greater loyalty to the teacher who, in addition to instructing them, was furnishing their living out of his own pocket; can wish that Agassiz had been less involved in the administrative and financial problems of his triple-phase career as teacher, naturalist, and museum director, so that he might have furthered more purposively the individual progress of each pupil.

The national issue was to be as sore a problem to Agassiz, in its disruption of his museum staff, as to each of the young men. Agassiz had been driven to America by revolution and internecine warfare in Europe twelve years earlier. Now civil war, breaking out in America, for the second time threatened interruption to scientific work. For Ned Morse the factors were even more complicated.

He was twenty-one, when youth is fiery, courageous, and adventurous; his natural impulse was to answer a call to arms with his comrades. He had been reared in Maine, in the doctrine that the Union of states should never be dissolved; the secession of the southern states outraged his fundamental ideas. Yet opposed to his desire to enlist were two powerful motives: his own keen interest in the pursuit of science and the insistence of his widowed mother and his fiancée that he could not be spared to enlist.

November 5 [1860], Monday: No lecture from Prof Agassiz today. At Nellie's this evening. We walked around and saw illuminations. I had wished to vote at the election because I wanted to contribute my mite to defeat [Anson] Burlingame. Many students had tried and were refused,[1] particularly if they were Bell-Everett men. I was assessed some time ago,[2] but my tax bill was unpaid. The principal mode of getting in was to go down to the City Hall and undergo an examination by the Mayor and aldermen. Of the twelve aldermen, only one was Bell-Everett; the greater number of them [were] Republicans and I knew if I went down there the Republicans would vote me out and swear I had no right to vote. I saw Waite, the Bell-Everett alderman, and he thought I might vote but said that if he should vouch for me, I would meet strong opposition and be rejected.

A happy idea struck me this noon. I borrowed a Lincoln and Hamlin medal, marched down to the clubroom of that party and stood in the doorway as if seeking for someone. A zealous Republican came along and, seeing Honest Abe peeking out from my waistcoat, supposed I was of the right stamp according to his idea; so he inquired if I was looking for anyone. I told him I was looking for an Alderman as I wished to vote. He asked for my tax bill, and I told him it was at home unpaid. He told me if I would go and get it, he would meet me at the Post Office and introduce me to Alder-

[1] On the grounds of not being residents of Cambridge.

[2] His status differed from that of a student in that he was a salaried assistant, not dependent on his parents.

man Chapman. Chapman, by the way, is one of the most rabid Republicans you ever saw.

I returned and met my zealous friend who introduced me to Chapman who took me confidentially by the arm, walked me down to their Clubroom and swore me; took my tax bill and said they would pay for it; and Chapman told me he would vouch for me and there was no need of my going down to the City Hall.

So far, so good. I now hunted up Waite and told him, if my name was vouched for by a Republican alderman, it was all right and that he must not find any objection. He acquiesced of course and this evening my name went onto the list with the unanimous consent of all the aldermen, for a wonder!

November 6, Tuesday: This morning went down to vote. There were four boxes to vote in, surrounded by Bell-Everetts and Republicans. Conspicuous in the crowd was Chapman, the rabid Republican. I secured my Bell-Everett votes [ballots], holding them closed in my hand; and called out my name and found they had got it down as Edward E. Morse. I turned round to Chapman and told him the mistake. He had it corrected and I then asked if everything was all right. He said it was. Then to his amazement, I cried, "All up, Bell-Everetts!" and deposited my votes wide open. Such merriment you never saw and Chapman was a picture to behold. . . .

Went down to the office this evening; large crowd there and I heard my job on the Alderman mentioned many times. Burlingame is defeated by a good round margin.

Morse showed Yankee ingenuity in making good his democratic right to cast a vote, and plenty of Yankee spunk in openly depositing votes for Bell and Everett in a Republican precinct. The secret "Australian" ballot was not introduced into the United States until after the presidential elections of 1884. In Ned Morse's day, one made his choice openly by asking for a ballot printed with his candidate's name, and deposited the ballots in the box under the eyes of a crowd. If one's candidate was unpopular and the crowd well seasoned with rowdies, voting became a hazardous undertaking.

Young Morse had put his money on the wrong horse in the election of 1860. Edward Everett was defeated in his own state of Massachusetts. The Bell-Everett ticket carried only three states, Kentucky, Tennessee, and Virginia, with a total of thirty-nine electoral votes. Although Ned Morse was pleased at Burlingame's defeat for re-election to the House of Representatives, the net result was a promotion for Burlingame, since President Lincoln appointed him minister to China in 1861.

November 8, Thursday: Prof's Monday morning lecture was delivered today. In it he stated that the so-called eyes of Radiates could not be homological with the eyes of the other classes. I don't believe it and shall write a refutation against it. Mr. Hansen gave me two tickets to the Boston Museum so I took Nellie in the evening.

November 9, Friday: We found out this morning that our position assigned to us in the Procession next Tuesday was a position dishonourable and mean, being in the very tail end of the line. We all [the student-assistants in Zoölogical Hall] held a meeting this afternoon and unanimously voted not to attend the procession at all. Professor Wyman said they had treated us shabby enough, and all who knew about it agree that it was a mean thing in the trustees thus to put us where they did. Felt sick to my stomach and went to bed early.

On November 7, when the journal states briefly, "At Nellie's this evening," there had been a regular meeting of the Natural History Society in Boston, with Rogers speaking on the Jurassic sandstones, followed by the usual debate between Rogers and Agassiz. At the close of the meeting the *Proceedings* state:

"Professor Agassiz invited the members of the Society to be present at the Inauguration of the Zoölogical Museum at Cambridge, on Tuesday, Nov. 13." This would be the occasion of the procession which had stirred up the combative faculties of the Zoölogical Hall boys.

Ned Morse could not join as ardently as usual with Shaler and Jim Foley and Putnam in upholding the rights and privileges and dignity of their status. His mind was preoccupied with a personal crisis which reached its climax on Sunday evening and left him in a mood in which processions were but trivial affairs.

November 11, Sunday: At Mrs. Richardson's to tea and spent the evening there. I have been with Nellie constantly since the night I called her sister. We understand each other. Our love is mutual and earnest. How grateful I am that I should now love and be loved in return. The only question I ask myself is "Am I worthy of her?" I will try to be.

November 12, Monday: We have had a row in regard to procession which after all may be a mistake. Putnam saw Prof today about it. Prof is very indignant and says the procession may go to the Devil. We are cleared up ready for tomorrow. With Nellie this evening.

November 13, Tuesday: Inauguration of the Museum today. As the papers made full reports of it, I shall stick the account in here which will do better than anything I might say. We all marched in the procession, distinguished from the rest by being dressed in black with black beavers and blue ribbons in our buttonholes. In the afternoon the Museum was overrun with visitors.

From this point in the journal, it may be assumed that the final entry each day is "evening with Nellie," unless otherwise indicated.

November 15, Thursday: Lecture from Professor Wyman this morning. Began copying the catalogue of alcoholic mollusca that I made last Winter. I have a table in the Students' room and find it very pleasant to sit down in one place after having been arranging the collection so much. Through the Museum today with Nellie and Mrs. Charles Richardson.

November 18, Sunday: Charles Robinson out here today; he stopped to dinner with me. I went all over the Museum with him, had nice social chat with him.

November 20, Tuesday: Splendid weather. Lecture from Professor Wyman. Row in College; Fresh[man] fired pistol at Soph[omore] this morning. Seven Sophs suspended.

Love makes alterations in a man's habits. Charles Robinson, coming out on Sunday for a good long religious discussion, must have been puzzled when Ned would go no further than "a nice social chat"; and here, with freshmen and sophomores indulging in pistol play a block distant, Ned Morse, who would run three miles to a fire or catch a train for Lawrence to see a mill that had collapsed, never even strolled across to the Yard to see the excitement. He did keep his promise to his mother to go home for the first Thanksgiving in the new home at Gorham. However, he had an ulterior motive. He called on Nellie's parents to gain their consent to his engagement to Nellie.

November 28, Wednesday [Portland]: Dined with Fred Morse at the office [Portland Company]; he has a little provision box and closet rigged up in his room and gets his own dinners. . . . This afternoon I went over to the Cape [Elizabeth]. . . . I called on Mrs. Owen, Nellie's mother. I like her very much; kind and gentle she appears. She welcomed me cordially and we talked quite a while. . . . Took the train for Gorham 5½ o'clock.

November 29, Thursday: THANKSGIVING. Nice breakfast and we fed as usual on such days, eating too much. . . . O, how different from last Thanksgiving is this day and yet I have an undefined idea that Father was with us.

December 5, Wednesday: Splendid lecture from Prof this afternoon on Glacier formation. . . . Attended meeting of Boston Society of Natural History this evening. Very interesting dissertation from Professor Rogers in which he stated his belief in the continuity of animals of the same species from one epoch to another and also the reappearance of animals after an elapse of epochs. Went down to the vestry of the Lee Street Church to escort Nellie home.

Agassiz did not debate with Rogers on this night, but his students contributed to the meeting. Addison Verrill "exhibited some specimens of minerals": tin ore from Mt. Mica, Paris, Maine; native arsenic from the eastern flank of Furlong Mountain, Greenwood, Maine; a crystal of chrysoberyl found on hill near Norway, Maine, in granite. Theodore Lyman "presented a description of a new star-fish."

Sunday, December 9: Went through an unusual performance this morning in the act of dressing up for church. How queer it seemed and how it carried me way back to times bygone when I was forced by a kind and indulgent Father who was doing what he considered right. How different were my feelings now! This morning attended Nellie to church; felt singular enough. The sermon was very interesting. . . . He said a skeptic was better who spoke his mind and expressed his doubts than he who avowed a belief and did not live up to [it]. He called it paralysis of the mind. . . .

December 10, Monday: Finished catalogue copying. We have 26,500 specimens of Mollusca in alcohol in the Museum, not including the cephalopods and *Naiades* and many still to be catalogued. These figures represent about 600 species, 400 of which are Gasteropods and 200 Conchifera. Snow and rain. Hard times in Boston.

December 11, Tuesday: Lecture by Professor Wyman. Put my pipe all lighted in my pocket just before I got to lecture and discovered myself to be on fire just before I got into the lecture room. Fire was extinguished without alarm.

December 12, Wednesday: We are having a long, uninterrupted spell of cold weather. Secession begins to look serious and from present appearance inevitable.

December 23, Sunday: Since last Tuesday I have been confined to my room with an attack similar to the one I had last Spring, a sore throat and backache. Nellie, the dear girl, has come every day to see me, bringing with her my daily food, preparing it in my room. . . . She has shown herself a true woman. . . . I cannot keep her out of my mind a moment; every thought and action seems only for her.

Through the Barry family, he had met relatives of theirs, the Bells, and several months earlier had changed his boarding place, taking his meals with the Bells.

December 25, CHRISTMAS: At Mr. Bell's all day. They had a Christmas tree and grand time. Mr. and Mrs. Barry and family, . . . Nellie Owen, were present. We had a grand supper, games, dances and a royal good time. It was one of the greatest pleasures of this world to see the little children when Uncle Joe [Bell], who was dressed up as Santa Claus, announced the presents from the tree.

December 28, Friday: Prof gave me Wednesday four species of *Unios* included in the genus *Melaptera* of Rafinesque and wished me to find out the generic features from the shell. I have been busily engaged on them since then; have made drawings of the hinge of *M. leptodon*, *Ohioensis*, *gracilis*, and *alata*. I find strong generic character specializing this genus though I am puzzled to know what to do with *Alasmodonta complaniotis* which shows many of the features of the genus *Melaptera*. . . . South Carolina seceded on the 20th of this month. News received states that Major Anderson has left Fort Moultrie and taken up position at Fort Sumter. Carolinians are in an open state of Rebellion.

December 31, Monday: The last day of this year; well, it has gone quick, very quick. My feelings through the year have been happy. My studies have progressed as well as I could wish though most of my labors in the Museum have been in arranging the collection of Shells and Mollusks. I have attended about one hundred scientific lectures and have taken ample notes. . . .

Our family circumstances are quite different. At the first of this year my Father was then on this Earth. . . . Now he has gone to a better and happier place. We lived in Portland. Now we live in Gorham, Maine, in a house of our own, neat and well-adapted to our wants. Two of the happiest things of my life transpired this year. The reconciliation with my only true friend [John Gould]. How horrible it seems to me now to imagine our position . . . at the first of the year. . . . Two months ago tonight I made my love known to Nellie Owen and was accepted. I have thus secured the one in whom I can seek comfort in time of sorrow, one who will strive for my happiness in the same way I shall strive for hers.

January 1, 1861, Tuesday: Another year before me and each year seems to pass quicker than the one preceding it. I start with a superficial knowledge of Geology and Zoölogy. By patient application this year I can gain a great deal in the more specialized aspect of the study. I am bound to do it. Must endeavor to read French a little every day if it is not more than a line. I start also with a dear girl at my side . . . ; I pray that she may enjoy good health and prosperity this year. My smoking habit, I must curtail. Liquor I have bid goodbye to. Coffee ditto. And I live very frugal and simple, boarding at Mrs. Bell's. . . . I have been for the past week engaged in the genus *Melaptera;* today I have been working on the same thing. Prof told me today that the strong line of growth noticeable a little way from the beaks of the Unionidae indicated their first spawning season. Spent the evening with Nellie.

January 2, Wednesday: Busy making up a report of Curator. Dr. [Augustus Addison] Gould was out here this morning. He told me the next month he should commence on the North American coast shells with the determination of publishing a second edition of his *Invertebrata of Massachusetts.* He looked at my marine collections and admired them very much. The fossils particularly pleased him.

In six years more Dr. Gould was to be dead of the Asiatic cholera; his project was never to be completed.

Lecture from Prof Agassiz today. He wished me to give him a list of the Maine *Helices,* which I promised to do. . . . N. Ferrand was out here this afternoon bringing with him 42 species of Cape Palmas shells. We exchanged with him, giving in return American *Helices.* Very interesting meeting at the Society of Natural History. Ordway read a paper showing the difference between *Paradoxides spinosa* and *Paradoxides harlani,* showing them to be distinct species while they had been considered identical by Rogers and Jackson. Poor Dr. Jackson was severely handled by Ordway and Prof much to our delight and his disgust.

Professor William Rogers and Dr. C. S. Jackson, vice-president of the Society of Natural History, might score in debates on vast nebulous theories of the origin of species; but when it came to minute, microscopical details, Agassiz and his brilliant young assistants were unbeatable, and for once Agassiz had pinned his opponents down on ground where he could triumph.

Both parties probably went home from the meeting in a high state of feeling and sat up in their own bailiwicks to discuss the question further, to judge from the journal's next entry.

January 3, Thursday: Tremendous snowstorm; the trees are all covered with it, giving a very beautiful appearance. Prof forgot all about lectures today.

January 4, Friday: Snowed more today. We have had sleighing now for some time. President's Fast Day today in consequence of the distracted state of the Union. Made out my report to Prof of the collections in my department since January, 1860. We have received about 4,000 species and 30,000 specimens.

James Buchanan, fifteenth president of the United States, might well seek some supernatural guidance on a situation that was completely out of his control. Jefferson Davis, senator from Mississippi and future president of the Confederate States of America, had stood up in the United States Senate to declare: "These are offenses no people can bear; it is far better, instead of attempting to preserve a forced and fruitless union, that we should peacefully part and each pursue his separate course." Andrew Johnson of Tennessee was proposing a Constitutional amendment whereby the president of the United States be chosen from a slave-holding state in 1864, and alternately thereafter from free and slave states. Mr. Noell of Missouri suggested abolishing the office of president in favor of an executive council of three.

On December 14 a majority of senators and representatives of the states of North Carolina, South Carolina, Georgia, Alabama, Mississippi, Louisiana, Florida, Texas, and Arkansas issued a public "Address to Our Constituents," proclaiming: "We are satisfied that the honor, safety and independence of the Southern People require the organization of a Southern Confederacy . . . that the primary object of each slave-holding State ought to be its speedy and absolute separation from a Union with hostile States."

Abraham Lincoln was not to be inducted into office as president until March 4, 1861. Meanwhile President Buchanan, timid of taking a decisive stand for either faction in the country he represented, laid his burdens on the Lord.[3]

Ned Morse, however, left the state of the country to President Buchanan and Providence and continued to concentrate on science.

January 5, Saturday: Examined a compound Ascidian this afternoon. The exterior looked like a lump of jelly. Found what I supposed to be the young.

January 6, Sunday: At church with Nellie this morning. . . . Went over to the Museum this afternoon and dissected Ascidian under microscope. Found the mass made up of numerous zoids [*sic*] and in the zoids I found the young, looking like a tadpole. Each contained four or five embryos.

January 7, Monday: Interesting meeting of the Club tonight, discussing the propriety of compulsory reading of papers. It was proposed to have a committee of three to select some subject for discussion and write it out in full on the blackboard to remain there a week. The meeting adjourned after choosing a committee to make arrangements in the by-laws to be conducive of good results in the whole.

The young zoölogists evidently were experiencing a phenomenon common to clubs of all kinds, in which certain of the more loquacious members tend to usurp the floor, and probably Professor Agassiz had suggested finding some device whereby the more retiring students might be encouraged.

January 8, Tuesday: Refilling bottles of Mollusca with alcohol this morning. . . . Nellie came down to the Museum this afternoon and I showed her examples in four departments of the Animal Kingdom. She shows a faculty of comprehending the different groups of

[3] See his Fast Day Proclamation, quoted in J. G. Nicolay and John Hay, *Abraham Lincoln,* II (1890), 435.

the Animal Kingdom and as soon as she gets fairly acquainted with the idea of the study, things will come easier to her.

The picture of gentle Nellie, in rustling hoopskirt sweeping the ground, shawl gracefully pinned about her shoulders, bonnet turned back from her pretty face, with rosebuds in the ruching and broad bonnet strings tied beneath her chin, gazing wonderingly at the immense skeleton of the Irish elk and the anomalous masses of mollusks in alcohol under the tutelage of her young scientist, is appealing. Throughout a long married life, she was to give him faith and coöperation.

If Ned's journal entries are any index of his subjects of conversation in the frequent "evening-with-Nellie" episodes, it is no wonder that she was driven to undertaking a study of natural history in order that the ideas might "come easier to her."

January 10, Thursday: Examined animals of *Melaptera Ohioensis*. The generic differences are obvious and striking in the animals as in the shell. The mantle and foot is at the posterior end and closes the gill cavity from the upper cavity. The outer gill reaches down and behind the inner gill. During impregnation the young are developed in the posterior part of this outer gill. The palpi are connected partially on the posterior edge. . . .

January 13, Sunday: This forenoon made drawing on wood of *Paradoxides* for Bertie [Ordway]. Went to church with Nellie, and Nellie and I went down to Mrs. Barry's this evening. . . . We stopped to tea and spent the evening with them; we sang most of the evening. Coming home we nearly froze, thermometer 9 below zero. . . . When I got home I found Verrill suffering severely with colic. Went after Dr. Wyman at one o'clock; thermometer 14 below zero, intensely cold. Hammered and rang away at the door for a long time till I nearly perished with the cold. Knocked and howled and raised all the neighbors in the immediate vicinity who told me to go to the Devil with my noise and I responded by telling them to go to Thunder. Went down to Mr. Wiley's and woke up Joe who directed me to Dr. Nichols, telling me that Dr. Wyman had moved up in town. After I got back to the Hall, I had to go out in the cold again . . . for medicine.

Alexander Graham Bell did not speak the historic first telephonic sentence, "Mr. Watson, come here. I want you," until 1876, and the first commercial telephone switchboard was not set up until 1878 at New Haven. Ad Verrill's colic sent Ned Morse out in the cold on one of the most familiar of New England errands, "going for the Doctor."

January 16, Wednesday: I have been working hard on drawing the animal of *Melaptera.* I am drawing every part of it under a glass [microscope] and enjoy the work very much. At Nellie's this evening. It makes me feel bad enough to think that she is going [back to Cape Elizabeth] so soon but it is for the best that she should leave, else I could not do much work. I am thinking of her continually.

January 17, Thursday: Still at work on same drawing. . . . Prof gave us a long lecture on Echinoderms. He did not present many things new to us; many things were almost word for word what he has given us before. However, I do not object to it as such repetitions only serve to strengthen the subject in our minds.

January 23, Wednesday: Received bundle of clothes from home; 12 handkerchiefs, 12 collars and two shirts of red flannel. How glad I am to get them. Made preparation of lingual ribbon and operculum of *Nassa trivittata.*

January 26, Saturday: Took out lingual ribbon of *Chiton ruber* and a very young *Buccinum undatum.* Find great difference existing between the teeth in young and old. Went into Boston and to the Depot with Nellie this afternoon. She went to Newburyport. What a lonely feeling came over me when I bid her goodbye. . . . Prof has been unwell for the past two or three days.

January 29, Tuesday: For the past two days we have had a snowstorm. . . . Club meeting postponed from last night till tonight. Ordway, Hyatt and I discussed the question "On the extinction and introduction of animals in different periods of time." Wrote again to Nellie tonight, told her I would come down to Newburyport next Saturday . . . ; I cannot bear to have her go home without seeing her again.

January 30, Wednesday: Am preparing an animal of Mr. Oliver's to draw from below with the gills spread. Prof told me this evening he should surely keep me at work, on the family Unionidae. Began reading Agassiz on Classification. Shall endeavour to read an hour every day before I go to bed.

The *Essay on Classification* was one of the most admired of Agassiz's works, comprising the principles of his lifetime of study. Alpheus Hyatt in the winter of 1860–61 actually learned the entire essay by heart.

January 31, Thursday: Short note from John Gould today with $8.00 enclosed. It comes just in the nick of time, the dear fellow. May God grant me a life to repay such kindness. Distributed a few Tertiary [fossil] shells in the cases this afternoon. . . . I have begun to study German and a systematic course of reading, beginning with Prof's Essay on Classification which I find to be very instructive and entertaining.

February 2, Saturday: . . . Took the train at 5 for Newburyport. The ride seemed very long and was tedious enough. Nellie's uncle Thomas Merrill was at the Depot for me and we went up to the house where my dear Nellie is stopping. . . . She immediately introduced me to her Grandmother and uncle and we went out to supper. . . . The folks live in an old-fashioned house; the furniture and everything is very antiquated. Nellie's uncle, or rather great uncle, is ninety years old and is blind and partially deaf. He is quite rich and spends all his income, about three or four thousand dollars, in the relief of the poor. He pays the weekly meat bills for twenty-five families in Newburyport and supports two students who are preparing for the ministry. Nellie's grandmother is a fine old lady, 70 years old, good-hearted and benevolent.

Nellie's innocent stratagem in stopping off at Newburyport on her way to Cape Elizabeth, inviting Ned to visit her in her great-uncle's home, is more apparent to the reader, eighty years after, than to Ned. His sturdy pride never allowed him to appeal for loans to any of the affluent men with whom his science associated him; and he listened to the account of Great-uncle Nat's philanthropies with an honest admiration untinged by self-interest.

February 3, Sunday: Very pleasant and cold. Uncle Thomas, Nellie and I went to church this morning at the Unitarian Church; very old-fashioned and the minister, though he spoke a good sermon, had no life or animation in him. Was glad when I got home. . . . Uncle Nat (that is the old gentleman) is an old-fashioned Christian. His Sabbath began Saturday night at 6 o'clock. He gives prayer every night, and tonight Nellie and I sang hymns, much to his delight.

Poor Nellie! Even the hymns did not compass her guileless little plot, for it never entered Ned's head to "make up" to Uncle Nat and retail his financial handicaps to the old philanthropist. Nellie, of course, thought her naturalist as deserving of aid in education as "two students preparing for the ministry," but Ned probably was independent enough to discuss Darwinism with an old-fashioned Christian who would have no hot food cooked in his house on the Sabbath.

February 4, Monday: Bid my Nellie farewell, and all the rest in the house and started for the Depot with Thomas. Nellie goes to Portland at 9 and I return to Cambridge. The ride up passed quickly and before I knew it I was in my seat in the Museum.

February 5, Tuesday: At the Museum hard at work. Putnam, Bertie [Ordway], and Verrill went to Washington yesterday to be gone some time.[4] Wrote a letter to Nellie of 28 pages; my fingers felt somewhat cramped by the time I got through.

February 6, Wednesday: Still at work on the *Melaptera.* Into meeting of the Boston Society of Natural History tonight. Carried in some of my drawings and Prof showed them to the assembly as illustrating the age of Unios; what had been commonly supposed to be a young shell he thought was at least seven or eight years old. Dr. Gould hardly believed it but Prof's reasons were apparently correct.

February 7, Thursday: Singular weather. Sun rose bright this morning; soon clouded up blowing a warm breeze. At 11 o'clock A.M. large drops of rain fell till 2 P.M. and the sun broke through the clouds. At 3 P.M. a most furious storm of wind and snow came upon us with the utmost violence; immense quantities of snow passed through the place, carried by the hurricane. At five o'clock the clouds passed away and the thermometer which had been at 42 degrees at 3 o'clock fell to zero at midnight and 4 degrees above zero at suppertime. I slept in my study. The wind continued to blow with the greatest fury all night. Should have frozen to death if I had slept in the bedroom.[5]

[4] For work at the Smithsonian Institution.

[5] By an odd coincidence this "singular weather" of 1860 was duplicated in every particular in 1939, as Ned Morse's journal was being transcribed, fourteen years after his death. On January 22, 1939, the day dawned sunny, shortly before noon became overcast with light rain. Just before 3 P.M., a wind of hurricane force came out of the west, with claps of thunder, driving icy snow pellets horizontally before it in impenetrable veils, and the thermometer dropped from 40° F. at 2:45 to 4° above zero by suppertime. The freakishness of New England weather has a consistency of its own.

February 8, Friday: Everything frozen up this morning; took me almost an hour to get ready to go to breakfast, so cold it was. Thermometer fell to 26 below zero this morning and was 11 below when I went to breakfast at 9 o'clock. . . . Made longitudinal section of *Melaptera;* traced the intestines out; laid open the organs of Bojanus and found out a lot of things I never knew before.

February 11, Monday: [Warm and pleasant]; the snow will go off very quick at this rate. Our bees came out of their hives in the entry today; retired again tonight. Made drawing of labial palpi of *Melaptera alatus* today.

Judging by the bees in the entry, Ad Verrill, in going to Washington, must have left his "zoo" behind to be looked after by the other students. Verrill, who had grown up in the wild country about Norway, Maine, seems always to have been fetching down from Norway live owls or snakes or various creatures for the museum.

February 27, Wednesday: Letter from Mr. John G. Anthony of Cincinnati wishing to exchange. Prof also gave me letter from Mr. Anthony which he wishes me to answer. Wrote to Fred about a suit of clothes. . . . Lent Putnam fifty cents.

February 28, Thursday: We are having splendid weather now. Heard robins singing and there are quite a number of birds now with us.

March 1, Friday: The first day of spring and a splendid day it is. I have made considerable progress in my investigations of the *Melaptera;* can make further advancement when I have compared the anatomy of the *Melaptera* with another allied genus. Thus far I have examined only the animal of *M. alatus*. . . . Received from Prof $25.00. $14.00 for board; $5 to Katie [the housekeeper at Zoölogical Hall] for wash; $1.00 for car tickets.

March 2, Saturday: Hat $2.50; boots $1.75; flowers .25; sundries .10 cents.

This was the first money Ned Morse had received from Agassiz since November 1, three months earlier. The other students seem to have been in the same situation, judging from the loan of fifty cents to Putnam.

March 7, Thursday: All the members of the Legislature paid a visit to the Museum this forenoon. Speeches were made by Prof and Governor Andrew and the visit was very pleasant. Prof wishes to get $20,000 from the State and I guess he will succeed. Into Boston this afternoon, went into Senate [in the State House] and saw Mr. Harries, reporter of [Boston] *Herald* who procured me the Trustees Report on the Museum [of Comparative Zoölogy]. Then he showed me all over the building and into every room. I was much pleased with my visit.

It would not have required much time to go into every room of the State House on Beacon Hill in 1861, as the addition at the rear, on the site of the old reservoir, had not yet been built, nor the white marble wings added facing Beacon Street.

The financial situation of the museum now began to disturb Ned, since he was anxious to achieve an income on which he could marry. He evidently had little satisfaction from querying "Prof," and had taken the step of surreptitiously procuring the official report of the trustees to the legislature. He was being dunned for the suit of clothes he had ordered the previous year.

March 9, Saturday: Letter from Fernald; wants me to pay a tailor's bill. . . . Wrote a long while to Nellie this evening.

March 16, Saturday: Letter from Mother; she likes Nellie very much. Attended Opera, *Ernani,* very nice. Saw John there. He got in about noon. . . . He brought me my new clothes and they were just in time.

March 17, Sunday: John is staying with me; we take our breakfasts in our room, bread and milk.

March 18, Monday: John and I went to the Opera this evening, *Un Balle en Muschovie*.[5a] Very peculiar Opera. I liked it very much.

March 19, Tuesday: At work on *Melaptera* as usual. Very interesting lecture from Professor Wyman. . . . My hand is very sore from a cut I got cold in. Attended Opera tonight with John. *Linda de Chamounix.* Miss [Clara Louise] Kellogg made her debut as Linda; she was enthusiastically received. She is beautiful and a charming singer.

March 20, Wednesday: Attended meeting of Boston Society of Natural History this evening, very interesting. Lecture from Prof this afternoon; stupid because we have heard it before. Finished letter of 12 pages to Nellie with much difficulty because I had to write it lefthanded. John went home this afternoon; I wish he could come up here and spend the time at lectures; I enjoyed his visit very much.

The facility with which Morse drew or wrote with his left hand later in life caused many of his friends to believe that the talent was natural; this journal entry demonstrates that it was acquired after he was twenty-one.

He had been absorbed for a fortnight in the study of a hitherto undetermined organ of a mollusk; now he solved the problem.

March 21, Thursday: Found the outlet to the cavity immediately above Bojanus' organ today; it is a posterior opening with top

[5a] So, at any rate, Morse seems to have written the title. The Boston *Evening Transcript* of March 18, 1861, shows that the opera was Verdi's *Un ballo in maschera.*

of cavity of inner site. The cavity is undoubtedly for the free ingress of water.

March 22, Friday: Terrible snowstorm last night. Horse railroad blocked up and everything covered with it.

March 23, Saturday: Ad Verrill got home from Washington this morning. His mission there has been attended with great success. He sent home from the Smithsonian 19 boxes of corals, birds, birds' eggs, etc. Opened some *Unio phosculens* for the purpose of making a drawing of the animal. The gills are filled in entirely different way from the *Melaptera*. In the *U. phosculens* the whole margin of the outer gill, or rather the lower half of the outer gills are filled with eggs and the gill takes on a beautiful appearance ruffled like a frill. Two lectures, one from Prof and one on cellular tissue from Clarke.

March 24, Sunday: Nat Shaler got home this morning. Our rooms [in Zoölogical Hall] are all filled up now.

March 27, Wednesday: Warm rain; snow going fast. Prof's lectures are getting uninteresting because he does not know what he is talking about. . . . Scudder came into the students' room today and told Prof in a hasty manner that Ordway had found a female crab with a male penis. Prof says, "Has he?" . . . I asked Scudder what a female penis would look like. They both had to laugh.

Rain, slush and mud, sticky. Letter from Charles Robinson with bill enclosed. O, how I wish I had the money to pay him. Dear me, dear me. Will it always be so?

March 28, Thursday: Met Charles Robinson at the Depot tonight. He laughed at me for worrying about the bill so much. We went to the Opera, *La Sonambula* [*sic*], Miss Kellogg taking that part. . . . a splendid opera and I was delighted with it. Charles came out with me and stopped with me tonight. Splendid lecture from Wyman today.

March 29, Friday: At the Opera with Charles again tonight, *La Juive,* the Jewess, and it is certainly a most brilliant opera. Miss

Colson as Rachel, Mrs. Hinckley as Eudoxia. The concerted pieces are truly magnificent. The Opera is heavy and grand. . . . Charles stopped with me again tonight. Made me a present of a nice pair of suspenders.

March 30, Saturday: Could not resist the temptation offered by the presentation of [the opera] *Martha*. Enjoyed it exceedingly; most of the airs are now familiar to me.

April 3, Wednesday: At meeting of the Society tonight; introduced Mr. Goodale, secretary of the Board of Agriculture of Maine, to Mr. [Jules] Marcou. Goodale is looking for a geologist to explore the State of Maine.

April 4, Thursday: For some time the Legislature and Senate have bothered over a bill to give the Museum $20,000 and yesterday afternoon it finally passed, much to the happiness and satisfaction of all of us connected with the Museum.

Agassiz, as soon as the funds were assured, resumed his extensive plans for collecting trips on behalf of the museum.

April 5, Friday: Prof told me yesterday that he should want me to go to Cincinnati this summer for the purpose of making drawings of the animals of Unios while yet alive.

Both *Unio* and *Limnea* are fresh-water mollusks inhabiting rivers. *Unio* and *Limnea* are widely distributed throughout the world. The Unionidae burrow in the soft mud of rivers and lakes. Their developmental history is unique among mollusks, as the young larva undergoes part of its life as a parasite on fish. The genus *Unio* was so common in the United States that its shells were used commercially in button-making. Its popular name is fresh-water mussel.

April 7, Sunday: Finished letter of 40 pages to Nellie. Commenced reading Lowe's Biographical History of Philosophy.

April 9, Tuesday: Splendid day. . . . First rate lecture from Professor Wyman. . . . Prof [Agassiz] is bound that I shall go West. Hurray!

Hurray, indeed! After eighteen months of lugging jars of alcohol, copying catalogues, preparing microscope sections, and making painstaking drawings, he was to have his turn at going to far places on "an expedition." He had toiled in the Cambridge heat all the previous summer, with no jealousy of the students who were sent to the Saskatchewan or the northern Maine forests; he had pegged away at *Melaptera* under the microscope while Verrill and Shaler went to the Smithsonian. Now his chance was promised him. Hurray!

Ned had been too absorbed in his scientific studies and his reading of German and philosophy to note developments in Major Anderson's long ordeal at Fort Sumter. The news one day later, which was to destroy his chance at the Midwest expedition, came as a complete surprise to him, and he saw no connection in it with his affairs.

April 10, Wednesday: Exciting news from the South. Lecture from Prof. At work on alcoholic Mollusca, putting them in glass-stoppered bottles [because they had found that the alcohol evaporated through corks].

April 11, Thursday: Interesting lecture from Professor Wyman. Sent letter to Nellie; she is at my house in Gorham.

April 12, Friday: Made estimate of jars wanted for Museum. . . . Prof talked with me on my trip out West.

April 13, Saturday: Lecture from Prof. He referred to the exciting news from the South and said that those who sought Nature always found relief if their minds were distressed. Received $40.00 from Prof. Went into Aquarial Gardens to make drawings of animal of

Apheris accentulis. . . . Terrible news from Charleston, South Carolina. The Forts [Confederate] had opened fire on Sumter, which replied. The fleet off the bar where it is preparing for action. God save the Union.

Paid out $14.00 board; $3 Katie, wash; $5.59 Bonell & Morre; $1 Post Office.

April 14, Sunday: Harry [Barry] up this afternoon; showed him over the cellar of the Museum. News of the surrender of Fort Sumter. Intense excitement.

April 15, Monday: Intensely exciting news this morning. All the previous news confirmed. Lincoln calls for 75,000 men. Saturday Prof after lecture told us what we should do. That we must work and show the people that amid such exciting times we should not go wild with the general excitement but work. That we must not read papers, etc. This morning Ordway was reading a paper and Prof. said we must remember what he said on Saturday, and then went out of the room. He immediately came back and said if we did not think as he did, he should forbid every paper from coming into the building. This was perfectly outrageous on Prof's part and we expressed our minds freely on the subject. Hansen spoke to him about it afterward, and Prof confessed that he was hasty in his expressions. . . . Received another letter from Nellie; she is very much afraid of my going to war. Club meeting tonight; I spoke of the differences existing between *Unio* species *Patella testitudinata,* supposed to be identical in Europe and New England. Bottling specimens today. All Northern States have got to add to their quota to make up the 75,000 and it will be done with the greatest alacrity.

April 16, Tuesday: Lecture from Wyman. . . . Letter from Sister Carrie this morning.[6] Troops responded to the call nobly, and Boston is just filling up with them. Every town contributed large sums of money for the support of the families of the soldiers while

[6] His mother, too, was afraid he would enlist.

absent. Ordway has joined the New England Guards. I never felt so excited about anything in my life.

April 17, Wednesday: All is excitement regarding the war news. I am terribly impatient to get off with them. Sent letter to Nellie and wrote one to Uncle Charles [Beckett].[7] Went into Boston this afternoon to see the troops off. . . . Saw the Sixth Regiment leave for Washington from Worcester Depot. Immense enthusiasm. Saw many affectionate partings. The troops look well in their new cloaks. They appeared jolly and hearty but I could detect an undertone of seriousness. We also saw the Fourth Regiment leave[8] for Fort Monroe, Virginia. Dear me, how I want to go!

April 18, Thursday: Professor Wyman delivered his last lecture for this term today. Nellie has requested me to try and leave off smoking; I will try for her dear sake, though she is ignorant regarding the effects of the use of it. I cannot tell how I can do it. . . . News of the secession of Virginia reached us tonight. Everyone feels blue and the signs are gloomy and bode no good. Wrote Nellie and Carrie.

April 19, Friday: Attended Cooke's lecture on chemistry; very interesting. Many troops have left Boston and many are still to leave. The most intense exciting news comes continually. The Massachusetts troops were attacked by a mob on their way through Baltimore and two of the soldiers were killed. It is a singular coincidence that this should have happened on the anniversary of the Battle of Lexington where, as in this case, Massachusetts blood was shed.

April 20, Saturday. Excitement. Excitement. I am nearly dead with impatience to be off and into the fight. I must go home and see Nellie first. Received a dear good letter from her this morning; do not feel much like answering it, for I am in a terrible state of turmoil and excitement.

[7] His mother's brother, to intercede with her.

[8] From the Providence station for Fall River, to embark on steamers.

April 21, Sunday: News of the capture of the "Star of the West" by the Secessionists reached us today. Also the lines, telegraph, are down between Washington and Baltimore. Troops are going to Washington by large masses and thousands more volunteering to go. Wrote letter to Nellie.

April 22, Monday: Still at work arranging alcoholic mollusca from corked bottles to glass-stoppered ones. Letters from Nellie and John Gould. . . . Club meeting tonight.

April 23, Tuesday: At Bell's this evening . . . Uncle Jim [Bell] drilled us in the manual [at arms].

April 24, Wednesday: John has enlisted in the Light Guards. If I was not engaged, I would enlist, too. I shall at any rate, when I am actually needed. Dear fellow, I would like so to be with him.

April 25, Thursday: Bertie left us today. His company has been ordered to Fort Independence. At Barry's this evening and Uncle James drilled Henry and myself. . . . At work helping Prof arrange the Unios. Prof looks rather cross today; he does not appear as if he liked Bertie's going away. If they knew in Boston how greedy Prof is of our time and how actually tyrannical he is at times and actually unpatriotic, it would kill the Museum.

Agassiz, however, was not a native American and for forty years had devoted himself with single-hearted purpose to science. He had had to abandon the museum he founded at Neuchâtel because of a European rebellion, and is entitled to sympathy in his distress at seeing the funds and staff of his new Museum of Comparative Zoölogy threatened by the outbreak of a war in the country where he had fled for peace to pursue his scientific work.

Morse at seventy recalled without emotion and prejudice the incidents of "Bertie" Ordway's going away to war. Speaking at the unveiling of the Agassiz tablet in the Hall of Fame, May 30, 1907, he paid tribute to Agassiz as a naturalist and a teacher and then described, more charitably than in 1861, Agassiz' farewell to Ordway:

> In the early part of the civil war [*sic*], one of our class [Bertie Ordway] enlisted and received an appointment as an officer of the line — the rest of us bought a fine sword and presented it to him. On showing the sword to Agassiz, he instantly threw himself into the attitude of a fencer and became absorbed in thrust and parry, utterly unconscious of our amazement at his earnestness and skill. We learned afterwards that as a student at Munich, he had not only fought a number of student duels in which he was always the victor, but on one occasion he had challenged a whole class. . . . After four had crossed swords with him and been vanquished the rest were quite ready to retire. . . .[9]

The journal continues:

> April 26, Friday. Attended Professor Cook's lecture on chemistry. Warlike news still comes. At Bell's this evening, drilling with Harry. Shall go home next Monday.
>
> April 27, Saturday: At work hard selecting series from Unios for collection and also educing genera.
>
> April 29, Monday: Started for home this afternoon. Received $20.00 from Prof.

Ned Morse went home to enlist in a Maine regiment. He thought he could gain his mother's consent, for she had always helped him in his dearest wishes. He expected that, after a few tears, Nellie would yield her compliance, and he would be off with John Gould.

There must have been some pathetic scenes. Years after-

[9] *Popular Science Monthly*, December 1907, p. 542.

ward, Morse cut from the diary the pages for the next fortnight. Evidently Jane Morse and Nellie Owen had agreed on their stand during Nellie's visit to Gorham. He could not persuade them, nor could he gainsay their cogent arguments. His mother was widowed, with three daughters and inadequate means. She undoubtedly could say with justice that it was not fair for him to leave the whole burden to big-hearted Fred, who was supporting the family, getting his own dinners in the office to economize, so that Ned could finish his courses with Agassiz. Ned finally surrendered. The journal resumes:

May 15, Wednesday: Bid goodbye to Nellie . . . and came to Boston this morning. . . . Maine Regiment, 2nd Company, came up today.

May 16, Thursday: At Museum. Boys all glad to see me back. Worked on Unios.

May 18, Saturday: Verrill went home today. . . . The trees are mostly leaved out, and the flowers are blooming. But our country, our poor country is in a sad state. War is rampant.

May 22, Wednesday: Stupid lecture this afternoon from Prof. I went to sleep. After lecture walked in with the boys to Aquarial Gardens; they received a whale this afternoon twelve feet long and we saw him in the big tank, cutting up at a terrible rate. . . . Prof was in there.

May 23, Thursday: At work as usual. Cannot make head or tail of the Alas. edentula. Shall leave them for a while for something else. . . . Bought Dalton's *Treatise on Human Physiology* $3.60; it is an excellent work and what I have been wanting for a long time.

May 24, Friday: Young Charles Hitchcock has got the situation of Geological Surveyor to the State of Maine. He came to the museum today wishing to get one of the Maine boys as a Zoölogical assistant. Prof told him that he would not let us go, that it would break up our studies, etc. After Hitchcock left, Prof came to me saying that it would be a good way for me to pass my summer vacation and I might go; so I went into Boston but could not find [Hitchcock]. Must write him about it. I would like to go very much as I might learn something in regard to Geology. . . . Called on Dr. Gould this afternoon; was introduced by him to Capt. Couthouy. He is a very pleasant man.

Captain Couthouy turned over his material from the Wilkes expedition to Dr. Gould after he had been invalided home from Honolulu. He served in the United States navy during the Civil War, being killed in action in the turret of the U.S.S. *Chillicothe* off Grand Ecore, Louisiana, March 3, 1864.

May 28, Tuesday: Letter from Mr. Hitchcock; he wants collector to go into the field next week to remain until September, so that chance is knocked in the head.

May 30, Thursday: Changing specimens from cork bottles to glass-stoppered ones. Prof showed me a lot of generic distinctions in the Unionidae. It will be splendid to work these all up and I hope to be able to do it.

Agassiz, with his magnetic personality, almost succeeded in making Ned Morse forget his inner turmoil over the war in a fresh field of research. Morse seems to have felt no bitterness against either his mother or Nellie; only a dull disappointment at a hard fate.

May 31, Friday: At work. Seems lonesome in the Museum without Bertie [who had joined the army] and the rest of the boys. . . .

June 1, Saturday: First Maine Regiment comes today. Went into town about noon. . . . Down to the Depot, train got in at 3. Found John Gould at once. He gave me his Journal written on thin paper since he has enlisted and he is to send it to me all the time he is away. Dear me, I wish I was going with him. Staid by him till the Regiment left. They drilled on the Common and I had a fine time with him. John Hudson, Ned Bicknell and a lot I knew were there. They went through drill on the Common. Dear boy, I wish I was with him and nothing but my darling Nellie prevents me from joining.

June 3, Monday: Changing bottles at the Museum. A corrupt hole and I wish I was out of it.

June 5, Wednesday: Prof looked awful blue today. He will not be able to humbug the people much longer. Gave Bertie a blowing up for being away with the New England Guards. Bertie said he was needed; then he spoke of taking a degree. Prof said there was no one in the Museum fit to take one. Bertie said he had been here three years and two-thirds of the time he had given to the Museum. Prof said such practical work was what we needed. Bertie thought that too much of a good thing was good for nothing.

June 7, Friday: I am getting the Mollusca in first rate condition. Told Prof yesterday that I was going to spend my vacation at home this summer and if he doesn't like it, he can lump it.

June 10, Monday: Shaler and Hyatt started for Anacostia [Anticosti] this morning; they are to wait for Verrill in Eastport.

It is obvious that Agassiz must have abandoned the Cincinnati expedition, on account of the armies in the field in the Ohio valley, and this, in addition to the family duty that kept him from enlisting to go to war with John Gould and Bertie Ordway, left Ned in a disgruntled state. His distaste for the museum sprang from the bitterness of seeing Shaler, Hyatt, and Verrill set out for Anticosti Island in the Gulf of St. Lawrence, where they were to collect fossils and marine animals. The three later worked in the vicinity of Grand Manan Island, off the coast, where Canada and Maine meet at Eastport. (Ned

Noritane Ninagawa

evidently confused the name of the Canadian island with that of the Maryland village near Washington.)

June 11, Tuesday: War, War, War. I don't write about it. Lies circulating about it in all the papers. It is singular that I don't write more about it, but I don't find time, though I am intensely interested in it. It is a luxury to know that all the Mollusca down cellar is finally safe from drying up.

June 13, Thursday: Kegged the Unios today. Verrill went this morning.

June 14, Friday: Received John's journal from Washington. Very interesting although some parts of it difficult to read.

June 15, Saturday: While in the Square this afternoon, the First Massachusetts Regiment came along on their way to Boston. How splendid the drums sounded! I marched way in with them to Boston and I believe I could have marched to Washington with them. Who could not fight with a drum corps?

June 18, Tuesday: My birthday; I am twenty-three years old. O dear, I wish I was twelve years younger. Letter from Ad Verrill from Eastport. Nellie will be here tomorrow.

June 21, Friday: Class Day. In the afternoon with Nellie to see the dancing and festivities of Class Day. . . . Beeman invited me into his room. We staid up there a little while refreshing ourselves with strawberries and cream. It rained and I got Nellie a cape and umbrella. Saw the students dance around the tree and it was a ludicrous sight to see them rush for the string of flowers nailed up on the tree about ten feet. Such a rough and tumble scrabble I never saw. With Nellie this evening at Mr. Richardson's . . . and we placed on each other's hands rings never to come off again.

June 22, Saturday: Worked hard at the Museum this morning. . . . Last lecture from Prof today.

June 24, Monday: Hard at work bottling Unios.

June 27, Thursday: Prof loaned me a nice microscope to carry down [to Maine] with me. Finished up work at the Museum. This evening Nellie and I took the boat for Portland. Prof gave me $50.00.

July 3, Wednesday: Received John's Journal and a letter from Prof Agassiz. He wishes me to collect alcoholic *Helices*. Has received a new work on the Genera of *Helices*. The *Helix pulchella* belongs to the family Verronia and is the only species known to the author.

July 16, Tuesday: Received John's journal from Washington. . . . Bertie Ordway called on him; he is in Washington trying to get an appointment to West Point. . . . Called on Dr. Tewksbury in town; talked with him on the origin of species; he is inclined [to be] Darwinian.

The first troops mustered in on Lincoln's call for 75,000 men were enlisted for three months and had to be officially discharged at the end of the time, most of them re-enlisting afterward.

August 3, Saturday: Nellie and I went in town today to see the return of the First [Maine] Regiment. They arrived about 2. . . . A harder, browner-looking set I never saw. I hardly knew John Gould with his whiskers and mustachios. They were marched around the city and then partook of a collation at the steamer wharf building. I tried to get in but could not. Went out on the end of the wharf and saw [Charles B.] Fuller in his boat. Went back and made out to get in the building; found John Gould and John Hudson, and Fuller gave us a splendid sail down the harbor. Stopped with John tonight. Received a splendid letter from Bertie Ordway who was at the battle of Bull Run and saw the whole of it.

Ned Morse loved his mother and Nellie. He was doing his duty by them with the best grace he could muster; but he meant to have no women folks around when he and John sat down to talk about soldiering. A sailboat in Portland Harbor, with their good friend Charles Fuller at the helm was a perfect man's sanctum, in which the two Johns, Gould and Hudson, might tell of the war.

Ned was dividing the summer vacation between his mother's home in Gorham and Mr. Owen's farm at Cape Elizabeth, lending a hand with the chores in both places, and collecting "alcoholic mollusca" and insects for the museum. His temper was sometimes short under the chafing inaction while all the boys he knew were going to war.

August 18, Sunday: Through my intense ugliness and obstinacy, I caused moments of pain to the dearest one to me on Earth. I suffered for it and had I been the only sufferer, I should not have cared; but Nellie, so good, so kind, forgave me and even attributed blame to herself, dear girl.

His Nellie understood him pretty well, and loved him enough to smooth the course of his daily life. A factor that contributed to his irritability was toothache. There is no mention of dentistry in the journal, and, although there were dentists practising in the mid-nineteenth century, their art was chiefly manifested in pulling teeth and fitting false teeth.

September 3, Tuesday: News reached us yesterday of a glorious victory of Federal navy at Hatteras inlet. Brought can of mussels from Fuller's.

September 4, Wednesday: Into Portland with Dr. Tewksbury. At Charles Fuller's a little while; his Aquarium is in splendid condition. Stopped with John all night. He is going back to the war as Sergeant Major in the Fifth Regiment. If I was free, how quick I would go.

September 10, Tuesday: Saw the Eighth Maine Regiment go through the city on the cars. Fine-looking boys.

September 13, Friday: Rode down to Fort Preble with Dr. Tewksbury. About 180 recruits there for the United States Regiment, and the toughest-looking fellows I ever saw.

September 17, Tuesday: Went into town today and made a cast of *Alasmodea arcuta;* then I sawed the cast in sections corresponding with the transverse cuts I made of the animal. I shall now be able to get an accurate outline of the interior of the shell. . . . Got a letter from Fred Putnam. Bertie Ordway is First Lieutenant and is recruiting in New Bedford. Jim Foley is captain of a Cavalry company.

September 18, Wednesday: Worked on drawing sections. . . . Ad Verrill out this afternoon. I showed Ad a species of Anemone on the shore which is a new species discovered by Stimpson at Grand Manan.

September 30, Monday: Rode in town [from Cape Elizabeth], went down to Portland Company and bid Fred goodbye; secured some valuable relics of Bateman's from Panama. Then [Nellie and I] drove to Fuller's shop and packed them with some other things to send off by express tonight. . . . Went out to the camp to find John Gould, but could not. Nellie packed my things so nice; we passed our last evening together.

October 1, Tuesday: Bid Nellie and all the rest goodbye and came in with Mr. Owen this morning. As soon as I got in the cars I buried myself in a newspaper and pipe . . . until I got to Boston. Brought some specimens of a new *Actinia* up with me alive. I carried one of these to the Aquarial Gardens and the rest brought to the Museum. The Museum looked natural enough. I missed Bertie [Ordway], Harry Rice, Nat [Shaler], Bowditch and Sam Scudder who are all absent. I felt so blue at leaving Nellie that I rushed over to Barry's in the hopes of singing away my pining, but it was no use. . . .

October 2, Wednesday: Prof was delighted with the collections I brought back which arrived safely. I commenced work this morning on the Unios.

October 4, Friday: At work hard on the Unios. . . . Attended Prof's lecture at Lowell Institute tonight. He regarded cats and dogs as belonging to the same family. Walked out.

October 6, Sunday: Went into Boston to see the Tenth Maine Regiment arrive. John Gould is Sergeant Major. I saw him and had a nice long talk. The Soldiers looked rugged and hearty. . . . Stimpson [of the Smithsonian Institution] was in this morning; he thinks that draughting would be a good line for me to earn my living, and I think so too. Wrote Nellie and Mother.

He had been back at the museum less than a week, but he knew he could not stand a winter of Unios, when every drop of blood in him cried out to join the soldiers, though every fiber of New England character constrained him to do his duty by the women-folks. He must have unburdened himself to Stimpson, or that earnest scientist and former pupil of Agassiz would not have been counseling him how to earn a living away from Agassiz. Ned put Stimpson's counsel into effect at once, as usual whenever a new idea had seized him.

October 8, Tuesday: Bertie Ordway, now First Lieutenant, in the Massachusetts 24th Regiment, came to the hall today; passed a very pleasant evening with him and Putnam in their room, talking over old times. Bertie gave me a cane and drawing tablet. Packard came today and is stopping in my room till Hansen goes.

October 10, Thursday: Went into Boston tonight to find a school of design; there are two at the Lowell Institute, one of models, blocks, etc; and the higher one, which I joined, of live models. There are about 20 in this school. I enjoyed it very much. Shall go every Monday and Thursday night.

November 1, Friday: Two years ago today I came to Cambridge; it hardly seems so long and yet at times it seems much longer. I have learnt a great deal since I have been here; also have learnt that there is a great deal to learn. I have attended many lectures and have ample notes of the same. I hope that before another two years are gone I shall be doing something to earn a living and also to be married. . . . Prof gave me $40.00.

November 15, Friday: We have had freezing weather the last two or three days. . . . Papers are brimming with accounts of the Federal victory at Beaufort, S. C. Our success there was a splendid one and if we can only hold it and make it a rendezvous it will be splendid.

November 19, Tuesday: News from Verrill [who had been called home]; his mother died last Thursday. Club meeting tonight. Burt Wilder is our president. I am at work now in my room preparing a paper to be read before the club showing my reasons for considering the articulates as a class assume a higher position than the mollusca.

December 1, Sunday: Have been in the house all day; made dissection and drawing of *Bulimus*. Stimpson in this forenoon; we had a jolly chat together.

December 2, Monday: At drawing school in evening. Drawing some things for Prof on wood. Received from Prof $35.00.

December 3, Tuesday: Paid $18 Bells [for board]; $3 Katie, wash; $6.00 boots; $2 [borrowed from] Packard; Post Office $1.

December 9, Monday: Drawing school this evening. Sick and miserable [sore throat]. Worked hard on Bryozoa and Terebratulina.

December 10, Tuesday: Club meeting tonight. Shaler read a paper which was full of mistakes which I shall correct.

December 17, Tuesday: Refuted Shaler's statements made at the last meeting and convinced the club that Shaler was wrong and I right. Chosen secretary pro tem. [Next day] Prof said I was perfectly correct in my remarks last night.

December 18, Wednesday: Alexander Agassiz took his degree this afternoon; we were all present. A few days ago Prof stated to Hyatt and myself that he should screw Alex harder than anyone and gave us to infer that he would be harder on Alex than he would be on us as he might be accused of partiality, but we were all astonished at the simple questions asked, not one of which we could not have answered; the examinations occupied only two hours and none asked questions but Prof.

December 20, Friday: Moved alcoholic collection of Unios in drawers to go down cellar.

December 25, Wednesday: The Bells had a Christmas supper but I missed my Nellie and did not enjoy it the way I did last year. The Barrys and Adams were there.

December 26, Thursday: Made up my mind to leave the Museum today. I know I can do better with my time at home than I can here. At Lowell Institute, took my drawing lesson. . . . Letter from Stimpson; Prof had carried it in his pocket a week.

December 27, Friday: Last night into Aquarial Gardens and saw a whale drag a boat around [the tank] with a girl in it.

December 30, Saturday: Letter from Mother and Nellie. Saw Prof this afternoon and told him I was going home. He appeared blue enough and when I asked him for money, he said he had not $10.00 in his pocket or the world.

December 31, Sunday: Prof came to me this morning and said he had misunderstood me last night and was astonished and indignant that I was going to leave on such short notice. He showed himself in his true light . . . ; as long as one will toady to him and be content to live on nothing, without hope of anything, so long will everything go smooth; but when one asserts his independence then is the man vexed and indignant. He insisted on returning the drawings I made of the Unios.[10] O, he was so wicked and unjust. Could get no money for my Cabinet and after paying me my month's wages, paying me in New York bills, proving that he told a lie last night, he bid me goodbye without wishing me any hope for the future. . . . Bid goodbye to all the folks and really felt sad at leaving them.

Another year has gone; what will the end of next year bring? Nothing of importance has happened to me personally, but the world has been startled by one of the greatest and most fearful rebellions on record and the world has witnessed a mighty uprising of a loyal people to crush it. I have suffered intensely in not being able to fight in the common cause, but Nellie would not hear to it; and John earnestly advised me not to go as I was not needed and they could get troops enough. . . .

I have learnt a good deal in Zoölogy this year. . . . The end of this terminates my connection with the Museum of Comparative Zoölogy where I have been for over two years. I have learnt everything there and feel indebted to Prof for such good as he has done me; but sincerely condemn his actions in other matters.

[10] For which Ned had expected to be paid.

CHAPTER SEVEN

HAVE LEARNED EVERYTHING 1862—1866

SO, in his twenty-third year, Ned Morse severed his connection with Agassiz at the Museum of Comparative Zoölogy. His manner of leaving, impetuous and boyishly brusque, undoubtedly offended Agassiz, but his reasons were sound and intelligent. With Yankee shrewdness, he had perceived that the disciples of a great master were eclipsed by the luminous center of their "solar system."

He made a clean break of it, evidently with the advice and approval of William Stimpson, who was also an old Agassiz student. Since there had been no written contract, he had no redress when Agassiz chose to consider the Morse cabinet of shells as a donation to the museum. Yet he had collected the shells before he joined Agassiz's staff. They represented the work of years. Visits made to view them in Portland by Stimpson, Cooper, and Carpenter were evidence of their value. Morse's bitterness at having his shells swallowed up in the Museum of Comparative Zoölogy, without recompense or credit, was reflected in his book, *Observations on the Terrestrial Pulmonifera of Maine,* published two years later: "Having most unfortunately parted with my own collections, the results of diligent labor of several years, I have depended greatly on the valuable collections of Adjutant John H. Gould and Mr. Charles Fuller of Portland, Me." It was, thus, fortunate that he had always shared "next bests" and duplicate shells with Gould and Fuller. Certainly he wrote the sentence above for the eye of Agassiz.

When Morse had paid his board bill to the Bells, his washing account to Katie, and bought a ticket to Portland, he had not five dollars in his pocket to show for two years of work at Cambridge. On the other hand, and he had justice enough to admit it at the time, Agassiz had given him an education in scientific principles and research that would underwrite the success of his future career.

Religion as a problem had ceased to disturb him, had practically faded from his mind. Earning a living was not a problem, for he now had an education, training in drawing, and the encouragement of Stimpson in setting up as a conchological draughtsman. He looked forward confidently to being able to get married shortly. The Civil War remained as the only unresolved conflict in his life. Until he achieved an inner peace, he could not settle down either to work or to marriage. The tug-of-war was there all day, all night as he tossed sleepless in bed. He wanted passionately to enlist; yet everyone dear to him said that he must not.

January 1, 1862: Had a pleasant trip down [to Portland]; sea calm and unruffled. . . . I commence the New Year independent, and poor as a church mouse. I must work hard in the profession I have chosen, that of Zoölogical draughting, and I shall try to lecture some. At any rate I shall make desperate attempts to earn money.

How splendid it was to see my darling Nellie. . . . We went to a dance at Lancaster Hall this evening. Met the little Freeman boy, an orphan that lived at Dan Hastings when I was there to school in 1850. I taught him to fife and he continued to practise; went off as fifer to the Fifth Maine Regiment. He was taken prisoner at Bulls Run and has just been sent home from Richmond. He entertained me with an account of his adventures.

January 4, Saturday: Received letter from Prime, who wishes me to draw some shells for him. Letter from Verrill.

January 5, Sunday: Corrected piece to read before the Society. Commenced writing a set of lectures.

It was probably William Stimpson who had given Morse's address to Temple Prime, merchant and amateur conchologist, of New York, as Prime hitherto has not been mentioned in the journal. Ten days later Prime sent Morse a box of shells to draw. Illustrations for books, magazines, pamphlets, or newspapers for a few years longer were still to be drawn by artists for reproduction by woodcut, and this was the profession Stimpson had suggested to Morse as a means of support while he continued independent scientific research. Its primary advantage was that the work could be carried on while he lived economically at his mother's home in Gorham.

January 6, Monday: Into town. This evening read a paper before the Portland Society of Natural History.

January 18, Saturday: Worked on drawings for Mr. Prime. Fred came out on the evening train and brought me the wood [blocks].

January 20, Monday: Finished drawings of Cyclas and the shell on wood.

January 22, Wednesday: Nellie and I went back to the Cape today.

January 23, Thursday: Hard at work all day chopping wood and getting ready for the party this evening. Mrs. Sarah Shaw has got up a surprise party and Silver Wedding to be given Mr. and Mrs. Owen tonight. I went in town twice to get things. The party came out late in the afternoon; fifty-five in all. We had dancing and singing. Several small sleighs and the large boat on runners brought the people to and from the house. Mrs. Owen was the recipient of very handsome silver presents from relatives and friends, a silver tea set; knife and fork; napkin ring, spoons, etc. Poor Nellie had a terrible headache all day. The supper was brought over during the day, consisting of turkeys, ice cream, ham; and the tables looked splendidly spread with cake and confections. I busied myself during the supper in serving out the coffee and it was quite a job to supply all their wants. Did not get to bed till after twelve, tired and sleepy enough.

Surprise parties from the neighbors were a feature of old New England life in rural communities to celebrate social occasions. Under the leadership of some bustling, executive housewife, a group of women in their own homes prepared in advance the turkeys, hams, ice cream, cakes, bread and butter, jellies, preserves, and custards for fifty to a hundred friends, to be smuggled into the house with the connivance of some member of the family. In this case, judging from the headache, Nellie was the accomplice, responsible for having the house swept and garnished, fires built in the best parlor (never opened except for funerals and parties), bedrooms tidied to serve as cloakrooms for the guests, and her parents warned in time to change slippers for best shoes, apron for best silk gown, and shirtsleeves for coat, collar, and tie.

February 1, Saturday: Made drawings of teeth of *Nassa observerta* for Prime. Fred came home tonight [from work at the Portland Company] bringing . . . a bundle from Sam Scudder. Enclosed I found a fine present from Sam in the shape of Savigny's Memoir on the *Ascidians,* a most valuable work which I wanted very much and I hardly know how to thank him for it. Also Stimpson's *Invertebrata of Grand Manan* from Stimpson; and Gould and Agassiz with two papers of Prof from Prof on the *Acephala* of Massachusetts Bay. . . . Scudder sent his brother's manuscript which I have read and think I can illustrate.

A branch of learning which men may spend a lifetime in studying without reaching proficiency is the science of reading the true intentions of friendship. As Paul of Tarsus wrote a couple of thousand years ago, "we see through a glass darkly" when we look at our friends, constrained by human limitations to view them through the glass of our own temperaments, their image often distorted by our emotions or prejudices.

A month earlier Morse, distraught, poverty-racked, yearning for marriage, money, and success in science, had burst precipitately into the study of an overwrought professor, harried with financial and administrative problems, weary for want of leisure to pursue his own researches. Brusquely he had announced that he was leaving at once. The ensuing interview had convinced Morse that Agassiz grudged him independence and success, Agassiz that his pupil lacked loyalty and courtesy. Now, in the tranquillity of country life at Gorham, Ned Morse received evidence that he had been mistaken. All the Cambridge friends in science loved him, wished him well, and were coöperating to further his career. At the Museum of Comparative Zoölogy they must have been speaking of him, planning how they could help, getting together a bundle of gifts from each one. "Prof" sent his new papers on the *Acephala* and *Principles of Zoölogy,* a textbook on which Dr. Augustus Addison Gould had collaborated; Stimpson contributed his work on Grand Manan; Scudder the Savigny work he knew Ned could not afford to buy, and a manuscript that Ned might illustrate for pay. Stimpson continued to advertise Morse as a zoölogical draughtsman to all the conchologists with whom he came in contact.

February 13, Thursday: Went into Portland soon after breakfast; Nellie did me up some dinner, ham, bread and butter and a lot of mince pie. I stopped at Fred Kimball's rooms; he made up some tea . . . and we had a jolly lunch. Went to Fuller's house and drew live Mollusca from his Aquarium. In the evening at John Hudson's; the Polymathic were all together. Made Tom and Jerry and had a jolly time. Played accompaniments [on the piano] to two violins, John Richardson and Carlton; guitar, Fred Kimball; trombone, Ned Bicknell. We made out quite an orchestra. Letter from Mr. [William G.] Binney today; he asks the price for 25 woodcuts of *Paludina*. . . .

Now that he had work, that he could see Nellie every few days, that he was buoyed up by his confidence of making a living, he settled easily and naturally into the community life of his old circle in Portland. He had forgotten that he had called it a country village — a dull, stupid place.

February 15, Saturday: Nellie and I rode into town today . . . into Back Cove where a thousand sleighs were congregated. A carnival; it was called Fantaisies; the Military Band and lively times generally. We enjoyed ourselves very much.

February 22, Saturday: Letter from Sam Scudder enclosing a letter from Stimpson to Put. Stimpson wants me to take the fluviatile and land shells while he takes the Marine. Good idea. Enclosed also was a shell from Dr. Gould; he wants me to draw six views of it.

February 28, Friday: Snowed very hard and blew a gale all day. Everything blocked up. Finished another drawing of Prime's. Shovelled path out tonight. Commenced reading Anthony Trollope's "Orley Farm" in Harper's [Magazine] tonight.

March 1, Saturday: It cleared off bright and beautiful this morning. The snow is piled up in high drifts around the house and looks like great waves just ready to dash on the house. Worked on drawings for Prime.

March 4, Tuesday: Raining hard all day. Received John's Journal [from the battlefront]. . . . One year ago today and Lincoln was inaugurated. Then the Union was upside down. Treason was everywhere. We had no troops in the field and the rebels were having it all their own way. Now we have 600,000 in the field, a victorious navy and victory, glorious victory in quick succession. May the good work go on till the Rebellion is among the things of the past.

March 5, Wednesday: Finished up all my drawing work. . . . Received Marcou's paper on Taconic system and Sam Scudder sent me Shaler's paper; I also got a letter from Shaler.

With his orders for drawings completed, Ned immediately resumed his own research work on land shells. The suggestion from Stimpson that he write a book on the land shells had crystallized into a project which would be completed in the course of the next two years. He had also been writing a series of lectures and was now looking for a chance to deliver them.

March 19, Wednesday: Letter from Mr. Prime with $20.00 in Government bills.

March 22, Saturday: Made preparations of lingual ribbons of *Buccinum undatum*, *Littorina radiata*, *Pecten testitudinalis* and *Crepidula plana*.

March 27, Thursday: This evening I called on Mr. Webb at the [Gorham] Seminary. He appears to be a first rate man and I had a nice two hours' chat with him. In the Cabinet room there is a very good collection of Shells and Minerals. He is going to get up a class before which I am going to lecture; at any rate make the attempt.

March 31, Monday: I am in hopes that I shall succeed at lecturing and if I only can I shall be glad enough, for I am feeling blue again and it seems awful that I should not be earning a good living.

April 1, Tuesday: Letters from John Gould and Ad Verrill. . . . John has moved on to Harper's Ferry.

April 2, Wednesday: Splendid weather though the ground is deeply covered with snow. . . . Sent my answer to Shaler's paper for the Boston Society of Natural History. Wrote to Bland [accepting work in drawing shells]. Fred brought me out [from Portland] today two packages from Mr. Bland, one containing Lithographer's stone and crayons; the other drawings and Bland's shells. So he has trusted to my ability to do them. Well, I will make a desperate attempt. Went into the Seminary tonight; had a nice time in conversation with Mr. Webb [about the lecture course]. . . .

April 9, Wednesday: Reading Boswell's "Life of Johnson"; very interesting.

April 10, Thursday: Letters from Bland and Prime. . . . Bland is in a hurry for the [lithographer's] Stone; Prime wishes seventeen figures of shells on wood. Also received John's Journal from Harper's Ferry.

April 14, Monday: Worked on Prime's things. Finished second volume of Life of Johnson. Letter from Binney this afternoon; wants forty shells drawn at 50 cents apiece. Very good.

April 15, Tuesday: Saw Ned Gould today; he says John [Gould] is Second Lieutenant. Hurray for John! I am glad he has got this promotion; it not only shows they appreciate John's fidelity to his duties but is pleasing to John. Fred brought out my blocks of wood.

April 17, Thursday: Received the rest of Bland's shells; shall draw them on the Stone at once.

April 22, Tuesday: John's Journal and a letter from Shaler . . . this afternoon; also box of shells to draw from Binney. Shaler is going to make defense of his article and wants me to supply him with notes of my paper.

April 23, Wednesday: Drew twelve *Paludinas* today without effort. . . . At Mr. Webb's this evening [arranging shells in his cabinet]. He told me he would put my name down in the catalogue of next term [at the Seminary] as Lecturer on Natural History. That will be good.

April 29, Tuesday: While eating chowder today got a fishbone in my throat and could not get it out. It pained and irritated me a good deal and I nearly stretched my jaws apart in trying to find it. Could see it but could not start it.

April 30, Wednesday: Went in town . . . to Dr. Wood who took the bone out of my throat, and it was such a relief. He said my throat looked bad and needed a regular treatment. Letter from Bland with proof of [drawings on lithographer's] Stone; he was much pleased with it and I was glad enough to know he liked it.

Morse had been suffering for four years from frequent attacks of "sore throat, shivery spells, fever, backache," etc. This was the first time any medical man had suggested "treatment" of the foci of infection. What he needed, evidently, was a tonsillectomy, which was still in the future as a surgical routine. He does not appear to have had the course of treatments suggested by Dr. Wood, probably for lack of funds. The omission was to cause the bitterest disappointment of his life a few months later.

May 10, Saturday: Drew seventeen of the hardest shells on wood today easily. As I get fifty cents for each one it is a profitable day's work. It seems strange that in this Journal I do not keep a record of passing events in the eventful things which are now transpiring. It is not because I do not take a deep interest in the welfare of our Union; but because reading the papers so much I could not commence without writing a good deal for which I have not the time. Be it known then that the Union Army is actually victorious everywhere. New Orleans is taken and almost every important point occupied by our troops. Even Yorktown where the Rebels were going to make a tremendous stand has been evacuated and the Rebels are in full retreat around Richmond where our Army will be soon. Certainly the Rebellion is on its last legs.

People still accepted General George B. McClellan's faith in his own abilities and destinies. The whole North believed that when McClellan finally moved on Richmond, the Rebellion would be crushed. Morse could not envision three more years of war, Lee crossing the Potomac into Maryland and Pennsylvania, the battles — Antietam, Gettysburg, Chickamauga, Chancellorsville, Vicksburg — that were to bleed South and North alike to pale, desperate brothers in a death grip. Morse continued to read optimistic prophecies of Union victory in the newspapers while he led a pastoral existence, planting potatoes in his mother's vegetable garden, collecting land shells as the Spring moved up the coast, working on the orders for woodcuts that came from his good friends.

May 19, Monday: Hard at work all day planting.

May 20, Tuesday: Worked hard all day planting.

May 24, Saturday: . . . out to the canal this morning. I got a lot of Unios and came to Gorham on foot. Stopped at Mrs. Cotton's for a drink of water and she insisted I should eat something, had splendid milk and fried parsnips and bread and butter. Got home about 2. Found letter from Binney with another shell enclosed. Feel tired all out and I am about used up.

May 26, Monday: Planted a lot of corn this morning. Drew a little but did not feel like work being all tired out.

May 27, Tuesday: Drew on my Unios. Letter from John. News yesterday that Banks [whom Ned met when, as governor of Massachusetts, he visited Agassiz's museum] was defeated and retreated through Winchester. John's regiment was engaged and I am anxious to hear from him. I hope to God he is safe.

May 31, Saturday: Up early and after breakfast started to walk to Portland and met Nellie at seven o'clock. It was splendid walking and I got within four and a half miles of Portland when I met my darling [driving a "horse-and-buggy"]. . . . Carried my circulars to Steve Berry to print and then went home [with Nellie] to the Cape.

June 1, Sunday: Got interested in the book "Thaddeus of Warsaw."

June 2, Monday: Got aggravating letter from Sam Scudder. He sent my manuscript down to have the letters put in plainer for the printer and it is lost. . . .

June 6, Friday: Came home [to Gorham] today in the cars expecting to deliver my first lecture tonight but found a note from Mr. Webb wishing to defer it till next Monday. I felt vexed at first, but found a letter from Sam Scudder asking me to rewrite the lost manuscript, as he would be held responsible [by the Boston Society of Natural History] for its loss. So after tea I sat down and rewrote the whole of it, not getting through till one o'clock. Letter from Bland with $13.00.

June 9, Monday: My first lecture came off this evening; what I had been dreading finally occurred. The graduating class, composed of five young ladies, Mr. Webb and two female assistants and Fred Morse composed the audience. Though a little embarrassed at first, I soon became interested and got along splendidly, feeling tiptop. I was perfectly happy when I got through at my unexpected success. Hurray! Wrote to Nellie tonight informing her of my good luck.

This first lecture, to five young ladies and two female assistants, was the beginning of a career in lecturing on natural history that Edward Morse was to carry on all through the United States and in Japan and Europe for nearly sixty years. He was a ready and entertaining speaker, graphic, with lightning sketches on the blackboard, executed with both hands working simultaneously; and he excelled at making scientific facts open enchanting vistas to large audiences. With no training in pedagogy, oratory, or dramatics, he combined all three instinctively in his lectures from the very first at Mr. Webb's Seminary in Gorham.

June 19, Thursday: Over to Natural History Society; came out in the evening train bringing with me specimens of live Starfish and one live *Actinia* which I am going to use tomorrow night at the lecture.

June 20, Friday: Painted three diagrams today [for his lecture]. Delivered third lecture tonight, on the Class of Echinoderms, and got along very easy. They all appeared much interested.

June 21, Saturday: Fred bought a cow this afternoon, $38.00; and I milked her.

In spite of the fact that he was busy and interested in what he was doing, the question of enlistment still troubled him.

August 20: After the repulse of McClellan, the President called for 300,000 men which has been generously responded to. Since then he has called for 300,000 more to be raised by Draft, and I have been excited enough. I have wanted to enlist, but Nellie and Mother made such a rumpus that I have not, hoping all the time that I might be drafted.

August 21, Thursday: Writing all day [on an article for the Portland *Transcript*]. Aching to enlist.

August 22, Friday: Worked around the house today. Restless and uneasy. Want to enlist. Began a letter to John Gould and before I finished it resolved to enlist.

So, after two years, he resolved his inner conflict. He had tried dutifully to accept the feelings of his mother and Nellie as to his responsibility towards them; he had endeavored earnestly to carry on his study of science in accordance with Agassiz's views on the relative importance of science and war; and he had resisted, with a minimum of complaint, the stirring appeal of patriotism, adventure, comradeship that gripped him as Bertie Ordway, John Gould, and thousands of other young men marched for war to the thrilling beat of drums.

August 23, Saturday: Went in early this morning. They have commenced raising a Young Men's Battalion. Talked with Nellie about enlisting and she, like a brave girl, said go, though the tears came to her eyes as she said it.

August 25, Monday: Signed my name to the enlisting paper of Company A, Twenty-First Regiment. This company is composed of young men of the city and will be an excellent company. I never felt better in my life than I did after I had enlisted. An event I have earnestly desired since the war began.

August 26, Tuesday: Went out to Gorham tonight.

As when his father died, Morse could not set down in his journal the depth of feeling, the intimate family scene, of the night when he went home to tell his mother he had enlisted. What Jane Morse said was written only in his heart.

August 27, Wednesday: Came in town this morning. With the company on parade in the streets this afternoon.

August 28, Thursday: Elected officers. Shaw is Captain; First Lieutenant, I forget his name; Second Lieutenant, George Abbott.

August 29, Friday: Marched around with the Company this afternoon. Nellie came out to Gorham with me tonight.

August 30, Saturday: Nellie and I looked over old letters and destroyed most of them.

Jane Morse hid the letters her son had written her, lest they too be destroyed; and many years later, when Jane Morse died, Ned found them and put them with the old journal.

September 1, Monday: Went up to the Examining Surgeon who examined me and I was rejected. I felt bad enough to cry. I had made up my mind to go, and then to be rejected. . . .

The chronic infection of his tonsils and the bad condition of his teeth had caused Morse to drop in weight from 148 pounds when he started work in Cambridge in 1860 to 131 pounds in the summer of 1862.

September 6, Saturday: Numerous disasters have befallen our army in Virginia. The rebels have crossed the Potomac and are in Maryland.

September 7, Sunday: Did not do much work today. Never felt so miserable in my life. Don't care a damn for anything; wish everything was in Hell, curse it and damn it. I do sincerely, truly and honestly believe that I am the laziest and most indolent fellow that ever lived. I sometimes regret that I ever gained the affection of Nellie, the dearest girl that ever lived. She should have had a smarter boy that I.

September 8, Monday: Hardly any work accomplished today. I seem to be completely upset mentally; cannot get my mind calmed down to a single thing. Letter from Fred Putnam in which he tells me that Hyatt, Wilder, Craigie, Bickmore, Luttrock, all members of the Agassiz Zoölogical Club, have enlisted. This, with Ordway, Foley, Bowditch and another, I forget his name, makes about the whole Club that have enlisted.

Morse's journal breaks off abruptly on the twentieth of September. Blank pages reflect the turmoil in his mind as he strove to adjust himself to circumstances. On the first of January, however, the old habit of summarizing the year just finished inspired him to write a résumé which gives the end of the story of his boyhood.

January 1, 1863: At the commencement of this year I returned from Cambridge to my home in Gorham with the firm resolve of making my way clear in view of marriage. As I was then situated in Cambridge, I received barely enough to keep me along and almost the only time I had to study was in the evening. Besides, the financial conditions of the Museum was very unsatisfactory. No hope for the future; small pay and the impending cloud of non-payment stared me in the face. On such uncertain ground I could not exist. At the commencement of the year I had about $10.00 in hand.

I received a few shells to draw for Mr. Prime and also some drawings on Stone which I did for Mr. Bland. I also issued a circular stating my occupation, that of Zoölogical Draughtsman, and sent it to quite a number. Had a large job of drawing to do for Binney and managed to lay by in the Bank in July the sum of $50 with all debts paid.

In the meantime I pursued my own studies; prepared a course of lectures in Zoölogy, with diagrams, colored, amounting to eighty-two figures. These lectures, seven in number, I delivered before the graduating class of the Gorham Seminary in the summer. Also prepared an article [on brachiopods] for the Boston Society of Natural History *Proceedings* which was published.

In July the work fell off. In August I had very little. In September I enlisted in the 25th Maine Regiment; when examination came I was rejected. About this time I commenced writing articles on Local Zoölogy for the [Portland] *Transcript.* So far I have written seven, four of which are illustrated.

September twenty-second I accepted a chance at the Portland Company at $9.00 a week, where I have been ever since. I have now, clear of all debts, about $75.00.

On his twenty-fifth birthday, June 18, 1863, Edward Sylvester Morse and Ellen Elizabeth Owen, the girl he had met and courted in Cambridge, were married at her home on Cape Elizabeth, near Portland.

In the year since he had resigned as Agassiz's assistant, to puzzle out for himself the truth or falsity of Darwin's theory of evolution, Morse earned exactly $256.41, by lectures and drawings of shells for conchologists. Nevertheless Nellie Owen and his family, alike, believed in his future as a scientist. Fred Morse was earning eighteen hundred dollars a year at the Portland Company works, humming with orders on account of the Civil War. The brothers had talked it over. Fred volunteered to be responsible for the support of their mother and the three younger girls, and, in addition, offered to enlarge the Gorham house to provide two rooms for Ned and his wife.

The naïveté of spirit with which Ned Morse assumed that the girl who loved him was prepared to share not only his scientific career but all other aspects of his life was never better illustrated than in the letters he wrote to his lifelong friend, John Gould, who was at Winchester with Sheridan and the Union army in the spring of 1863.

On March 15 he wrote about his plans:

You must be here, John, when we are married in June. . . . As everything is so infernally high, I have concluded not to go to keeping house at present but to board. Nellie is not strong enough, neither, to take care of a house and I shall not be able to keep a girl. At the Portland Co. we are full of work; we have finished rebuilding the engine of a steamer called the *Harvest Moon*; then we are building a beam engine for a steamer being built at Bath, . . . besides two big engines for U. S. gunboats.

Two weeks later he continues:

March 29, 1863

Fred Morse is going to have the back part of the house run up two stories, making a large room over the dining room and another over the kitchen. . . . The two front rooms on the second story are to be occupied by Mr. and Mrs. Edward Morse, the one over the parlor being our bedroom and the one over the sitting room to be my sanctum. . . .

John, this summer I am going to make a descriptive catalogue of the Fluviatile and Terrestrial Mollusks of Maine. I am going to get up a big thing on the anatomy of all the fresh-water mussels of the state. . . . I am very ambitious on it, as it will be quite an original thing.

But the surprising bit is this paragraph:

John, don't you think you can go off with us on our marriage tour? I think of going up to Bethel to stay a week or ten days, and I want you to go very much. What a delightful time we should have!

Fortunately, the battle of Gettysburg prevented John Gould from spending Nellie's honeymoon at Bethel hunting for land snails. Instead he sent a wedding present of twenty dollars, which, Ned wrote, "shall be devoted to the purpose of going to Bethel; we expect to go as soon as some of their boarders leave. If I go to Brunswick I hope to be better off financially than for years."

The reference to Brunswick has to do with Morse's application for a professorship at Bowdoin College, Brunswick, Maine. Jeffries Wyman, the Harvard anatomist, wrote a letter of recommendation for him. But Bowdoin College had its own financial difficulties, and it was several years before funds were available for the professorship in zoölogy.

During that summer, Alpheus Hyatt, engaged in his studies on fresh-water Polyzoa, spent a few weeks at Morse's home in Maine. Absorbed in science, with a philosophical bent, he was very different in temperament from the sophisticated Shaler and the versatile, boisterous Morse.

> So absorbed would he become at times [recalled Morse] that he appeared to be in a dream. . . . He was specially working on certain principles of classification, the outlines of which were then dawning on him. He sat down at a table opposite me to write out some views we had been discussing. The table was covered with books, the inkstand being on my side of the table . . . [and] he was forced to reach his pen over a few of the books to get at the ink. Noticing this, I began to hedge the inkstand in with books more completely; book after book was added to the impedimenta. Without once noticing the obstructive ramparts I was building up, he had partially to rise from his chair in order to fill his pen. Finally I perched a huge dictionary on top of the pile, and then only, when he had fairly to stand up in order to get at the ink, he broke out with, "Confound it Edward, what have you been doing!" [1]

Morse himself held to the scientific career he had planned with all the tenacity of a barnacle attached to a wharf piling. Although he went in on the train with Fred to do his draughtsman's work at the Portland Company four days a week, two days a week he worked in his study at shell drawings. It was tedious work. These were not new shells, nor animals that he could dissect for his own study and publications. Wealthy ama-

[1] *Proceedings of the Boston Society of Natural History*, XXX (1901), 416–417.

teurs of conchology like Temple Prime and Thomas Bland, New York merchants, paid him fifty cents apiece for illustrations for their papers. W. G. Binney commissioned him to make the drawings for studies on the shells of North America which the Smithsonian Institution was to publish.

Nevertheless, he worked away unflaggingly. Nellie, for her part, helped Jane Morse and the girls with the housework and cooking in the large Victorian cottage on Church Street, Gorham, near the old academy on the hilltop where Ned had given his first lectures on natural history. At the end of a year and a half, however, their responsibilities were increased; their first child, Edith, was born December 9, 1864, in a yellow frame cottage around the corner from Jane Morse's, which had proved too crowded for all the family.

Agassiz's anger at the arguments over evolution had melted rapidly in generous recognition of Morse's independent work, and in 1864 he forwarded $250.00 in belated payment for the collection of shells which Deacon Morse had packed and shipped down to the Museum of Comparative Zoölogy in 1860. The sum brought Morse's earnings for 1864 to $494.49. It was not enough. On July 5, 1865, Morse, taking stock, counted $285.00 in outstanding bills, with assets of $5.00 in hand. The Smithsonian owed him $330.00, but, he wrote to John Gould:

> There has been trouble between the Institution and Mr. Binney, the former charging Binney with having more work done than was authorized and while they are settling the affair, I have to suffer . . . with my family cares, my studies and these bills hanging over me like a black cloud, it [is] worrying me to pieces. Our little daughter Edith is healthy and beautiful; she is seven months old today.

By October he was writing: "My only embarrassment financially has been not receiving my pay from the Smithsonian, which still hangs fire owing to the destruction of their papers, etc., by fire." The Bowdoin professorship, too, was still uncertain.

In March 1864 Morse had published his first book, issued from the press of his old friend Stephen Berry, in Portland, *Observations on the Terrestrial Pulmonifera of Maine, Including a Catalogue of All the Species of Terrestrial and Fluviatile Mollusca Known to Inhabit the State.* He had illustrated it himself, with ten plates from nature and one hundred and four small line drawings. As an original scientific work it was a success. William Stimpson wrote from the Smithsonian Institution: "With its great elaboration and full illustration it beats anything yet done in Europe on the same subject." Isaac Lea, the outstanding expert of the country on mollusks, endorsed it as "a credit to the Science of the country." The thin brown volume sold for $1.50. In the first year the sales amounted to $34.25; in 1865, to $17.00.

Catching at almost anything that would enable him to earn money and at the same time keep on with his scientific work, Morse moved his wife and the baby into Portland in the fall of 1865 to board with the Owens at eight dollars a week, while he sought lecture engagements.

I have settled nearly $400 of bills this month [he wrote to John Gould, December 10, 1865], having received the money due me from the Smithsonian, and am only in debt about $150. It is evident that to get along . . . I must find some other source of revenue. This I hope to do in lecturing. . . . It is rather humiliating that I in the prime of life cannot earn more, and yet I feel that I must stick to Natural History. . . . I hope the time may come when I shall spend my whole time in scientific research.

To augment his imperfect English, Agassiz had illustrated his lectures by means of blackboard sketches. Morse, a superb and practised draughtsman, adopted this device and developed it into a tour de force in which he sketched in chalk with both hands simultaneously. This ambidexterity was fortuitous. In 1861, as the journal shows, he had septicemia of the right hand and taught himself to use his left hand for lecture notes at the Museum and twenty-four-page letters to Nellie at Cape Elizabeth. In later years, under the pressure of carrying on three or four careers at once, and the spell of public admiration, Morse honestly forgot the origin of his ability. When the Wistar Institute in Philadelphia wished to examine his brain, he bequeathed it to them, expecting himself that an anatomical factor of ambidexterity might be discovered in the autopsy.

The usual price of admission to lectures was fifty cents, and it was customary for the lecturer's fee to depend on the "gate money." By May he could tell John Gould:

May 15, 1866

I lectured every night in the week [for two weeks at Norway, Paris, and Paris Hill]. It was very rainy and muddy most of the time and in two of the places I had small audiences. Everybody personally expressed their great interest in the lectures . . . I made about $50.00 above expenses. . . . A course of lectures is engaged for Westbrook Seminary for $50.00 cash. I commence a course at Gorham tomorrow night. In about two weeks I expect to commence a strictly scientific course of eight lectures before the Doctors and Medical School in Portland.

He had to lecture for a living, but nothing could prevent him from carrying on his researches. John Gould was still "reconstructing the South" with his regiment as provost marshal at Darlington, S. C., and Morse asks him to "collect land shells, as I am writing a monograph on the *Pupae* of North America." Then he had a flash of the same scientific opportunism that had envisioned including John on a shell-collecting honeymoon: "Another thing I would like very much, though I suppose you could not get it, and that is a skull of an African Negro. If you should hear of one and can get it, please do." Gould, a uniformed major of the United States army, declined to rob graves or otherwise procure an African's skull. It is one of the few things he refused to do in seventy years of friendship.

Morse had every intention of spending his life in Portland. His old friends Dr. William Wood, Charles Bolles, and Charles B. Fuller had undertaken to raise a fund of five hundred dollars to put the collection and rooms of the Society of Natural History in order — a task which Morse had been engaged to perform. At a special meeting of the society Morse was voted general curator, and plans were afoot to put the position on a permanent basis.

Then came Portland's great fire of 1866. Smoke and flames rolled over the city. Grandmother Abigail Beckett, with the family of Charles Beckett and Nellie Owen Morse and her baby girl, took refuge in the cemetery on the hill, and the old lady was credited with saving everyone present by ordering them to lie down, faces to the ground, to avoid suffocation. Morse worked heroically at the Natural History rooms. He gathered all the portable specimens and files and books into "cradles" extemporized from the long benches in the lecture hall, and organized volunteers to carry them to safety. When the roof had begun to blaze, he was the last to leave the building, carrying in his arms the heavy, gilt-framed oil portrait of Humboldt presented to the society by Henry Wadsworth Longfellow.

After the fire, when meetings were resumed, the members of the Natural History Society spread on the records a fulsome resolution of thanks to their curator for his efforts in saving the collections. But the fire had ended hope of raising a permanent fund to employ a curator on a salary. Portland had enough to do for the time being to rebuild the city. Ned Morse was out of a job.

The implications of the Portland fire as they affected Morse were not in the mind of Dr. Augustus A. Gould when he read in the Boston newspaper of the conflagration. With a truly scientific detachment, he dashed off a note to Morse to learn the fate of the *Teredo*, or shipworm, which he had commissioned Morse to draw for him.

Boston, July 8, 1866

Dear Morse: I have been anxiously expecting you for two weeks, aye three, with your drawings of Teredo. I have been very anxious about it, because I cannot move a step until I have them. The printer has had the matter in type for a month but cannot print. And now, alas, I fear that all is lost,—drawings, specimens and all. Now, is it? I can only hope that as the Natural History Collection [in Portland] is again destroyed, you may be able to come to Boston and do up my work; and your wood engraver, too, may be glad to come and do likewise. Pray inform me at once how it is. . . .

Across the Atlantic, however, another New Englander who had known poverty and handicaps in boyhood was about to create new jobs for a number of scholarly Americans.

George Peabody was one of the romantic figures of the nineteenth century. Victor Hugo said, "John Brown and George Peabody are America's characteristic contributions to the historic figures of this age." A poor boy of South Danvers, Massachusetts (which subsequently changed its corporate name to Peabody in his honor), he began business as a grocer's clerk at the age of eleven; at forty-two he was the rival of the Barings and the Rothschilds in London as a dealer in foreign exchange and securities. A confirmed bachelor, whose tastes were so simple that he spent but three thousand dollars a year in his London household, he accumulated an immense fortune and became one of the first great public benefactors. He antedated Franklin Roosevelt's housing program by giving to the city of London in 1865 two model apartment buildings, renting at two shillings sixpence per week per room, "for that section of the laborious poor who occupy a position above the pauper and to assist them by furnishing comfortable tenements at reasonable rates in healthy locations." In recognition of this philanthropy, the Prince of Wales unveiled a statue of George Peabody on the east side of the Royal Exchange, London, in 1869.

In 1854, Peabody, who had taken Junius Spencer Morgan into partnership, made a visit to America and inaugurated a series of munificent gifts to his native land. The Peabody Institute in Peabody, Massachusetts, completed in 1854, was the first. It is an endowed building for public education where there may still be seen Peabody's portrait by Healy, as well as the miniature of Queen Victoria presented to him in royal recognition of his $1,500,000 housing gift to London. The Peabody Institute received an endowment of $250,000 for a library and free lectures.

With virtually no formal education himself, Peabody directed his benefactions chiefly towards providing education for adults. To Baltimore, where part of his early business career had been spent, he gave $1,500,000 for the Peabody Institute in that city, which includes a free library, an endowment for lectures, an academy of music, and an art gallery. A visit to the South, originally intended to permit him to take the cure at White Sulphur Springs, led him to set up the $3,500,000 Peabody Education Fund for schooling in the South. To Yale and Harvard he gave equal sums — $150,000 — endowing at Harvard the Peabody Museum of Archaeology and Ethnology, at Yale the Peabody Museum of Natural History.

His only sister, a Mrs. Daniels, lived in Georgetown, near Salem, and on his visits to America Peabody passed much of his time in his native Essex County, of which Salem was the social center and county seat. Young Fred Putnam, who had been a student under Agassiz with Edward Morse, and who came of a family prominent in Essex County circles, had a chance to meet the silver-haired old bachelor, with deep-lined face and simple manners, a chance to pour out enthusiastically his dream of a scientific center of the new Darwinian school to be developed around the old Essex Institute of Salem. Peabody must have given a hint of what to expect, because in the spring of 1866 Fred Putnam was writing to his former classmates, Morse, Hyatt, and Packard, to make no commitments, as by fall they could all be employed in Salem.

The Portland fire made up Ned Morse's mind for him. Nellie and the baby went to Gorham temporarily, and by September 1866 Morse was at Mrs. Baker's boardinghouse in Salem, financed by a commission to make the drawings for Hyatt's forthcoming study of hydroids.

A sketch inscribed "Patented Sept. 12, 1866, by E. S. Morse" shows Morse with a mustache (he had not yet grown his famous beard) seated naked in a china wash-basin, knees doubled and feet in another basin, with pitcher handy. The title is "Proposed Plan of an Economical Bathtub in Mrs. Baker's Boarding House." He soon left, however, because of what in a letter home he describes as "strange news. Mr. Baker died suddenly last night. Bicknell and I had just gone upstairs when Mrs. Baker called us frantically. Down we went just as his heart stopped beating. I have all the more reason to leave the house now."

He and Hyatt took a house together, in 1867 bringing their families to join them, but the arrangement was a temporary one, and he wrote to John Gould on January 27, 1868:

> [I have] no alternative but to build or board, and boarding is frightful as to comfort and price; besides, if I boarded I should lose one of the best Irish girls; besides I should have to send Grandmother [Owen] back to Portland. . . . I have bought the land! Price $800, paid on it so far, $400. Good chance of selling my diagrams for $300, time to pay balance two years. Lot is 300 feet deep, 67½ feet wide and cleared of every stone and weed.

He built a house, which he was to call home for fifty-seven years, at 12 Linden Street, Salem.

CHAPTER EIGHT

FOUR INSANE YOUNG MEN 1867—1876

GEORGE PEABODY'S gift of $140,000 "for the promotion of science and useful knowledge in the county of Essex" was placed in the hands of the trustees of the Essex Institute by an instrument of trust dated February 26, 1867. It provided, among other things, for the purchase of the East India Marine Hall, owned by the East India Marine Society, and for the establishment of a museum "for the benefit and instruction of the public." Arrangements were made with the East India Marine Society for the permanent deposit of its collections of "natural and artificial curiosities, particularly such as are to be found beyond Cape of Good Hope or Cape Horn," and with the Essex Institute for the deposit of its natural history collections. Thus the Peabody Academy of Science came into being.

By May 1, 1867, the formalities had been completed, and on May 16, 1867, four of Agassiz's former students — Morse, Hyatt, Packard, and Caleb Cooke — were appointed members of the staff, of which Frederic Ward Putnam had already been made director. Though the others were to leave the Academy before many years had passed, Morse was to serve the interests of the institution for most of the rest of his life. He held a curatorship from 1868 to 1871 and the directorship from 1880 to 1914; from 1914 to 1925 he was director emeritus.

The East India Marine Hall, a brick building with a granite front on Essex Street, Salem, had been built in 1824–25 to house the curiosities that were brought home with faithful regularity by Salem's sea captains from the time of the establishment in 1799 of the East India Marine Society. Some alterations had to be made in it when the Peabody Academy took it over, and the work of taking down and packing the old collections of the East India Society, and the new arrangement of the cases when the building was ready, must have reminded the young men vividly of their first months with Agassiz, assembling the collections of the Museum of Comparative Zoölogy.

While the dismantling of the old East India Marine Hall went on, Morse was engaged in making plans for the new cases and the proposed alterations. In the course of this work, he invented an "adjustable bracket for shelves," which was patented February 25, 1868. He further, "for the sum of Five Dollars," granted and assigned to a firm composed of Edward S. Morse and Aaron Goldthwaite, Jr., the title to this invention. Perhaps Goldthwaite was not the ideal manufacturing partner; at any rate, the bracket had small sale. The only place outside the Peabody Museum at Salem where Morse ever saw one of his brackets in use was at Hamburg, Germany, in 1888.

For the most part Agassiz's old students worked in their chosen fields. Putnam took vertebrates; Packard, entomology; Hyatt, geology and paleontology; and Morse had his shells — Malacology, as it was called. Verrill was there briefly but was soon called to Yale.

All were in their early twenties. Putnam had been married in June 1864 to Adelaide M. Edmands; his babies of the early Salem days, Eben and F. L. Putnam, were later to become a well-known historian and a Boston banker. Hyatt married, on January 7, 1867, Ardella Beebe of Kinderhook, N. Y. The name of one of his daughters is carved above the entrance of a New York museum; she was Anna Hyatt Huntington, sculptor of note. Alpheus Packard, too, was married in 1867, in October, to a girl with the old Salem name of Elizabeth Derby Walcott.

George Peabody lived to see the new museum dedicated in the summer of 1869 at a service in Tabernacle Church, Salem, during the session of the American Association for the Advancement of Science. Coming from London for the occasion, he entertained the trustees of all his American benefactions, as well as personal friends, at the Peabody Institute in the town of Peabody. Dr. Oliver Wendell Holmes read a poem, and an "elegant lunch" was served in the library, to which Morse, with his appreciation of victuals, doubtless did justice.

A few months later Morse, Putnam, Hyatt, and Packard went to Peabody again for the funeral of the man whose generous thought had given them the beginning of their careers. Peabody had barely returned to England when he died, November 4, 1869. His funeral was held in Westminster Abbey, and his "honored remains" were sent back to America on the British warship *Monarch*, which docked at Portland, Maine. The Eastern Railroad, of which Peabody had been one of the original incorporators, fitted out a special train to carry the body to Peabody. Minute guns tolled in Essex County as the body was taken from the train to a funeral car, eleven feet long and seven feet wide, draped in black velvet with silver lace and fringe, studded with stars. Winged silver cherubs bore up the casket. Six black horses drew the car through immense crowds.

Among the few papers from these early years that Morse kept

until his death was a sheet in his own handwriting, an agreement dated November 22, 1866, four months before Peabody's $140,000 became an actuality. The sheet bears an orange two-cent revenue stamp, for Civil War debts still pressed heavily on the country. Scrawled in ink on the stamp are the initials of Edward S. Morse, A. S. Packard, Jr., Alpheus Hyatt, and Frederic W. Putnam. For three weeks previous to this, the four had been discussing an exciting idea which they now put into execution. They founded a magazine.

The "Agreement of the Editors of the *American Naturalist*" [1] begins:

> The following expenses are to be paid first and in the following succession: 1. Paper. 2. Printing. 3. Engraving. 4. Binding. 5. Incidentals. 6. Advertising. 7. Drawing. 8. Chief Editor's actual expenses for the year, board, clothing, travelling. Should any money remain after the above expenses are paid, this amount will be divided among the Editors according to the amount of their contributions pro rata.

This was a curiously unoptimistic prospectus, but American business had not yet recovered from the Civil War, and none of the four editors came from wealthy homes. They all knew what it was to scratch for a dollar. Packard oversaw the arrangement of the literary matter; Putnam looked after the publishing details; Hyatt kept the books, and Morse, as secretary, had charge of the mailing list and subscriptions. Their first printing was of two hundred and fifty copies.

The magazine was published until 1878. It never made its editors rich, but it was the first publication to combine science

[1] It is now in the possession of the Museum of Comparative Zoölogy at Harvard University, the gift of Morse's daughter, Mrs. Russell Robb.

and popular interest and to use artistic illustrations for scientific articles. Packard wrote most of the articles, and Morse made most of the drawings. His first sketch for the magazine, naturally enough, made use of a shell, but it was enlivened with grace and fantasy; it showed a snail rearing its shell up to apply its horned eyes to the eyepieces of a microscope. Putnam, Morse recalled in 1915 at the time of Putnam's death, was indefatigable in his work on the *Naturalist*, attending to the business part of it, correcting proof, and finally establishing a printing office known as the Salem Press.

One of the out-of-print publications of Putnam's press was *Today*, of which the Boston *Evening Transcript* said:

> The Salem Oratorio and Essex Institute Fair opened last night.... Two of its productions are before us: the initial number of "Today," the paper that is to be issued daily, and a pamphlet. . . . The latter, bearing the title of "Old Houses of Salem," is exquisitely funny, the very cover showing genius for sly and delicate humor which further manifests itself in comical silhouettes that embellish the witty letter-press.[2]

The cover was the work of Morse, a cartoon directed against Salem's sentimentalists. It shows a jumping-jack on a gallows a-top a hill, with a Spanish caravel unloading at the wharves next a wood-burning locomotive and train, and, amid street advertisements of lager beer and "Machine Stitchers Wanted," a card: "Six Witches will be hung today. Admission 12½ cents."

The pseudo history of the "witty letter-press" is a candid account of Agassiz's four youngsters, who had by 1870 been four years in Salem:

[2] Clipping in the Peabody Museum, Salem; dated October 31, 1870, by Margaretta Brooks.

Many years ago an eccentric genius took forcible possession of the church [in the rear of Plummer Hall] and converted it into a menagerie. Besides a lot of wild beasts that he confined in the pews, there were four insane young men who ran wild on bugs, rocks, fishes and shells. . . . One was guilty of the heterodox idea that stone walls preached sermons. . . . Another one pulled snails' teeth, and the other one tortured insects by sticking pins into them. . . . Their mania consisted in hoarding up every conceivable object that they could lay their hands on. . . . It seems these insane men annoyed the animals by calling them the most outrageous names, and not content with crushing them under the incubus of one name, they would give each one a dozen. . . . [The sketch closes with a scene in which the insects rebel.] This load of scathing insults was beyond endurance, and a frantic fight ensued. Having a good many legs, [*Scolopocryptops exspinosa*] kicked right and left, and set everything a going. The insane men brought their biggest words to bear . . . taunted them with their homologies and morphologies; . . . and the bugs flew, and the rocks flew, the fishes scaled impregnable positions, and the snails shelled them without mercy. . . . So great was the rumpus that the old church was shaken from its foundations. . . .

But this was mere fooling. During the five years that he was associated with the organizing of the Peabody Museum, Morse displayed in more serious matters the perpetually amazing energy of youth. He was drawing plans and doing the manual work of arranging specimens at the museum, illustrating articles and writing contributions for the *American Naturalist*, supervising the magazine's mailing list and subscriptions, building a house and rearing a family (his son John was born August 2, 1870), lecturing on zoölogy at Bowdoin College and Maine State College, and traveling from Maine to Mississippi delivering popular lectures on science five months of the year. Nevertheless, he still found hours to spend at his microscope and time to write scientific papers.

His publications from 1863 to 1870 were descriptive, rather than analytical or argumentative. His lectures, frankly a means of livelihood, were popular in character. The titles included "Sketches of Animal Life," "Ancient Glaciers of the Continent," "Flowers and Their Friends," "Glimpses of Insect Life." Various contemporary newspaper comments were quoted in the circulars issued to advertise the lectures. Said the *Daily Despatch* of Columbus, Ohio:

He draws outlines of mountains, valleys, crevasses in glaciers, insects . . . with remarkable celerity and accuracy. He seldom breaks the thread of his remarks while drawing but keeps on in a rapid and distinct tone of voice, saying more in one hour and saying it better than a very large majority of speakers. . . . He talks in a friendly and pleasant way to the people, and not to the ceiling or to a manuscript or to some invisible thing way beyond the audience. . . ,

while the Worcester *Spy* reported:

His expertness in drawing with both hands called forth enthusiastic applause, as the butterfly with its delicate wings spread for flight, or the frog, with a naturalness that would seem to give him power to jump from the board to the audience, grew in an instant under his rapid manipulation of the chalk.[3]

Of one of these lectures Morse wrote to John Gould:

January 9, 1870

The [American Institute] committee [in New York City] had prepared expressly for me a blackboard four feet wide and forty feet long. On this I spread myself for an hour and fifteen minutes, having the whole stage to myself, as the blackboard crowded off the accustomed dignitaries, who sat below, Horace Greeley among the rest. . . . Two gentlemen from the new Normal School, New York City, intended to be the biggest dump of the kind in the country,

[3] The quotations are from a pamphlet advertising Morse's lectures.

came on to Salem last Friday and asked if I would accept the chair of Natural Sciences, salary $3,500. . . . I think now I shall not accept, but still the offer tickles me extremely.

The years were busy, peaceful, happy. At the old home in Gorham everything went smoothly, too. Caroline Morse was married in June, and Mary in November, of 1868; Nellie Morse in June 1870. With the last sister off his hands, Fred could indulge in matrimony himself and promptly did so. His wedding to Rhobie Pierce Gilman of Portland took place in September 1870. Jane Morse continued to live in the white cottage, and Grandma Owen, who since she was widowed had spent her winters in Salem with Ned and Nellie, now began spending part of her summers in Gorham with Jane Morse.

Although very different in temperament, Fred and Ned Morse were astonishingly similar in appearance. About this time both grew luxuriant brown beards and became so twinlike that they even deceived themselves. Ned Morse once, entering a room, shouted "Fred, old boy!" and rushed through an amused crowd to try to shake hands with himself in a long mirror. Another time in Boston, seeing "Fred" in a cab, he opened the door and jumped in to greet his brother, only to find the cab empty. He had been deceived by his own reflection in the glass of the window.

The difference between the two brothers was in point of view. Ned lost himself in enthusiasm; the pursuit of science was imperative for him. His earnestness in this disarmed criticism; all his life his family, friends, and associates cheerfully sacrificed themselves to further his work. Fred Morse, on the other hand, had a deep sense of responsibility to others. He was highly talented as an artist but relinquished any hope of training or practice in art to go to work at sixteen in the Portland Company plant in order that he might contribute to the support of the family. After the death of his father, he not only assumed the support of his mother and three younger sisters but heartily encouraged Ned to stick to science. He had complete faith in Ned's genius for science, and the relations between the brothers were unmarred by criticism. Fred Morse's reward for his early sacrifices was a success in business that elevated him to the head of the Portland Company and permitted him to retire at fifty-five years of age, to devote his leisure to painting landscapes in oils.

Forty years of going to the same office in the same town every day would have been the death of Edward Morse. There was always something new just around the corner, and he had to go look for it. At the end of four years, the Peabody Museum had been reduced to order and established in a routine. It no longer commanded Morse's entire enthusiasm, and he had had enough success in lecturing to believe that five months of hard winter work would earn enough to leave him seven months for research. He resigned from the staff of the museum in 1871. Putnam stayed as director until 1873, when he succeeded Jeffries Wyman as curator at the Peabody Museum of Archaeology and Ethnology at Harvard. Packard remained at Salem until 1878; Brown University called him then as professor of zoölogy. Hyatt had been the first of the quartet to leave, going to Boston in 1870 as custodian of the Museum of Natural History, professor of zoölogy and paleontology at the Massachusetts Institute of Technology, and curator of fossil cephalopods at the Museum of Comparative Zoölogy at Harvard.

By 1870 Morse was earning up to five thousand dollars a year from his lectures alone, his itinerary ranging from Portland, Maine, to California, from Chicago to Baltimore. He had no manager, was affiliated with no "Bureau." He and Nellie sent out the circulars, made the engagements. He had no fixed fee, but talked for the sum a community could afford. On October 25, 1872, he spoke at Ipswich, Massachusetts, for ten dollars, and on January 21, 1873, delivered the same address at Troy, New York, for one hundred.

Lecturing was something he did to earn a living. He was burdened with interest on the mortgage on the Linden Street house, with the need of paying a Swedish or Irish "girl" to spare Nellie's frail strength, with all the expenses of his own household and contributions to the support of Jane Morse in Gorham and Nellie's widowed mother. Most of all, he needed money, above the necessities of his daily life, for the pursuit of his scientific studies.

He knew that he had made a genuine success in lecturing when he was invited to deliver the winter course at the Lowell Institute, Boston, in 1871–72. He, the pupil, was to stand on the platform where the master, Agassiz, had launched an American career twenty-five years earlier. The public stood in line for tickets to the course of twelve lectures on "Natural History, with Special Reference to the Lower Animals of New England," by Professor Edward S. Morse. "My first Lowell lecture came off last night," he wrote to John Gould on December 7, 1871. "The audience was large and attentive. Hope I can keep them so through the rest. The tickets were applied for in a rush, 1,400 being distributed in little over an hour."

Meanwhile, his scientific studies on brachiopods and birds had been leading him in the direction of Darwinism and away from the position of Agassiz.

Morse had been too young, Louis Agassiz too old, at the time of their meeting to achieve a happy relationship. Morse's first impulse to respond to Agassiz's personality by installing him as a father in place of the Baptist deacon was frustrated by the two issues of Darwinism and the Civil War. At twenty-three, Morse had no tolerance for a mature scientist's reluctance to abandon the work of a lifetime because an Englishman had written a book or for the inability of a Swiss to immerse himself in American political questions. The Civil War over, Agassiz, to whom it had never been important, easily forgot that Morse had left him to enlist and readily understood the compulsion that had urged Morse into marriage at the cost of a career at the Museum of Comparative Zoölogy. He renewed the friendship and made overtures to his brilliant pupil with the belated payment in 1864 for Morse's collection of shells. Perhaps he hoped that Morse would become an ally in research to refute Darwin, for at that time Morse had not declared himself definitely in print on the subject.

Now in 1873, at a field meeting of the Essex Institute at Salem, Morse proclaimed himself a Darwinian. Doubting Agassiz's teaching, he had set himself either to prove or disprove it, and his ten years of research had convinced him that Darwin was right.

He had selected the Brachiopoda for study not only because they were shellfish in his own field but because they were the most readily available forms which could be studied in series from the fossils of the earliest strata to living specimens on the American coast. In the course of collecting living brachiopods at Beaufort, North Carolina, at Eastport, Maine, and in the St. Lawrence River, and poring over specimens in his home laboratory, he perceived an error in the current classification of them as mollusks. As a result, he wrote and illustrated a pamphlet, *The Brachiopoda, A Division of Annelida* (1870), in which he proved that these creatures of the ocean's muddy shore were not shellfish but worms; not properly classified with the Mollusca but with the Annelida. This was Morse's outstanding contribution to scientific knowledge.

His expeditions were made at his own expense. No museum, no university, paid for his journeys or his laboratory. He thought he worked out of a pure devotion to science, but subconsciously he was carrying on the ten-year-long argument with Agassiz. In a long letter to John Gould on October 25, 1873, he wrote:

> Agassiz still believes that the Brachiopod is like a clam, in having the shells upon the sides, and two years ago he confessed to me that he knew nothing about the structure of Brachiopods. Still less does he know about the leading features of worms, yet . . . he says in acknowledging the receipt of my pamphlet that he "shall not leave it until he ascertains whether my position is tenable or not!"

It was gratifying, in the face of Agassiz's doubt and disparagement, to receive from England Charles Darwin's comment on the same pamphlet, which Morse quoted in a letter to John Gould on October 25, 1873:

> I must have the pleasure of thanking you for your kindness in sending me your Essay on the Brachiopoda. I have just read it with the greatest interest, and you seem to me (though I am not a competent judge) to make out with remarkable clearness an extremely strong case. What a wonderful change it is to an old naturalist to

have to look on these "shells" as "worms," but, as you truly say, as far as external appearance is concerned, the case is not more wonderful than that of cirripedes. . . . With the most sincere respect for your admirable labours, I remain, dear Sir, Yours faithfully, . . . P. S. Your woodcut of Lingula is most skilfully introduced to compel one to see its likeness to an annelid.[4]

The pamphlet on the brachiopods had been published in 1870. In 1872 Morse unlimbered another gun on Agassiz's entrenchments, in *On the Tarsus and Carpus of Birds,* in which he took as his premise a casual observation made by Agassiz himself and carried it through to an argument for Darwinism. It was based on a study of the embryos of birds — the bank swallow, the eave swallow, the kingbird, the crow blackbird, the bluebird, and the chipping sparrow. "At an early stage of the embryo the leg and wing are almost exactly alike," wrote Morse, and added, "This similarity was noticed by Agassiz a long time ago, as well as the fact that in the embryo robin the toes are webbed."

Poverty and pressure of time were to drive Morse throughout his life to find the universe in his immediate environment. He found birds for his study of evolutional anatomy in the orchard behind his house on Linden Street, Salem — source material that is in great contrast with that of the men engaged in similar studies under university sponsorship. When Morse

[4] Included in *More Letters of Charles Darwin* (I, 350), as is also another letter to Morse, called forth evidently by Morse's *Remarks on the Adaptive Coloration of the Mollusca* (Boston, 1871): "I am much obliged to you for having sent me your two interesting papers, and for the kind writing on the cover. I am very glad to have my error corrected about the protective colouring of shells. It is no excuse for my broad statement; but I had in my mind the species which are brightly or beautifully coloured, and I can as yet hardly think that the colouring in such cases is protective." *Ibid.*, II, 95.

corresponded with Dr. Jeffries Wyman, the Harvard anatomist, Wyman wanted to check his findings with the bones of the extinct Great Auk; and Elliott Coues insisted on dragging into the discussions a penguin from the Antarctic.

The acclaim that greeted *On the Tarsus and Carpus of Birds* reconciled master and pupil. Morse, triumphant, was no longer in conflict with his old teacher; Agassiz, mollified by this acknowledgment of his early teaching, no longer denied the validity of his pupil's reasoning. He invited Morse to be one of the faculty of the first Summer School of Science at Penikese Island in Buzzards Bay in 1873. They had a last pleasant, affectionate period together. Four months later, on December 14, 1873, Agassiz died.

John Gould, however, at his banking counter in Portland, closely affiliated with the church in that provincial community, worried lest Ned Morse by his open avowal of Darwinism should sacrifice public support for his popular lectures. In a long letter dated October 25, 1873, Morse stated his views:

> . . . I know your letter is prompted by the solicitude you have always felt for me . . . and your advice I will follow as far as keeping out of bitter controversy is concerned; but as for pursuing a neutral course of life through policy and for any promised success that might attend it, I should rather come down to one meal a day and lumber along in debt than follow so humiliating a path. . . . The public demand further knowledge concerning Darwinism, of which they are intensely ignorant. . . . My chief care must be to avoid that "rigidity of mind" that prevents one from remodelling his opinions; there is nothing [more] glorious to my mind than the graceful abandoning of one's position if it be false, as Lyell, Gray, Bentham, Hooker and a whole host of followers of Darwin who were thirteen years ago his stoutest opponents. . . .

Surely, John, it is a long way for the Evangelical Church to jump before it says Amen to such views; and yet my impression is that we shall both live to see it. . . . [Clergymen] discuss Darwinism and approve of H[enry] W[ard] Beecher's utterance that he isn't [so] concerned where he comes from as where he is going to, failing to see that his origin from lower forms of life knocks in the head Adam and Eve, hence Original Sin, hence the necessity for vicarious atonement, hence everything that savors of the bad place. . . .

Next year, the year after and so on, new modifications will arise, and your only refuge will ultimately be [to] enter Unitarianism or the Catholic Church. . . . Catholics and Jews are much like certain ancient forms of animal life which . . . have survived by adapting themselves to the greatest number of conditions. . . .

Here, John, I have spent a whole precious forenoon when I had forty other things to do, to write this; but your letter stirred me up, and I could only settle myself by writing this screed. . . .

Earlier in the same letter he had said:

Don't be afraid, my boy, that I will lose my head. A leopard cannot change his spots (save through Natural Selection) nor can Morse get his head turned as long as he carries on in the undignified way so characteristic of him. As to my being a loon, I will admit; but when you say I cannot reason, that hurts me.

My classification of Mollusks, of the Tarsus and Carpus of Birds, and my Systematic Position of Brachiopods called out the most delicious letters from Dana, Wyman, Leidy, Gegenbaur and Darwin.

The character of the proximal tarsus series I reasoned out long before I had ever seen an embryo bird and the greatest anatomist in the country admitted the validity of my reasoning. On the strength of one Brachiopod alone I reasoned out its position in the Animal Kingdom months before I had ever seen a living *Lingula*, *Rhychonella* and *Discina*. It isn't every day that a prominent class of animals is walked out of one branch and into another. I will confess that when I got Darwin's letter I felt for the first time amply repaid for all the hard study I had spent on it.

This ebullient letter to his closest friend was Morse's secret shout of joy at resolving the mental conflict dating from his days at Harvard under Agassiz. He had taken thirteen years to make his choice between the theology of Deacon Johnathan's teaching and Darwinism. Under the spell of Agassiz, buttressing his early home-training, he had once written to the same John Gould:

Cambridge, Zoölogical Hall
March 11, 1860

Received your letter and read it with interest because in it you said that you had been reading Darwin's work and you hoped it would prove to be a correct theory. Now, John, stop and think of it for a moment and now don't you perceive that if his theory were true it would leave one without a God? Most certainly it would, for the origin of species according to his idea would be simply chance and nothing else. . . . It is simply chance and, then again, Darwin does not support his theory by a single fact in nature. . . . You will always hear it said that such and such things have run wild, in speaking of breeds where they have degenerated, in other words, returned to its first form which God created. . . .

I wish you was here, John, that I might talk with you on this subject, or better, that you might hear Prof [Agassiz] talk on it. . . . In my next letter I will give you more but I will say one more thing; and that is that Darwin's chapter on the struggle for existence smacks too strongly of Calvinism and yet Darwin's picture looks a good deal like Calvin's drawing of us poor worms. . . .

A week later he was writing: "If you could hear such men as Wyman and Agassiz talk on Darwin's theory you would soon be convinced of the fallacy of it."

He had been sincere in writing to John Gould that he would be willing "to come down to one meal a day." He had completed his researches for his *Land Shells of Maine* on an income of five hundred dollars a year, wearing a second-hand overcoat given him by a friend. He was content to make a brilliant and original contribution to comparative anatomy by robbing birds' nests in a Salem apple orchard, while his friends headed sponsored expeditions to hunt for the bones of the Great Auk.

He never minded looking shabby, and he exercised his ingenuity to be comfortable. In 1875 he "invented" a garment which forty years later was to be glorified in olive drab and horizon blue for army officers' wear in the field. He described it in a letter to his brother, May 25, 1875:

> Another wrinkle I have . . . seems good. I have had my pants slit up nearly to the knee and a gore taken out; a series of buttons bring the legs tightly around the calf of the leg as follows [sketch], the knee left baggy and loose. Have also bought a pair of stout boots for $2.25. These were high-laced boots and I have had them cut down . . . [with tabs] to button each side of the leg.

The sketch in Morse's letter is a neat forecast of the military breeches or jodhpurs of the next generation.

In addition to his duties at the Peabody Academy, his writings and drawings for the *American Naturalist,* his itinerant lectures on natural history all over the country, and his researches and publications on brachiopods and birds, he carried for four years a heavy program of university teaching. He lectured in 1870–71 on comparative anatomy and zoölogy at Maine State College in Orono, Maine, and from 1870 to 1873 at Bowdoin College, Brunswick, Maine; at Harvard College in 1872–73; and at the Summer School on Penikese Island, Buzzards Bay, Massachusetts, in the summers of 1873 and 1874.

He was successful enough at teaching to have several flattering offers. One of these was from the "new Normal School" in New York City, mentioned in his letter of January 9, 1870, to John Gould. Another opportunity was the invitation to found a department of science at Princeton, for which he was recommended by Asa Gray, the great Harvard botanist. There is no record of why Morse declined this opening, but undoubtedly the heavy debt incurred in building his house in Salem shackled him to the locality; moreover, he had worked out a way of life that suited him. He had a convenient home for his family, companionship with Putnam at the Peabody Academy, affiliations with the Boston Society of Natural History, and such friends there and at Harvard as Scudder, Shaler, and Alexander Agassiz. His lectures and the three-hundred-dollar honorarium for his courses in zoölogy and comparative anatomy at Bowdoin afforded both income and leisure for his private researches. Yet he gratefully preserved the memento of the Princeton offer:

Botanic Gardens, Cambridge, Massachusetts
December 17, 1872

My dear Morse:

President McCosh of Princeton was here. He wants to start a Natural History School at Princeton, — wants specially a man who will devote himself to research, and offers opportunities — offers no great pay, certainly, to begin with. But, as I have told him, you are just the man he needs. I hope you will take his offer, if made, into serious consideration.

Ever yours truly,
A. Gray

Bowdoin, with its old red-brick buildings in a setting of green Maine pine trees, awarded him his first honorary degree — that of Ph.D. *honoris causa* — in June 1871. The Bowdoin archives fail to show any record of the reasons that impelled the trustees to vote such an honor to a young man who had earned no university degree. Perhaps it was through pride in a son of Maine who, in his *Observations on the Terrestrial Pulmonifera of Maine* and *The Brachiopoda, A Division of Annelida,* had produced original and valuable scientific studies. Bowdoin indirectly gave Morse another gift — a lifelong friendship with George Lincoln Goodale, professor of botany at Bowdoin, 1868–1873, and professor of botany at Harvard afterwards, who was responsible for procuring for Harvard the famous Blaschka "glass flowers." Goodale, graduated from Amherst in 1860, had taken his M.D. degree at Bowdoin in 1863 but, like old Dr. Nathaniel True of Bethel, preferred teaching science to practising medicine. Morse himself expressed his happy remembrance of old days at Brunswick in his bequest of a thousand dollars for a book fund to the library of Bowdoin.

Various other honors fell to him. A lively interest in all research and a genial nature led him throughout his life to take an active part in the meetings of learned societies. In 1869 he was made a fellow of the American Academy of Arts and Sciences; in 1876 he was elected to the National Academy of Sciences. In the latter year he was also chosen a vice-president of the American Association for the Advancement of Science; ten years later, 1886, he was the first of Agassiz's students to become its president.

Morse's lecture tour in the late spring of 1874 took him across the continent to California, although he had to hurry back for his second summer of teaching at Penikese. In San Francisco he heard for the first time of the abundance of varieties of

Brachiopoda in Japan. The news was the soft sounding of a horn introducing the second dominant motif in his life. He resolved that his next scientific expedition should be to Japan to collect brachiopods. First, however, to obtain the money for the journey, he must complete some work that was in process. "Have put off my Japan trip till September," he wrote on February 20, 1875, to John Gould, "as I cannot finish the little book for two or three months yet. It is an awful job. So I shall see you in Portland on my way to Eastport in June." Postponement never diverted Morse from something he had resolved upon. He envisaged a delay of "two or three months" in starting for Japan; actually he endured a delay of three years.

The little book was his *First Book of Zoölogy,* published by Appleton in 1875. The Boston *Daily Globe,* joining a chorus of flattering press notices, recognized Morse's aim: "We have been struck with the happy manner in which the author has popularized the subjects which he treats, and the simplicity and clearness of his descriptions. . . . The volume is an admirable companion for a person visiting the country or seashore as well as the pupils for whom it is especially designed."

Morse's own paucity of formal education perhaps aided him in evolving in 1875 a pedagogic method that in 1940 would be styled "progressive." The Chicago *Tribune* for September 18, 1875, describes it:

> At last a textbook of zoölogy has been prepared for the use of schools which displays the rare quality of common-sense in the construction of its plan. Its author has not gone to books or art but to Nature for a key to the best method of instructing children. He does not forget that he was once a boy himself, with a fondness for watching spiders spinning their webs and ants digging their holes and

snails crawling about with their houses on their backs; and with a longing to know, without a deal of discouraging study, the life-history of these odd little creatures. The book follows the pupil, rather than leads him; relying upon it that when Nature inspires a child to study her works, the instincts are to be developed instead of being repressed.

Morse's children, Edith and John, were perhaps the models on which he learned to "follow rather than lead" his pupils in nature study. They were always free to play about his laboratory, and they and their mother accompanied him to Penikese Island for the summer school sessions with Agassiz.[5]

The book went into two editions and was eventually translated into German and Japanese. It was adopted as a textbook by the state departments of education in New Jersey and Ohio, and by many individual schools. It contained a hundred and eighty-eight pages of lucid prose and nearly two hundred charming drawings by Morse. Stamped in gold on the brown cloth covers were his drawings of the Maine land shells of his own boyhood.

Despite the numerous demands on his time, Morse was

[5] Morse had a favorite anecdote about his young son at Penikese: "My little boy, four years old, was fond of wandering through the laboratory and curiously examining the contents of the dissecting pans. At one table Miss White, a teacher from New Bedford, was studying the gross anatomy of a cat. The creature was eviscerated, and Miss White was hard at work on the body with her dissecting scissors and needles. The boy came along and looked over the disemboweled creature, and the teacher, expecting some exclamation of disgust, leaned back in her chair curious to know what observation the child would make. Finally after a critical examination of the remains and the dissecting implements, he turned to her and asked, 'Miss White, are you trying to mend that kitten?' Certainly a most natural question, as the cat was evidently in need of extensive repairs." ("Agassiz and the School at Penikese," *Science*, Oct. 12, 1923, p. 274.)

generous in his efforts to help his friends. He encouraged John Gould through several years of arduous work in the preparation of his *History of the First — Tenth — Twenty-ninth Maine Regiment* (Portland, 1871) in the Civil War. He not only made drawings to illustrate the book, from rough sketches penciled by Gould, but undertook to find a publisher:

October 7, 1875

Got your roll of drawings last night and have just drawn them out. . . . Add any lettering you want on the drawings or any correction. . . . Return drawings at once and I will put them on wood [for engraving] when I have the leisure. Shall go to New York the last of this month so have your manuscript ready if possible.

November 17, 1875

. . . Your book is still under consideration at Appleton's. I have blown your horn to the highest extent of my abilities. The fact is that Appleton's are blue at present. They have an immense stock on hand of Dickens, etc., which are not selling at all on account of hard times. But should they decline it, I shall see Harpers and Scribners personally.

The panic of 1873 had made hard times all through the country. Not only Appleton's was "blue." In the year that Morse attained the dignity of election as a vice-president of the American Association for the Advancement of Science, his finances were so straitened that his cook could nearly cause a moratorium:

November 7, 1876

Dear John: . . . will send you fifty dollars the last of the month for interest on my note. Newburyport lecture for $75 has been delayed a month; should have been tonight. Greatest calamity of all: Mathilda, our cook, goes to Sweden tomorrow and I had to raise $160 to pay back wages.

Still the great green brachiopods of Japan gleamed in his mind's eye. If he had to "come down to one meal a day," he meant to go hunt for them. In the spring of 1877 he planned a trip of three months, a summer's expedition, to collect brachiopods.

When Philip Carpenter from England came to see a boy's land shells at Portland, Morse had known only that the visitor "got his feet all over mud" exploring the landslide; it was hidden from his vision that, through Carpenter, he would go to Harvard as assistant to Agassiz. Now again, in Salem, focusing his aim on brachiopods, he had not even a premonition of how the journey to Japan was to mould his future life.

CHAPTER NINE

SCARLET LACQUER AND SHELLS
1877–1879

EDWARD MORSE was endowed with an observant eye, an amazingly retentive memory for form and fact, an analytical mind. He had, however, virtually no aesthetic feeling. Arriving on his fortieth birthday in Japan, the most picturesque country of his day, he wrote barely two paragraphs of description of its landscape in his thousand-page journal during the next two years. Privileged to visit a land only ten years removed from actual medieval feudalism, he overlooked the poetry and romance of that incredible anachronism, missed that transcendent opportunity to look backward through three centuries. Anything beyond the evidence of Morse's senses was beyond his interest. Japan's religions, her philosophy, history, or government, did not impinge on his consciousness, but he could spend hours in a street sketching the novel tools of a carpenter at work.

Morse was one of the first foreigners to visit the stately groves of dark-green cryptomeria trees on Nikko's mountain slope, brooding above the scarlet lacquer mausoleums of the Tokugawa shoguns. Of the beauty of these jewel-buildings and cathedral trees he wrote next to nothing. Nikko to him was significant because of the land and fresh-water shells he found on Nantai mountain or Chuzenji's shore.

No foreigner was ever more alien to Japan than Morse; yet no foreigner was more beloved, and none left a deeper mark on the country. The reason was that Morse never tried to analyze or understand the Japanese. He accepted them as human beings and equals instead of considering them curiosities or barbarians. He made them partners at sight in his own pursuit of science.

Morse had planned for three years to spend a summer in Japan collecting brachiopods. Banker Gould was called on to finance the trip.

May 18, 1877

Your draft for $150 came this morning. Should I get a [chance to] lecture in San Francisco, will cancel part of it within two weeks. I leave tonight at six o'clock and it makes me groan to leave the wife and children so long.

He sailed from San Francisco, May 29, 1877, in the steamer *Tokio* and landed at Yokohama in June. One of the things he took with him was a letter of introduction to Dr. David Murray, who, through acquaintance with Japanese students sent to Rutgers College in New Jersey, had been appointed in 1873 superintendent of the Mombusho (Department of Education). A Japanese mission to study western methods of education, headed by Prince Iwakura, had returned in 1873 after visiting America and Europe. Viscount Fujimaro Tanaka of the Iwakura mission became vice-minister of education and recommended the adoption of the American system and the appointment of Dr. Murray. Now, early in the morning of the day after his arrival in Japan, Morse took the newly constructed railroad to Tokyo, eighteen miles distant, to call on Murray and ask his assistance in the red tape restricting the movements of foreigners outside the treaty ports.

A shell, five thousand years old, that lay beside the railroad track on that June day changed the whole course of Morse's life. As the train passed the village of Omori, Morse's camera eye perceived on the raw earth of an embankment a quantity of bleached shells of *Arca granosa.* As the result of his field work in Maine and North Carolina, he identified the formation instantly as a prehistoric kitchen midden. Writing about this incident forty years later, in the same unconscious enhancement of himself which he displayed about his ambidexterity (he attributed it to unique talent, forgetting the blood-poisoning at Harvard that forced him to use his left hand), Morse solemnly declared that "Dr. Murray . . . was the only one to whom I told the nature of the deposit."[1] This agrees neither with the recollections of his Japanese pupils nor with the implications of a letter written to John Gould in July 1877. Furthermore, it is against all the ebullient character of Morse himself.

He admittedly told Dr. Murray, and, while Murray took him to the new university to meet Professor Toyama, Cornell graduate, and Vice-Minister Tanaka, Morse must inevitably have bubbled over to them about his new discovery. He found the students using Gray's *Manual of Botany* and fascinated the Japanese with his anecdotes of intimate acquaintance with the author; but when he spoke of the ancient shell mounds at Omori, he astounded them. Archaeology, until that moment, had not existed in Japan. Zoölogy, particularly marine biology, was also a book with uncut pages for a race emerging from

[1] *Japan Day by Day*, I, 288.

three centuries of isolation. Morse's enthusiastic talk of brachiopods and their importance in evolution impressed his hearers deeply.

Dr. Murray, who had planned a journey to Nikko, invited Morse to join the party. Hurrying back to the hotel for his luggage, Morse reported for the expedition with eighty pounds of impedimenta, chiefly "insect net, shell scoop, collecting bottles." He found *Corbicula, Pisidium,* and *Lymnea* near Lake Chuzenji, and ignored the elaborate art of the Tokugawa mausoleums.

Meanwhile, in Tokyo, the Japanese authorities were holding hurried, exigent conferences. The bluff, bearded, exuberant professor from Salem had appealed to them as no other foreigner they had met. They liked him. They could learn from him. They must keep him in Japan. Before Murray and Morse returned from Nikko, the decision was made, the proposal drawn up. Morse had applied for a permit to spend a month on the coast, dredging for brachiopods; the answer was an invitation to organize a department of zoölogy at the university and to found a museum of natural history.

Morse, for the first and last time in his life, indulged in some unintentional Yankee bargaining over this affair. He was loath to give up his work in America. The Japanese, more accustomed to dealing with foreign merchants than scientists, interpreted his reluctance as a desire for more money. Since they wanted Morse badly, they raised the offer to a sum Morse could not resist. Then it became expedient for both parties to the bargain to keep it secret. Both Morse and the Mombusho would have been mobbed by protesting foreigners had it been published that he was receiving more money than any other foreign teacher. Only to John Gould did Morse reveal the figures:

Tokyo, July 11, 1877

I write to tell you confidentially, for even my wife does not know it and so I implore you to keep it safely to yourself. I have written Nellie that the college has engaged me for a year, that I am to return in November to America and come back the first of March. . . . Now what I have not told Nellie is this: They want me so bad for the second year that they have made an arrangement with the Educational Minister to pay me $800 as an advisory assistant, to sort of tell them how to do things. The first year they pay me $4,200 gold and furnish me with a house and stable or in lieu of house and stable $30 a month, and pay my expenses home in November, which is $400 and home again whenever I go. I told them I would not stay out unless I could bring my family or a salary of $5,000, and as they could not pay me that, since $4,200 is the highest salary now paid, they made this arrangement. . . .

I shall bring out Nellie, Edie and John. Dr. Murray assures me that I ought to lay up out of this for the two years at least $4,000 and with the hopes of clearing my house of the $3,500 [mortgage] on it, I am going to do it. I shall be home this winter for two months lecturing and it will be only the winter after next that I shall be away from my lecture business. I hope, however, to get up lectures on Japan which will go like hot cakes, but above all I will have unparalleled facilities to study tropical life and shall make some important scientific contributions. Then I hope to write a book on Japan . . . and with sundry articles for *Popular Science Monthly* I hope to get finally my head above water. I don't want Nellie to know about this till I get home. . . . They pay me $350 and the pay comes every month. My time commences today, and so I am now under pay and shall have no occasion to draw on my letter of credit.

Thus, three weeks after he had landed in Japan, Morse had been offered and had accepted a two-year contract to establish a department of zoölogy at the new Imperial University of Tokyo. He had stipulated for time to collect the brachiopods which had lured him to the Orient, and the university offered to furnish him a laboratory and staff for the summer vacation months. On July 21, 1877, he was established at Enoshima, founding the first marine biological laboratory in the Pacific regions, the beginnings of the splendid research institution still functioning at Misaki.

Morse's discovery of the Omori shell mounds made the Japanese appreciative of his ability. It also deeply disappointed young David Starr Jordan. Jordan had been a student under Agassiz and Morse of the Penikese summer school in 1873. He had put in his application for the chair of zoölogy at Tokyo Imperial University and thought he had the inside track through his friendship at Cornell with Professor Yatabe. By sheer chance, the appointment was diverted to Morse, who had not sought it, did not want it, and had to be cajoled into accepting it by a secret arrangement of a higher salary than that of any other foreign teacher. Unintentionally, Morse shunted Jordan away from Japan and gave him to the Leland Stanford Junior University as its great president. The weatherbeaten shell of *Arca granosa* by the railroad tracks in Omori thus had an influence like that of Aladdin's lamp in building two great universities on opposite shores of the Pacific.

Enoshima, where Morse's laboratory was established, is one of the scenes beloved in Japanese art, its twelfth-century lacquered shrines visited annually by a million pilgrims. It is a rocky, tree-crowned island rising from the blue water off Kamakura, with a superb vista of Fuji's snowy peak in the sky across the bay. A village of fifty fishermen's families clusters in the only cove that breaks the sweep of cliff. A single narrow street of steps hewn in the rock leads upward to the Shinto shrines. Shops line the street, displaying souvenirs made from shells for the endless throngs of pilgrims.

The university rented a fisherman's hut on the shore for Morse's laboratory at thirty sen a month. Fishermen brought him the spoil of their nets, or took him out with dragnet and dredge on the long swells of the Pacific. The summer days went by in a whirl: "Busy making plans for a summer laboratory, arranging a course of study for a class of ninety students, planning the foundation of a museum of natural history."

Morse was too busy to inquire into the history of the red lacquer shrines above on the hilltop, dedicated to the Shinto goddesses Tagitsu Hime, Ichikishima Hime, Tagiri Hime. He never knew, apparently, of the pilgrimage made to the shrines by Yoritomo, great shogun of the thirteenth century, or that of the Emperor Meiji in 1875. He profited, however, by being quartered in the special apartments at Tachibana Inn built for the Emperor's stay. He did row around the island one day to the cave dedicated as a shrine to Benten, goddess protector of seafarers; but instead of studying the rock carvings of the cave, he "scanned the walls closely for evidence of twilight insects and to my delight found two little spiders, two very small sow-bugs and two cave crickets."

His fame was spreading through the country. A youth walked two hundred miles from Kaga province to beg Morse to accept him as semi-student, semi-servant, requiring nothing but three or four hours of Morse's time every week, any food to live on, and any space to abide in. Japanese scholars came to the inn and turned his room of an evening into a seminar. "Mr.

William Sturgis Bigelow and Ernest Fenollosa in Japan

Ikkoto brought in the work he is engaged in, that of translating 'What the Ancients Thought of the World' from the 'Popular Science Monthly.' Mr. Matsumura is studying my little 'Textbook of Zoölogy'; Professor Toyama an analytical table of the animal kingdom, and he was hard at work upon it." [2]

It was July 30 before Morse had a moment to look for

> the objects that first induced me to come to Japan; namely Brachiopods. . . . Conceive my astonishment and delight when the first haul brought up twenty small *Lingula*. . . . Were it not for my assistants I could never get through the work of sorting the material dredged. The sea bottom is very rich in marine life, and while I am studying the precious *Lingula* my men are at work separating the different groups: shells, sea urchins, starfish, and the like.[3]

He had only a month for the brachiopods before the university opened in Tokyo. He and his assistants moved the marine collection and themselves in a fifty-mile trip by jinrikisha, each passenger holding in his lap a basket of fragile coral, glass sponges, and shells, and "finally unloaded at the University . . . the nucleus of a museum — the first zoölogical museum in Japan."

The Imperial University of Tokyo had been established on April 12, 1877, three months before Morse arrived, unheralded, in Japan. The medical school was housed at Hongo, in the former estate of the feudal daimyo of Kaga. The colleges of law, science, and literature were for that period in the Kaisei school buildings at Hitotsubashi.

Japanese teachers with whom Morse was closely associated

[2] *Japan Day by Day*, I, 176.
[3] *Ibid.*, I, 183, 190.

included Masakazu Tonoyama, born in 1848, who, with twelve-year-old Tairoku Kikuchi, had been sent to England to study at the age of sixteen, returning to Japan in 1869. The next year, with Arinori Mori, he went to America, where in 1872 he entered the University of Michigan and there heard Morse lecture on evolution. He was professor of English and psychology at Hitotsubashi from 1876. Ryokichi Yatabe, who studied botany at Cornell (1871–1876), in 1877 was superintendent of the Koishikawa Botanical Garden. One of his assistants was Ninzo Matsumura, a *Koshinsei* (student educated by his clan), of the Matsuoka clan who had dispensation for special studies mornings at the university under Morse. Yatabe assigned Matsumura as assistant to Morse in the marine biological laboratory at Enoshima from June to September 1877. Hideo Takamine (1854–1910), an Aizu clansman, was graduated in 1878 from the Oswego Normal School, Syracuse, New York. He met Morse on the latter's second journey to Japan, on the steamer from San Francisco, and through him Morse became acquainted with a brilliant student of English, Tsunejiro Miyaoka, who boarded at the Takamine home while attending school in Tokyo. Among the students in Morse's first classes in zoölogy were Chujiro Sasaki, Isao Iijima, Yutaro Iwakawa, Aikitsu Tanakadate, and Chiomatsu Ishikawa. All became leaders in science in Japan.

When the university resumed sessions in September 1877, Morse inaugurated the first course of lectures on zoölogy ever given in the country, began the excavation of the shell mounds at Omori, and established Japan's first scientific societies on the international model.

He reached Japan only two years after Japanese samurai, by

imperial edict, had given up wearing their traditional badge of rank, the two swords in the girdle. Many mature Japanese men still wore their hair in a waxed queue at the back of the head, designed to fit the lacquered helmet of a feudal warrior. Near the university classrooms at Hitotsubashi, Morse passed daily the old Tokugawa castle, now become the palace of the Emperor Meiji. In the moat below massive granite walls blossomed giant pink lotus, and in the straggling pine branches above the walls perched hundreds of cormorants. The railroad and the telegraph were the first visible signs that Tokyo as a city, Japan as a nation, were in transition from placid medieval culture to modern mechanistic civilization.

Yet Morse lectured in English and in his classroom needed no interpreter. The foresight of the Japanese authorities had decreed the establishment of an English preparatory school preceding the establishment of a university. So earnest was the application of the students, so efficient the instruction, that the twelve-year-old playmate of young John Morse, Tsunejiro Miyaoka, was capable of translating Morse's lectures on evolution extemporaneously during a trip Morse made to explore archaeological remains in an inland village.[4]

In November Morse sailed for America, on leave of absence

[4] Miyaoka entered the preparatory school in Tokyo at the age of ten. Morse ultimately procured for him the post of interpreter to Percival Lowell, special envoy to the Korean court. Miyaoka accompanied the Korean embassy to Washington, and with Lowell visited Brookline, Massachusetts, where he carried in his arms Lowell's young sister, Amy Lowell, the future poet. He married a lady-in-waiting of the Princess Asoka and had a distinguished career as counsellor in international law at the Japanese embassies in Berlin, London, and Washington. Because of his command of international law, he was named by President Coolidge as the neutral member of the five-man commission to settle the United States–Uruguay treaty.

from the university, in order to fulfill his lecture engagements for the winter season in America.

His family in Salem were thrilled with excitement when he revealed the secret of his two-year contract to teach in Tokyo. Arrangements were made for a neighbor, Henry M. Brooks, to manage Morse's property in his absence, and Brooks's daughter, Margaretta, who had already done secretarial work for Morse, undertook to attend to his correspondence. The family left Salem in March for Japan. Morse described the trip to John Gould:

April 18, 1878

It was delightful to see the children enjoy the trip across the continent, the scenery, the Indians, everything, and the four days in San Francisco with a ride to the Cliff House . . . and then the grand sail across [the Pacific] put back by head winds and fearful seas, water dashing into the cabin, and the vessel rocking. . . . [Then] our arrival in Japan, . . . their first experience with the Japanese, their jinrickisha and finally a ride of five miles through Tokio. . . . I found my house ready and in the best of order, repainted, repapered and re-strawcarpeted throughout. The servants are engaged and nothing to do but buy the things for housekeeping . . . most fearful job. It is impossible to find furniture, dishes and the things we require, and has nearly drove [*sic*] me frantic. At the College all seemed glad to see me.

Morse's house address in Tokyo for the next two years, "Kagayashiki Gobankan," was an epitome of the three centuries of history preceding. "Gobankan" means "house number five"; and "Kagayashiki," "castle barracks of the Lord of Kaga province." The Marquis Maeda, daimyo of Kaga in western Japan, donated his hereditary Tokyo castle to the government as a site for the first Imperial University. To this day the students' nickname for the university is "Akamon" (Red Gate), recalling the historic red lacquer portals erected in 1828 for the marriage procession of a shogun's daughter to a Maeda heir.

In 1636, disillusioned by commercial, political, and religious strife between Dutch, English, Portuguese, and Spanish voyagers, the first shoguns of the Tokugawa regime, whose mausoleums glint like rubies among the cryptomeria trees of Nikko, decreed and enforced the expulsion of aliens; total isolation for Japan. Except for an annual trade exchange with the Dutch, for more than two hundred years no foreigners entered Japan, no Japanese left the island empire. It was a seclusion of choice, not ignorance. Japanese officials, through the Dutch, kept informed of the progress of western imperialism in the Asian hemisphere.

They were aware of the development of a mechanistic culture and of its inexorable victory wherever it penetrated. They knew of the domination of India, Burma, and the Malay peninsula by Great Britain; of the expansion of the Russian empire across Siberia to the Pacific coast; and of the unsuccessful war fought by China in 1839 to maintain the Celestial Kingdom's isolation. The Japanese knew in the middle of the nineteenth century that Great Britain, France, Russia, and the United States were engaged in rivalry to establish trading privileges in Japan.

A fundamental expression of Japanese racial philosophy is the principle of *wa,* which may be translated as "victory by assimilation," or "acceptance of new conditions without surrendering fixed principles." It was first defined in writing in 594 A.D. by Shotoku Taishi; it was still operating in the statesmanship of Prince Konoe in 1940. It dictated the decision of the Japanese in 1853, when Commodore Matthew Calbraith Perry brought his United States frigates into Tokyo Bay. The Japanese perceived in advance the futility of opposing feudal armies equipped with bow and arrow to warships and troops armed with guns. They resolved, nevertheless, to avoid foreign domination such as had occurred in India, China, the East Indies, the Philippines. This nation, beleaguered by enemies bent on exploiting its natural resources, dedicated itself almost as one man to maintaining independence by mastering the technique that had made its enemies powerful.

The accession of the Emperor Meiji in 1868, sometimes described by foreigners as a revolution, might more properly be termed an evolution of government. The feudalistic regime that had sufficed under the Tokugawa isolation was voluntarily terminated. Feudal chiefs laid their estates and their powers at the feet of the Emperor for the good of the nation. Hundreds of picked Japanese were sent abroad to study western civilization, and returned to instruct their countrymen in the techniques of civil administration, educational methods, industrial development.

The barrier of language prevented western nations at first from comprehending Japanese philosophy and psychology. Merchants expected to find it as easy to cheat the natives as it had been to despoil the blacks of Africa. Missionaries came to convert ignorant idolators to the precepts of Christianity. The Japanese understood both motives and tolerated in silence an invasion they could not oppose effectively by force. They were deliberately practising *wa.*

Morse came into this atmosphere under unique circumstances. He arrived in Japan with no mission save to collect brachiopods. Having no professional preconceptions, he accepted the Japanese on a plane of equality. He included them with himself as seekers after scientific truth. He observed their civilization and culture factually, with a scientist's detachment, and accorded it validity in its own environment. This was the explanation of the instantaneous response of the Japanese, individually and officially, to his personality.

Then he began to teach the Darwinian doctrine of evolution. It was still new, not yet generally accepted in the western world. It remained, in the Christian countries, a controversial question that was to be debated until in 1925, at Dayton, Tennessee, it was argued by such representative American citizens as William Jennings Bryan and Clarence Darrow. Christian civilization had pinned its faith to a tenet of individual creation. Louis Agassiz, the greatest scientist of his day, son of a Swiss Calvinist pastor, illustrated in his refusal to accept Darwinism the tremendous traditional strength of denominational opposition to evolution.

The Japanese to whom Morse brought the doctrine of evolution had no Christian bias. The strongest social force in their lives was Shinto, that deep racial bond, neither a religion nor a political philosophy, yet including the fanatic self-abnegation and social solidarity of both. Shinto is the mystic union of ancestors, living citizens, and posterity; its expression includes veneration for ancestors, coöperation of citizens, provision for the future generations. The missionaries had sought to overthrow, to disparage Shintoism. Identifying it with idolatry and paganism, they called on converts to replace allegiance to the living representative of the Sun-goddess by dedication to a Christian God.

In disclosing the scientific and philosophic doctrine of evolution, Morse, all unconsciously, reconciled for the Japanese the acceptance of western civilization and the concept of loyalty to ancestors. Evolution was the endorsement of ancestor-veneration. Evolution taught the continuity of the life of the past with the life of the future on earth. Evolution endowed even an *Arca granosa* with ancestors.

Vaguely the missionaries in Japan in the 1870's sensed this. In a denominational publication entitled *The Heathen Woman's Friend* [5] for February 1879, a missionary wrote from Japan:

> Prof. Morse is untiring in his efforts to sow scepticism. His peculiar socialistic views find a ready lodgment in the hearts of the Japanese and it has looked sometimes as though he were going to raze all that the missionaries are building. But there are some earnest Christian men in the government schools. . . . Dr. Faulds feels that he is called of God to answer Prof. Morse's arguments and is publishing the most powerful replies to his lectures. At the Conference it was recommended that Prof. Morse be made a special subject of prayer with us all.

Yoichi Honda, Bishop of the Japanese Methodist Church, left a record of this controversy from the convert's point of view:

> In Tokyo the learned members of the Christian Church assisted by publishing a magazine, called *Rikugo Zasshi,* in 1887. . . . [It] entered more and more eagerly into a struggle: first with sceptics, empiricists as well as utilitarians, and then with the University thinkers. . . . Dr. H. P. Faulds, founder of the Tsukiji Hospital, was a fair antagonist of Professor E. S. Morse. The former, presenting himself wherever the latter lectured, argued against the evolutionary theory as by no means wholly proven. But other [Japanese] Chris-

[5] In the collection of the Portland Society of Natural History.

tians found the drift of the theory to be irresistible, and changed their logic, claiming theism to be consistent with evolution. . . .[6]

A Japanese scientist, Yujiro Miyake, has recorded:

By far the greatest impression produced upon the thinking public of Japan was the advent of Professor E. S. Morse, who spared no pains to introduce the theories advanced by Darwin and Huxley. The Darwinian theory of man's descent from a monkey was in itself enough to surprise the Japanese students, and Professor Morse's eloquent discourses, accompanied by skilful figures on the black-board not only made a great impression on students, but also had a great influence on the public.[7]

Morse, with his combative, frank nature, responded to the missionaries' attacks upon his science by a vigorous hostility to the attackers. Only the prudent influence of John Gould prevented him from waging open warfare on Dr. Faulds *et al.* He wrote to Gould:

April 18, 1878

On the boat coming over [from San Francisco] was a disagreeable-looking man, one whom you might suspect would fawn on you and steal your pocketbook. . . . After a while we noticed that his little boy, not quite three years old, a beautiful little child, was often missing from the table; then he was heard whipping the child fearfully in his stateroom, and one of the passengers yelled to him to stop it. Two days after another passenger was forced to speak to him, and if a man ever came near being thrown overboard, he did. The little child was forced to carry in its hands a Bible, and, if it mislaid it, the child showed the greatest alarm. In the course of a week we found this man was a Presbyterian missionary! Going to Formosa where they are kind to their children, polite and generous and where they have no Idols! He was interruped by the passenger four times for beating [his child]. . . .

[6] Shigenobu Okuma, *Fifty Years of New Japan*, II, 88.

[7] *Ibid.*, II, 230.

July 14, 1878

Don't be alarmed about my publishing. I don't intend to go into print at all on the missionary record but I intend to have an array of facts showing that the expenses of the missionaries are entirely out of proportion to the good they accomplish and that the teaching of peculiar religious views are [*sic*] an obstacle in attempts to change a people. When the people change their habits, it is from imitation or because experience shows it to be more successful, and the logical absurdity of doctrine for these people arises when the city which contains the greatest number of representatives of Christian nations is notoriously the most licentious, profane, indecent and corrupt city in Japan. The banks are gouges, the merchants are horse jockeys and the whole foreign settlement given up to debauchery. Now, please don't publish this, but I assure you it is a fact without the slightest exaggeration. Yokohama is a bad place and for that reason I scrupulously avoid going there more than I can help. The Japs know this and they don't want the religion of the [Yokohama] people. . . . Understand there are some delightful people there among the missionaries but as a town it is reckless and absolutely indecent.

Meanwhile he had been pursuing his own interests. On April 18, 1878, he had told John Gould: "Since I have been away, my two special students have repeatedly visited the shell heaps and among the stuff is an immense amount of new material. I hope to make a large illustrated memoir on it."

In this sentence is revealed the genesis of the Tokyo University Press. In the course of sketching and describing the wealth of prehistoric pottery, implements, and artifacts from the Omori shell mounds, Morse discovered that among the university supplies were a number of lithographic stones. Through his early training as an engraver he possessed the skill to transfer his drawings to the stones for printing, and he induced the officials to bring out his "memoir" as the first publication of the Imperial University of Tokyo.

The late Baron Hiroyuki Kato, first president of the university, used to say that he always dreaded to see Morse enter his office. "I am charged with the administration of university funds," he would explain. "Usually I can review projects advanced and exercise my judgment for the good of the institution. Professor Morse was so eloquent, so irresistible in his magnetic personality, that I recognized I was helpless to oppose any ideas he should advance!" But Morse was never egoistic in his projects. In his few months in America he had collected and brought back to Baron Kato twenty-five hundred books and scientific brochures as a nucleus for the library of the Imperial University of Tokyo.

His *Shell Mounds of Omori,* under the university imprint, elicited a spontaneous tribute from Charles Darwin that was highly valued by the Japanese. Darwin wrote:

> Down, Oct. 21st, 1879
>
> . . . your paper . . . has interested me greatly. The increase in the number of ridges in the three species of *Arca* seems to be a very noteworthy fact. . . . What a constant state of fluctuation the whole organic world seems to be in! . . . Of all the wonders in the world, the progress of Japan, in which you have been aiding, seems to me about the most wonderful.[8]

There was a kinetic quality about Morse. He was a powerful instigator of action. Whetting the desire of his students and his Japanese associates for scientific knowledge, he threw off sparks of action and inspiration as unconsciously as a grindstone sharpening an instrument or tool. He had no sense of his own importance, no insistence on his own dignity. For example, he wrote in his journal of his public lectures:

[8] *More Letters of Charles Darwin*, II, 383–384.

The lectures were given in a large hall, and the audience averaged from six to eight hundred. . . . It was interesting to see the auditors squatting on the matted floor, — not closely jammed together, — attentive and evidently eager to understand the lectures on evolution in religion, in the solar system, and in the animal kingdom. The platform was only slightly raised above the level of the matted floor. At times it was so cold that I had to wear my thick winter ulster while lecturing. Compelled to be in my stockinged feet, I endeavored in vain to stand in one place, but by the end of the lecture my feet were very cold. At the end of the lecture many of the auditors would rise to exchange greetings with some friends in other parts of the hall. I used to watch the place where some corpulent auditor was sitting, and if he rose, I would find the hot spot on the mats where he had sat and warm my feet till the lectures proceeded.[9]

His university lectures, in Kai sei gaku, a bare wooden building, unheated and unfurnished, at Hitotsubashi, before the new brick buildings had been constructed at Kagayashiki, were to smaller audiences.

The class is divided into two divisions of forty-five pupils each, so each lecture has to be given twice, which is somewhat exhausting. I am in love with my students already; it is a delight to teach such good boys, all greedy to learn. . . . I anticipate a delightful experience in presenting Darwinism pure and simple. Especially noticeable is their alert recognition of my drawings of various animals on the blackboard. These young men are sons of samurai, some of great wealth, others poor, but all modest and polite to one another and very quiet and attentive.[10]

Almost every pupil in that first class was, in effect, a fellowship holder. Daimyos of the great feudal clans personally selected the most brilliant youths in their provinces and subsidized their university education. Morse's students without

[9] *Japan Day by Day*, II, 427–428. [10] *Ibid.*, I, 284.

exception developed into leaders in their own land. What Agassiz's students became in the university world of America, Morse's pupils were to be in the educational field in Japan. Thus it is accurate to say that Japan's progress in the fields of medicine, biological, botanical or agricultural research, physics and archaeology, stems from that group of ninety promising students under a brilliant teacher in 1877.

These talented students took down faithful notes of Morse's lectures, almost verbatim; and when he made a facile sketch on the blackboard, they duplicated it, with the Japanese innate skill in drawing, in their notebooks. The first book on zoölogy published in Japanese, in 1883, was entitled *Theory of Animal Evolution as Given by the American Professor, Edward S. Morse, and Taken Down by His Pupil Chiomatsu Ishikawa.*

Morse established a warm relationship with his pupils and the Japanese professors with whom he was associated, a relationship that endured through the forty-odd years of his subsequent residence in America. Typical is the story of Matsuura, who was a special student of Morse's in the first summer at Enoshima. Morse wrote in his journal an account of a visit he made to Matsuura's quarters at Hitotsubashi:

> The dormitories are two stories in height. A single row of rooms on each story opens on a broad veranda. Each two rooms accommodate seven students, the one below being the study room and the one above the bedroom. . . . Cold and barren, they have neither the interest nor comfort of the native house nor the coziness of a student's room at home. A narrow crib for one person only, and seven of these scattered without order about the room, no pictures . . . ; no furniture. . . . The rooms were very cold, but stoves are being put in.[11]

[11] *Japan Day by Day*, I, 361.

When Matsuura developed beriberi, Morse visited him often in the hospital and at his death attended the funeral.

> [The coffin] was borne on the shoulders of four men. . . . We followed the body in an irregular manner, but with sober and orderly demeanour. All the students were in native dress . . . and their clogs made a curious clatter on the hard road bed. A walk of nearly a mile and a half brought us to the cemetery. . . . Soon a Buddhist priest with clean-shaven head came out bearing sprigs of leaves which he leaned against the side of the coffin . . . ; he then put on his rich brocade robes, lighted the candle, knelt down beside the coffin, and began mumbling a prayer, occasionally tapping a little bell which he had with him. . . . After the coffin had been lowered [into the grave] many of the students pushed in a little earth with their [paper] umbrellas; others took up a handful of earth and tossed it in.[12]

Morse contributed to the expenses for burial and a monument, and, at the request of the students, composed an epitaph in English which was engraved on the reverse of the stone at the cemetery in Uyeno, Tokyo. As late as 1940 it was the custom for surviving pupils of Morse to visit this grave and maintain its care. William James, Harvard psychologist and student of religious manifestations, hearing Morse tell the story at a Lowell Institute lecture, was so deeply impressed that he asked Morse for copies of the epitaphs and made notes on the incident. Even the great earthquake of 1923 at Tokyo did not throw down the gravestone of Saiyoshiku Matsuura, who died July 5, 1878. Carved on it, in the alien tongue in which he studied science, signed with his teacher's name, are these words: "A faithful student, a sincere friend, a lover of nature, holding the belief that in moral as well as in physical questions 'the

[12] *Japan Day by Day*, I, 408–410.

ultimate court of appeal is observation and experiment and not authority.' Such was Matsuura. Edward S. Morse."

Morse's journal records that in October 1877 "the students interested in botany and zoölogy came together the other day at my suggestion and formed a biological society. The members are exclusively Japanese."[13] The Tokyo *Times* for December 7, 1878, notices an early meeting of the Biological Society of the Tokyo Dai Gaku held on Sunday, December 1:

> Prof. Morse communicated some facts in relation to the *Limacidae* or slugs, a family he said was world-wide in its distribution. . . . Prof. Yatabe made a communication on the fructification of the Algae. . . . Mr. Matsubara exhibited specimens of living *Hydrae* which he had found near Uyeno. . . . At the next meeting of the society Mr. Takamine will exhibit specimens of *Amphioxus* and microscopic sections of the same.

Along with the theories of Darwin, Morse had introduced to the Japanese the western laboratory technique of dissection and the making of sections and slides for microscopic study of plants and animals. He had succeeded in transplanting to Tokyo the Natural History Society meetings which were the most familiar and formative of his own experiences in Portland and Boston.

Even before his leave of absence in America, he had been drafted by the Japanese to extend his lectures to public audiences. On October 6, 1877, he delivered the first western-style public lecture ever sponsored in Japan, one of a series of four for which tickets were sold and advertising posters displayed. "It was delightful," he remarks, "to explain the Darwinian theory without running up against theological prejudice as I

[13] *Japan Day by Day*, I, 340.

often did at home. . . . One of the Japanese professors told me that this was the first [public] lecture ever given in Japan on Darwinism or Evolution."[14]

The importance attached by the Japanese to Morse's discoveries in archaeology is illustrated in the visit made in December 1877, while Morse was in America, by the Emperor Mutsuhito in person to an exhibit of the antiquities excavated by Morse and his students from the Omori shell mounds. The exhibit had been arranged in the science department of Imperial University at Hitotsubashi.

Imperial patronage in Japan had no flavor of the commercialism found in such phrases as "Bootmaker to His Majesty the King." The Emperor and Empress of Japan accorded personal recognition only to educational, charitable, or patriotic undertakings, and the Japanese public regarded imperial patronage as a moral guide to conduct. In Japanese opinion, the Emperor's study of the artifacts from Omori placed the imperial cachet on the pursuit of science as an occupation. Without the imperial approval, Morse could not have succeeded in his object of educating the Japanese in the spirit of international scholastic coöperation. Undoubtedly the Foreign Office had been consulted (though Morse was not aware of this) before the university consented to entrusting him with the three collections of duplicate specimens from Omori which he took home with him in 1877 as "exchanges" for the Smithsonian, Yale, and Salem museums, the choice of American museums being dictated by Morse's personal friendship with Putnam at Salem, Verrill at Yale, and Henry at the Smithsonian.

It was due to imperial interest also that Morse later was in-

[14] *Japan Day by Day*, I, 339–340.

vited by Count Tachibana, director of the Peers' School, to give a course of four lectures on evolution, attended by members of the imperial family and the great nobles of the land. At the end of the series, a banquet attended by three hundred and fifty aristocrats was given in Morse's honor. The Department of Education arranged another series of almost equal impressiveness. Morse wrote to John Gould of the latter:

January 26, 1879

> Besides many other things I have been invited by the Educational Department to give three lectures on Evolution at the Educational Hall, the Princes, Prime Minister, Chief Councillors and a host of other high dignitaries were invited. The Department made an appropriation of one thousand dollars to defray the expenses, as after each lecture they give refreshments in the shape of cakes and wine.
>
> Temporary sheds were built to accommodate the horses of the officers and their escorts, and the whole affair was very swell. I gave the first lecture yesterday afternoon and the audience was of the choicest description and very intelligent and attentive. Prof. Yatabe interpreted for me.

Meanwhile, Morse had suddenly developed a new interest in a subject far removed from science — Japanese pottery. To his contemporaries it must have seemed inexplicable. The pottery of Japan is not readily appreciated by westerners. It is a product sophisticated almost to the point of decadence, a branch of art requiring of its devotees a high degree of culture and artistic education. In the Imperial Household Museum at Uyeno Park, Tokyo, for instance, is a pottery tea caddy of rough brownish ware, about two inches long by one inch wide. It is valued at forty thousand dollars. But to an untutored tourist it looks like something the cook might keep cloves in, after acquiring it with premium stamps from a corner grocery.

When Edward Morse first went to Japan, he had had no aesthetic training. His idea of poetry was a good rollicking limerick. His favorite pictures were big bouncing horses by Rosa Bonheur. As for music, he doted on the "Pilgrims' Chorus" performed by a brass band. How did Morse, of all men, happen to achieve mastery in this cult of super-simplicity, of ultra-refinement, of the negation of bigness, color, and flamboyancy? How could this marine biologist, who was not a professional artist, not a creative craftsman, not a student of the fine arts, become the supreme authority of his day on Japanese pottery, consultant to the British Museum, the Royal Museum in Dresden, the Freer Art Gallery, the Boston Museum of Fine Arts? It is an odd story, made up of a series of coincidences.

In the late autumn of 1878 in Tokyo, Morse's nerves and digestion began to bother him. For the previous ten years in America he had lived under a taxing routine of university courses and lecture tours, with their train journeys, restaurant meals, hotel beds. The psychiatry of his day had not evolved the nomenclature to tell him that he was the victim of an inner conflict, but it is plain to see in his journal and correspondence that he drove himself to lecture in order to finance the private research that was paramount to him. His Tokyo doctor evidently perceived pragmatically the results of this inner conflict, for his prescription for indigestion was a five-mile daily walk. When Morse rebelled at the chore of tramping through the endless miles of one-story monotonous Japanese houses which surrounded Kagayashiki, the doctor employed therapeutic psychology. "Find some hobby, some interest to pursue during your walks," he said.

A few days later Morse came home with something to show

the household. It was a byword, Morse's naïve inclusion of the nearest human being in his current enthusiasm. The family loved to tell of his dashing into John Gould's house in Portland, Maine, brimming with excitement so irrepressible that, finding only six-year-old Lizzie in the room, he informed the youngster breathlessly: "I have just discovered that the Brachiopoda resemble the Annelida instead of the Mollusca!" So now in Tokyo, he told twelve-year-old Miyaoka that he had found in a china shop a small saucer that was an exact replica of a pecten shell.

It was this chance discovery of a saucer resembling a shell that first turned Morse's attention to Japanese pottery. Daily as he walked through the streets, past the open-fronted Japanese shops, he looked for and found other examples of pottery modeled on shells. When he showed his shell-saucers to his Japanese friends, they were not enthusiastic. Politely, gently, they explained that these saucers were not old, that they were not from famed potteries. Morse became aware that there was a cult of pottery in Japan and that good pottery was to be recognized by an incised potter's mark.

Once again chance played its role in making a man into an expert. Morse's camera eye, his uncanny ability to carry an image in his mind with complete fidelity of detail, enabled him to recognize at a glance the unfamiliar hieroglyphs of the various potters' marks. He could not read the Japanese written language, but he could carry in his memory and identify unerringly hundreds of potters' signatures in Japanese characters.

The third factor that made a ceramics connoisseur of Edward Morse was his passion for collecting. All his life he had had a compulsion to collect, catalogue, and classify objects, beginning with the Maine land shells of his boyhood, and this natural impulse had been disciplined and encouraged by the great collector Agassiz.

His Japanese friends, distressed at the cheap and common bits of pottery praised by Morse, took him to a gathering of connoisseurs and made him ambitious to excel at the game they played of rivalry in identifying rare specimens. For hours they would kneel in a circle on the pale gold straw *tatami,* fortified with tiny thimblefuls of clear tea or *sake* and small silver-bowled Japanese pipes of tobacco, studying pottery. Tea bowls or tea caddies would be unwrapped from brocade covers and passed from hand to hand in grave silence. Each guest had inkstone and brush and a new cup of white pottery in which to write secretly his judgment of the origin, date, and potter's name to be ascribed to the exhibits produced by the host. There would be a prize for the greatest number of correct attributions.

This game, Morse perceived, was nothing more nor less than an exact science, a classification of objects according to certain physical criteria. He felt the need of a textbook on the subject, but he found that there was only one, and it had never been published. It was the work and property of Noritane Ninagawa, student of ceramics.

Soon Morse had Ninagawa as his tutor, with every Sunday afternoon devoted to the lessons. Ninagawa, with his scanty gray scholar's beard, kimono of thin gray silk, stiff black silk *hakama,* like a divided skirt, short over-kimono or *haori,* on which his family crest was embroidered in white, would sit uncomfortably on the foreign chairs in Morse's library, and finally slip comfortably to kneel on his heels on the straw matting, surrounded by the pottery Morse had purchased in his walks of the week. Here he taught Morse the esoterics of Japanese pottery, instructed him in the "feel" of clays from various provinces, pointed out the tricks of firing the different glazes, sketched for him the identifying marks of famous potters.

They became great friends. Ninagawa allowed Morse to make notes from his manuscript on dates, provinces, hereditary families of potters. "One day I actually abducted him," wrote Morse in his journal, "carrying him in my jinrikisha, against his protestations, to a photographer, and had his picture made, the first and only one he ever had."[15] Their acquaintance was timed as though by Fate's schedule. Ninagawa transmitted to the American lore gathered through a lifetime — and died within a year. Morse attended his funeral. But Ninagawa had handed on a torch to the only man who wanted to receive it.

Japan at the moment, young Japan, had no leisure to preserve that lore or cherish its traditional heirlooms. Sons of the old samurai, in hard straits financially because of surrendering feudal hereditary incomes and lands to the Throne, were selling off family collections to the first bidder for cash to live on, cash to study the new western learning, cash to establish businesses. The ancient cult of the tea-ceremony was forgotten for a quarter of a century while young Japan was getting on its feet. When the pendulum swung back, the Japanese found that their old treasures had been scattered to the four winds. It was during that interregnum of Japanese culture that Morse fortuitously began collecting pottery. Viscount Fujimaro Tanaka, vice-minister of education under whom Morse had taught at Imperial University, grumbled in the 1890's when he

[15] *Japan Day by Day*, II, 190.

began to collect pottery himself: "That rascal Morse has carried away every decent bit of old pottery in Japan and put it behind glass in the Boston museum."

The Japanese were astounded at the quickness Morse displayed in mastering the fine points of old pottery. His enthusiasm and expert appreciation of their treasures won their hearts. Count Kentaro Kaneko, a native of the province of Chikuzen in western Japan, had a casual chat with Morse in which the latter eagerly asked about the Takatori pottery from Chikuzen. Like Baron Kato of Imperial University, Kaneko found himself irresistibly moved to aid Morse. He wrote home for a wonderful collection of Takatori ware and presented it to Morse. Count Okuma, founder of Waseda University, once invited Morse for tea and showed his pottery. Morse's recognition and appreciation of each piece so touched the Count that he sent the whole collection to Morse's home as a gift.

Morse, be it remembered, was not discovering Japanese pottery; he was discovering that the western world was ignorant of a highly developed art in old Japan. The barriers of language, the prohibition of travel in the interior of the country, and racial differences presented to the first foreigners, after Perry opened Japan to the world, obstacles as real to the understanding of its culture as the layers of dust and debris to excavators of an Egyptian tomb or Cretan city. The first diplomats, merchants, and missionaries to reside in Japan's treaty ports had no contact with the homes of the people, no entree to the educated, cultivated society of Japan, no key to the art or literature of the country. Morse, however, through his post as advisor to the government in science, had official freedom to travel widely through the island empire; his scientific habit of observation disposed him to note well new objects; and his very lack of western culture left his mind open to that of the Orient.

In his capacity as advisor, Morse headed a number of official survey expeditions that took him, under government auspices, through sections of the country never before visited by foreigners. On July 13, 1878, with his assistants Taneda and Sasaki and his associate in botany, Professor Yatabe, Morse left Tokyo by steamer for the Hokkaido, the largest northern island of the Japanese archipelago. At Hakodate he established a marine biological station and museum, commandeering space in the customhouse. He went dredging in the Straits of Tsugaru and, at the first cast, "brought up beautiful Brachiopods, one a Terebratulina light red in color." He rode horseback over densely wooded mountain ranges, frequented by bears and populated by a primitive race, the hairy Ainus. The journal of this period is full of adventures, including culinary ones.

> I have finally resigned myself to the Japanese food of this region and to consider my stomach as a dietetic laboratory which will assimilate the necessary nutritive elements from the material offered. . . . I had for supper marine worms . . . [probably of] the genus *Sabella.* They were eaten raw and the taste was precisely as seaweed smells at low tide. I ate a large plateful and slept soundly.[16]

He went hunting for land shells near Sapporo, crawling about on hands and knees, overturning damp leaves and bits of bark.

> I had been searching for these little objects for some time when I heard a number of shouts, as if of warning. Looking up I saw, at a distance of fifty or seventy-five yards, a number of hairy Ainus, in a row, shouting at me and gesticulating. I waved my hand in recog-

[16] *Japan Day by Day,* I, 439–440.

nition of their call . . . , whereupon they became more violent in their gestures and one pulled his bow and arrow in a series of jerks in what seemed to be a threatening manner. Then it suddenly occurred to me that they thought I was hunting for their graves, which they defend even to the extent of murder, and recalling the deadly poison of the arrow tips I reluctantly got up and walked away. [Later] they explained that one of their men had been killed and eaten by a bear a few days before, and that they had set a bear trap with a huge poison arrow, and they were warning me that I might get shot if I did not get out. . . . They were afraid of coming nearer, not knowing quite where the string was which would spring the bow; and I on my hands and knees crawling about like a bear with the hidden trap ready to shoot me! [17]

Morse, who had never ridden before, had more actual trouble with horses than with bears, trotting along in a cavalcade of eight horses, piloted by an Ainu guide.

Suddenly, for no apparent reason, three of our horses ran away, and I was on one of them. . . . Everything portable was shed, first hats; then . . . tin botanical boxes, bags, and packages. . . . I managed to hold on to everything: my pith sun hat, my colored eyeglasses, and a cigar-holder with lighted cigar were undisturbed.[18]

Later he was not so lucky.

I had been trying to fix the end of my umbrella in my shoe as an easier way of carrying it, but the joggling of the horse prevented me. Leaning over to see the shoe I again attempted rather impatiently to jab the point in the shoe, when, in some way, I missed the mark and the umbrella hit the horse under the belly. He instantly shied and I was thrown to the ground striking on my head and shoulder. . . . Looking up, I found Sasaki on the ground also. . . . Our horses went tearing down the road and we after them.[19]

[17] *Japan Day by Day*, II, 12.
[18] *Ibid.*, II, 31.
[19] *Ibid.*, II, 42.

His training in the field converted him to a Japanese diet at his university office:

. . . though I have to eat it [his lunch] on a table laden with snakes, worms, and skulls, I find my appetite is not affected by the surroundings. The wooden bucket [sketch] contains boiled rice. . . . There is also a large piece of broiled fish . . . tender and delicious; another dish contains a slice of salted ginger, radish, and a bunch of green leaves of something. Having acquired the use of chopsticks, I recommend them as the most simple and economical device ever invented by man.[20]

The next spring, leaving on May 8, 1879, Morse headed a university expedition to dredge in Kagoshima Gulf, off Nagasaki and off Kobe, for semitropical fauna. On this long exploring trip the leitmotif of shells was reintroduced. At Takahashi in Satsuma province, Morse and his assistants took a fishing-boat down-river to the sea.

Imagine my amazement at finding a large number of the shells of the large green *Lingula anatina*! . . . I soon learned that they were dug at low tide and were a common article of food. Here was the creature that alone had brought me first to Japan and for the moment I felt like abandoning everything to devote my whole attention to this ancient worm.[21]

There were adventures in this hinterland of Japan. He spent a hot, sultry day collecting

many fine specimens of carrion-eating snails in the refuse piles. . . . The stench was dreadful. . . . On our way back we . . . saw . . . a warning notice which read, "Cholera is now prevalent; be careful!" . . . I felt uncomfortable the whole day knowing how

[20] Letter to Gould.
[21] *Japan Day by Day*, II, 144.

I had exposed myself overhauling garbage heaps . . . and so thirsty all the time I had to drink water once in a while.[22]

Out at sea, dredging, the breeze died down and made Morse late to dinner with the governor of Satsuma:

The Governor had invited me to dinner at six o'clock, but it was nine o'clock before we got back to the landing-place. . . . As I walked into his room the Governor . . . greeted me cordially, and I did not detect the slightest impatience in his manner. . . . He had a wonderful collection of chrysanthemums in his garden, and these were illuminated by hundreds of lanterns. He then showed me a number of old Satsuma and other pieces [of pottery]. . . . I showed my appreciation [of the dinner] by eating heartily. The only mouthful I had had since four o'clock in the morning was two sweet potatoes with coarse and dirty salt. . . . I showed them a number of tricks on the floor that I had learned as a boy in a country academy, and what with saké and the games we had a delightful time. The Governor gave me the Satsuma bottle he drank from.[23]

The bottle was a collector's item, and is now in the Boston Museum of Fine Arts, "purchased" with a country boy's antics learned at Bethel Academy.

Morse's most thrilling adventure, told to shudders and exclamations in lecture halls in America for years to come, occurred when the governor of Yatsushiro at Onamura told him of shell heaps and a cave filled with ancient pottery. Morse's immediate enthusiasm made light of the tropical typhoon that was blowing: "Though it was raining in torrents we waded through the mud up a steep incline for nearly half a mile, when . . . they stopped and pointed over the precipitous side

[22] *Japan Day by Day*, II, 153.
[23] *Ibid.*, II, 157–160.

of the road where ten feet below was an opening out of which the muddy water was pouring like a sluice, and that was the entrance!" The governor ordered the peasants, gathering to view the bearded foreigner, to dam the gutter, diverting the stream until a hole two feet in diameter was exposed in the roof of the cave.

A dozen or more villagers had collected, and generous payment was offered to any one who would allow himself to be lowered into the cave, but superstitious fear at entering a sepulchral vault was so great that not one of them was willing to go down . . . and as my assistant did not volunteer there was nothing to do but go myself. . . . I got two jinrikisha men to grab my hands and lower me. It was as dark as a pocket, and the little light from a rainy sky was cut off by the curious and awe-stricken crowd that shaded the hole. I stretched out my legs in vain to find something to touch, and finally jerked my hands from the grasp of the men and dropped into the water nearly up to my middle. There was a momentary silence and then shouts of horror. . . . I called back to my assistant that I was all right, when in agitated tones he told me that great poisonous centipedes were crawling out of the opening! I had on my wide-brimmed hat and a slippery rubber coat, and what I had supposed to be crumbs of earth and pebbles tumbling from the sides of the ragged hole were huge centipedes dropping on me! I stood literally in a cascade of the venomous creatures. . . . As I got accustomed to the dim light I saw them by hundreds floating in the water, and after waiting till the current had drained them off I groped around in the sand for the pottery. . . . It was a hideous experience, but my slippery raincoat and broad-brimmed hat saved me, for the creatures could not retain a hold and tumbled into the water as fast as they struck me. Had I not been so excited over the pottery, my loathsome position in this dark and noisome cave, crouching in a cascade of centipedes, would have horrified me. I got three specimens of the creatures for the Museum, made a sketch of the wall of the cave toward the opening, and then had a rope lowered to me and was pulled up.[24]

At luncheon, before the expedition, Morse had sketched for the governor probable pottery forms to be found. "The Governor drew out the sketch and I heard him speak to the natives in wonder that I, a foreigner, who had come ten thousand *li* [miles] across the seas, should describe precisely the shape of the vessels to be found, which they had never seen. The natives looked at me as a foreign devil, indeed. . . ."[25]

The return to Tokyo was made overland, by jinrikisha, visiting out-of-the-way provinces off the railroad and tarrying in the principal cities to visit potteries. Morse made voluminous notes into the history of potter families, sketched thousands of impressions of potters' marks. At Nara his lack of training in the history of art caused him to take casually his extraordinary luck in being the first foreigner to see the collection in the Shoso-in:

> A marvelous old wooden storehouse . . . built a thousand years ago to preserve objects belonging to the emperor of that time. . . . To appreciate the marvelous character of the collection one must imagine a similar storehouse in England which should contain the household objects belonging to King Alfred. . . .[26]

[24] *Japan Day by Day*, II, 170–172.

[25] *Ibid.*, II, 172. The word "li" for "ri" is probably a typographical error. Although Morse never learned to speak Japanese fluently, he had a keen and accurate ear and excellent memory. Japanese visitors at his Salem home, forty years after Morse had left Japan, recorded with wonder Morse's flawless rendering in Japanese chant of long poetical passages from the Noh dramas. There is no letter "l" in the Japanese alphabet; therefore what Morse undoubtedly heard the governor of Yatsushiro say was "ri," a Japanese measure of distance corresponding roughly to two and a half English miles. In stating that Morse had come ten thousand *ri*, the governor was using poetic license, like Puck, who "put a girdle round the earth."

[26] *Ibid.*, II, 182–183.

He sketched the old pottery there, and it never occurred to him, apparently, that anything except pottery was worth cataloguing, describing, or sketching. It never occurred to him that his official position, his travels, his friendships, were affording him an unparalleled opportunity to make a study of an ancient civilization that must inevitably be altered and forgotten in the near future. His mind was set on continuing his researches on brachiopods and filled with nostalgia for his homeland. He wrote to Gould on January 26, 1879: "I am looking forward with the keenest delight to getting home again. . . . Perhaps we can do Mahoosac this fall, say September." Three months later he wrote to Gould again, perhaps in response to advice that he stay in Japan:

May 2, 1879

Your advice is good, but think of the ten years I have given to the Brachiopods. That work cannot wait any longer. . . . This is the prime reason or one of them. I want to get out my Japanese book while the vim is on me. I want to get out my Second Book of Zoölogy while I am young. I want to publish a popular work on Evolution while it is still remembered, for if I wait much longer it would be like working on a book trying to prove that the Earth revolves, etc. This, however, is all selfish; Edith is growing up, so is John. They are literally alone in having no playmates and they have no competition in their studies. They must have that stimulus. I need the same. The Boston Society of Natural History meetings I miss. So I come home and face starvation, as you say, with a courage only equalled by my paganism. Thanks all the same for your good advice.

Six months later he carried out his decision, refusing to renew his contract at Imperial University, and packed up his pottery, his brachiopods, and his family for removal to Salem.[27]

[27] See Appendix for an account of the honors paid Morse in Japan after his death.

One of the last events of his stay in Japan was the arrival of General Ulysses S. Grant with his wife, son, and secretary on a tour of the world. Morse's conception of Grant was founded on soldiers' grumblings and the gossip of Civil War days. He subscribed to the dinner and reception arranged for Grant at Uyeno Park, Tokyo, by the American colony, "but had no special desire to go, having no time for such affairs. . . . Reluctantly I attended the dinner . . . and despite my prejudices admired the quiet, dignified, yet easy tone of his [Grant's] voice." He had taken his attractive young daughter Edith, now sixteen years old, to the dinner. General Grant won his heart by taking her hand and introducing her to his son, "with some witty remarks about his little boy," who was a robust six-foot fellow.

> My prejudices, due to the infernal slanders of our newspapers, were promptly swept away when I watched the man. . . . I hurried to Kaga Yashiki . . . , and had my boy, nine years old, awakened from a sound sleep, dressed, and hurried to the reception, that he might remember in after years he had seen the great General. At the dinner General Grant did not touch a drop of wine of any kind, and the stories of his intemperate habits, I was told, were gross exaggerations.[28]

On the steamer from Yokohama to San Francisco in September 1879 Morse and his family were fellow passengers with the Grant party. The famous soldier and former president of the United States spent most of the three-week voyage teaching nine-year-old John Morse to play chess.

[28] *Japan Day by Day*, II, 200–201.

CHAPTER TEN

THINGS COMING IN THICK
1879–1882

RETURNING to the Linden Street house in Salem, in October 1879, Morse expected to pick up life in America where he had left it, two years earlier, in quest of the great green brachiopods. He was not introspective by nature and he had failed to note the subtle spell that the Orient weaves. His materialism, adopted in reaction to the religious conflict with his father, armored him against yielding himself as completely to Japan as some of his contemporaries were to do — men like Ernest Fenollosa, Bigelow, Lafcadio Hearn. Nevertheless a bewitchment which he did not define impeded wholehearted return to his old activities in Salem.

It was the culture of Japan that he missed, but his life in America had been so barren of culture that he was slow to realize why the Salem home no longer seemed homelike. His original standard of interior decoration is suggested by a photograph taken at Salem in 1878, just before he sailed for the second time to Japan. It shows Morse seated in his study before a plain deal table heaped with pottery, shells, and papers. A drawing board is hinged conveniently at his left. Above his head is a gas chandelier with four opaque white globes. Under his feet is a torn and wrinkled strip of cocoanut fiber matting. Two folding chairs of the type ordinarily associated with funerals flank the table. Wooden shelves along the wall are crammed with pamphlets and scientific books; cardboard filing cases are piled in the corner. The windows are uncurtained, the plaster of the ceiling stained. The ornaments include a picture of his curly-headed youngster John, a steel engraving of a scene in some European city, a horseshoe crab mounted on a board, and an arrangement of the caudal bones of some vertebrate. Morse, his brown beard flowing over his breast, cigar in his left hand, bends his attention to the open pages of a large book.

The luxuriant beard, fashionable in the era after the Civil War, expressed authority and benignity; it set off his noble head and balanced the composition of his figure as he stood above audience or class on the platform. It also curtained his shirtfront from his own view, so that his Japanese students for years remembered as the most characteristic campus scene at Imperial University a morning ritual at Kagayashiki. Running down the steps and jumping into the waiting jinrikisha would come the bearded professor, arms waving as he put on his overcoat. Up would go the shafts as he seated himself, and off would lope the *kurumaya* in mushroom straw hat, blue cotton tights, and straw sandals. Far behind would be seen the panting figure of a curly-headed small boy, one arm outstretched desperately waving the forgotten cravat, piping gustily, "Papa! Your tie!"

The insidious passion for collecting pottery which Morse contracted in Japan proved even more expensive than the pursuit of brachiopods. For a dozen years he spent every cent he could earn or borrow, above minimum household expenses, on rare and exquisite Japanese pottery and elaborate provisions for safeguarding and displaying it. Meanwhile wallpaper peeled, carpets wore into ragged holes, the furnace failed to work at 12 Linden Street, Salem. His wife wept about the wallpaper, but Morse assured her it was not important. The house was so cold that custards froze in the pantry, and she had to pretend loyally to her children that she had been making "ice cream"; but Morse borrowed money in 1883 to build an annex for the one hundred and five barrels of pottery waiting to be unpacked.

Ellen Morse's loyalty and love endured through it all, but hers was not an easy existence. She was not sustained by the burning excitement over shells and ancient shards that compensated her husband for discomforts. Morse had been too busy teaching university students, pursuing his own researches, and writing scientific pamphlets to remember his intention of their courtship days to "teach Nellie the classification of species." It was simpler and quicker to accept the instrument he found ready to his hand, the tool all busy men of multiple interests must have, the perfect secretary. Morse found his secretary just across the garden, in Margaretta Brooks, daughter of Henry Mason Brooks, his associate at the Essex Institute. Brooks was secretary of the Institute, in which Morse was a curator.

Brooks was a gentle, companionable soul. Sometimes of a morning the young Morses at breakfast would hear footsteps in the hall and Brooks's voice by the hatrack, murmuring apologetically, "I'm afraid Mr. Morse took my hat again last night." He loved music, was one of the founders of the Salem Oratorio Society, and published, in 1888, a book, *Olden-Time Music,* for which Morse wrote a preface. Morse enjoyed evenings around the piano with Brooks, letting out his pleasant, though untrained, tenor voice. Brooks admired Morse, with a touch of awe at his achievements in science and his independence of manner. On catastrophic occasions, when polite endurance was forced to the last ditch, Henry Brooks would explode in the one objurgation permitted to pass his lips, "Well, well — as Mr. Morse would say, By Thunder!"

A scrap of envelope, in Morse's handwriting, addressed to "Miss Margarette W. Brooks, care of Henry M. Brooks, Esq., Salem, Massachusetts, U.S.A." and postmarked June 18, 1878, from Tokyo, is the earliest evidence of the association of scientist and secretary that was to endure fifty years. During his absence in Japan, Morse entrusted to her the care of his correspondence. He also mailed back to her for copying and filing the pages of his daily journal.

The journal, handwritten with pen and ink, profusely illustrated with sketches, was a catch-all for Morse's daily thoughts and occupations. Margaretta was charged with the task of making it do duty as diary, notes for future books, and correspondence with Morse's mother and friends.

Written with candor, keen observation, and considerable detail, it had charm and wide general interest at a time when Japan was still a land of mystery to most of the western world. Evidently there were requests for a wider circulation than in the intimate family circle, for Morse wrote, suddenly, to John Gould:

March 23, 1879

It alarms me to think of my Journals going to Lowell, Longfellow and others. I am so afraid they will be lost and if they are I can never recall the things I have written. On the whole, I think I must say

No. After you have seen them, I should rather you would return them at once to Maggie Brooks.

Margaretta (as she spelled her name in later years) was one of four Brooks sisters, of whom three were closely connected with the Morse family. Alice Brooks was for twenty-five years in the business office of the school of the Museum of Fine Arts in Boston. Jennie Brooks, of artistic ability, was the playmate and friend through life of Morse's daughter Edith. Margaretta as a girl was interested in natural history, making collections of shells along the Salem beaches. Perhaps that first attracted Morse's notice to her.

She was by temperament neat, painstaking, orderly, as evidenced by the Peabody Museum files. The card catalogue of Morse's five thousand specimens of Japanese pottery is in her clear handwriting. The manuscript of every book Morse wrote was typed by her. She had that estimable faculty in a secretary of admiring and approving all that Morse wrote, said, or did. She fitted her tasks around his eccentricities, achieving efficiency despite the dusty chaos that Morse insisted on in his library. Josephine, the faithful family housekeeper, once said, "I never saw the top of Professor Morse's study table until after he was dead." A deluge of shells, clippings, notes, tea jars, and microscope slides gradually inundated three rooms in the Salem house.

For her imperturbability and calm, Morse at first nicknamed his young secretary "the Saxon"; as he realized more fully her value in his work, and became acquainted with the Japanese language, he evolved a punning tribute. Margaretta means "pearl" and a jewel in Japanese is *tama*. The quiet, dark-haired shadow by Morse's desk soon was referred to and addressed as *O Tama San* — Miss Jewel.

Margaretta Brooks never married. Doubtless she loved Edward Morse. Most people found it impossible not to love that radiant mind and magnetic personality. It would have been difficult for her to serve and aid him with fidelity and understanding through a lifetime unless she had loved him. But there was nothing more. Both came of old New England Puritan stock and lived in an era when Freud had not preached the danger of warping the libido by repression. Unconsciously and innocently, Morse and Margaretta Brooks "sublimated" any attraction they mutually felt in a happy, companionable absorption in scientific and literary occupations.

On his return to America, Morse plunged eagerly into the old routine for which he had been homesick in the Orient. "Lecture engagements and other things are coming in thick," he wrote to John Gould, October 2, 1879.

He received an invitation to give a course of lectures on Japan at the Lowell Institute, whose fee of one thousand dollars was a coveted plum. Said the Boston *Evening Transcript* of "Professor Morse's unique course of lectures" at the Lowell Institute, "no other of the several winter courses has been so thronged and no other has given such apparent delight. The audiences have surged up to the very platform steps . . . following this course with the help of the blackboard has been like making a tour of the insular empire."

Morse's lectures on natural history were still popular in cities all over the country — Washington, Milwaukee, Minneapolis, St. Louis. The Chicago *Times* portrays their atmosphere:

Professor Morse is a natural humorist without a taint of vulgarity. He possesses a fine vein of the ludicrous and while he was busy as a beaver drawing skeletons, he would continually let slip something that kept the audience in the best of humor. Whether he desires to represent a protoplasm or a fashionably dressed lady, he does both with about five chalkmarks apparently promiscuously dashed upon the board, but there is no mistaking what it is intended to represent.

Another contemporary wrote of Morse, "His chalk talks on the subject of Evolution without doubt did more than anyone else to bring the great mass of thinking people of America to the acceptance of Darwin's work." [1]

Morse had come home to complete his research on brachiopods and to write a book on Japan for Appleton's and a Second Book of Zoölogy, but he found a new preoccupation awaiting him at Salem. His colleagues of the days under Agassiz and the founding of the Peabody Museum at Salem had all been advancing in their own notable careers. Putnam had been called to Cambridge and New York, at the American Museum of Natural History; Hyatt was in Boston at the New England Museum of Natural History; Verrill was at Yale; Packard at Brown University. Salem wanted Morse to carry on, and in 1880 he accepted the post of director of the Peabody Museum in Salem which he was to hold until his death.

Nothing ever became an exclusive interest with him, however. He found time in the midst of country-wide lecturing, research, and the direction of the Peabody Museum to make up for the old humiliation of being rejected for the Union army. In Japan he had played soldier with his son John and young Miyaoka and had put on a miniature battle for the Offi-

[1] From undated clippings at the Peabody Museum.

cers' Club of the Japanese army. Now in May of 1880 he organized his boyhood cronies in Portland for a battle on a scale so elaborate that they had to hire a hall.

May 15, 1880

Get some small hall [he writes to John Gould] with a floor about thirty or thirty-five feet square. We ought to have at least eight men on a side. . . . With that number the game becomes intensely interesting. I am having more cannon made and I wish we had more soldiers.

May 31, 1880

. . . Have all [the matches] cut the right length [to represent toy soldiers], sharpened, and the blocks on which they stand, one thousand of them, are finished. . . . Don't make town talk of it or have a single spectator. . . . I will finish soldiers, guns, and everything. . . . See about hall and have it stocked. Wear old pants.

A year later they were at it again.

May 16, 1881

Could we not get the Army and Navy Hall or is it too large and expensive? At all events we want a good large floor and no skulking behind scenery.

Morse often combined these affrays of working out military tactics and strategy, crawling over the floor in old pants, pushing toy soldiers and artillery into position, with a mountain-climbing expedition, such as the trip to Mahoosac or Carrigain which he had written of from Japan. He was so gregarious by nature that he threw off impromptu clubs as a pinwheel does sparks. His White Mountain Club was an amateur forerunner of the Appalachian Club, and his associates recognized his as the cohesive force, as is shown in a printed ballad, "Song of Mt. Carrigain," which begins,

This mountain has a mossy top
And Morsey men would climb it
And to them all that did befall
I now propose to rhyme it.

In July 1880 Ellen Morse had an attack, described, in pre-appendectomy nomenclature, as inflammation of the bowels. Apparently she endured the tortures of acute appendicitis, rupture, and peritonitis. The sickness that restricted her activities and caused her suffering until her death in 1911 was probably due to resulting adhesions and ulceration. Morse was in despair when the doctor told him his wife was probably dying, but, when the crisis passed, his spirits rose quickly and he returned to his passion for land shells. In six weeks he was off to the mountains again, looking for *Helices.* A short time later (September 18) he wrote to Gould: "I fear there will be no mountains for me this Fall. I am swamped with work."

His trips to the mountains had been in connection with a paper on "the gradual dispersion of certain mollusks in New England" which he contributed to the *Bulletin* of the Essex Institute. With his usual diversity of interests, he is reported in the *Bulletin* [2] as contributing remarks on the mud wasp, barn-swallows, and the Japanese people at another meeting, and later in the winter he got up to make "remarks on the Japanese language, glacial phenomena and *mollusca.*" Meantime, for the Boston Society of Natural History, he prepared a memoir on "the identity of the ascending process of the astralgus in birds with the intermedium"; for the American Association for the Advancement of Science he explored shell heaps in New England, with interpolations about his observations on the shell

[2] XII (1880), 136.

mounds of Omori in Japan; wrote a paper on "Changes in Mya and Lunatia since the Deposition of New England Shell Heaps"; and simultaneously tossed off a popular article on "Prehistoric Man in America" for the *North American Review*. He also prepared a schedule for the Essex County Summer School of Biology, initiated by Putnam during his directorship, in imitation of Agassiz's summer school at Penikese. Morse gave lectures on invertebrata in the Essex County school, which was "designed expressly for teachers of public and private schools . . . class limited to twenty-four; tickets ten dollars."

He took a final month in the summer of 1881 at Tuckernuck Island, off Nantucket, as the guest of Dr. William Sturgis Bigelow. He emerged from this vacation with a new compulsion working at high pressure. In Japan he had longed for the meetings of the Boston Society of Natural History, for research on his brachiopods, and now he found he could not settle down to them. He could not compose himself to write the "Second Book of Zoölogy"; he could not get started on a book on evolution, nor one on Japan. He had a bad case of nostalgia for Japan, and in talking with Bigelow at Tuckernuck he had convinced himself that his notes needed amplification, his pottery collection additions, and the Peabody Museum an ethnological collection. When Edward Morse had persuaded himself of anything, he never had the slightest difficulty in persuading others.

In a few months he had persuaded his wife that, instead of papering the front hall, they should add to the pottery exhibits; he had converted the trustees of the Peabody Museum to financing an ethnological expedition to the Orient; and he had induced Dr. Bigelow to visit Japan.

CHAPTER ELEVEN

FLOWER VASE OF BRONZE
1882—1883

WITH Dr. Bigelow, Morse began his third journey to Japan in 1882, leaving Boston the first week in May. Their steamer's sailing was postponed to May 16, giving them a week in San Francisco. They visited Chinatown, rode out to view the giant redwoods of California, and had a day at the Presidio with Major Sanger. To Morse's militaristic joy, the Major "had his light battery of four guns out. . . . Galloping away like thunder at the sound of a bugle, they would unlimber and fire thirty shots before a minute had elapsed." Either the United States field artillery set a record on that May day or else Morse had no stop watch.

When Bigelow and Morse arrived in Japan, Ernest Fenollosa joined them at Tokyo for a journey through the great centers of Japan's art and antiquities, hitherto inaccessible to most foreigners. They journeyed by jinrikisha or boat, for the railroad had not yet been extended through the empire, and stayed at native inns because tourist hotels of foreign architecture and furnishing had not yet been built.

Down the old Tokkaido they traveled, the pine-shaded dirt highway worn to a track four feet below the bases of the lofty trees by the feet of horses and men of the daimyos' processions, bringing tribute to the Shogun at Yedo for three centuries.

They climbed the steep Hakone mountain passes, past cliffs carved with gigantic stone Buddhas; they took ferries across pebble-choked rivers; they rode through the emerald-green patchwork of rice fields, embroidered with the pink of lotus blossoms or the white of cranes wading in the mud. At each large city where they tarried — Nagoya, Nara, Kyoto — they separated on their several searches. Morse was hunting for old pottery, Bigelow for sword guards, Fenollosa for screens and painted scrolls.

At night, by candlelight in the inns, shut behind the seclusion of the translucent paper *shoji,* sprawled on the soft straw mats, they would show one another the treasures found that day, and Okakura,[1] their interpreter, eager to perfect his English, would sit quietly listening. Bigelow would have a rare old pair of swords, perhaps; Fenollosa a *kakemono,* Morse a tea jar or bowl. One night Morse suddenly fell silent. It was unprecedented, and the others asked him what was wrong.

"Many fine things of Japanese art are now on the market, like those we are buying. It is like the lifeblood of Japan seeping from a hidden wound. They do not know how sad it is to let their beautiful treasures leave their country," answered Morse.

His words went like a sword into the heart of the young Japanese sitting in the shadows. In that moment Okakura conceived a mission. On his return to Tokyo he went to officials and important citizens and never rested until they, too, recog-

[1] Kakuzo Okakura (1862–1913) in 1906 came to Boston, first as advisor to, later as curator of, the Department of Chinese and Japanese Art at the Museum of Fine Arts. His published works include *The Book of Tea, The Awakening of Japan, The Ideals of the East, The Heart of Heaven* (a posthumous collection of his essays).

nized the situation. Morse's words were the seed, flowering in the mind of Okakura, that produced the law of *Koko Ho* (National Treasures) by which, in 1884, all remaining objects of ancient art were registered and restricted from export. Japan's museums in Tokyo, Nara, and Kyoto owe their collections, in great degree, to the sincerity of Edward Morse and the quick percipience of Okakura.[2]

The story of Morse's peregrinations with Bigelow and Fenollosa has been told in detail in his book *Japan Day by Day,* and his acquisitions are listed in his *Catalogue . . . of Japanese Pottery.* A letter from Bigelow, written after Morse left Japan to go on to Europe, re-creates informally the flavor of their summer together in 1882:

Myanoshita, Japan
September 3, 1883

My dear old Woolly: It is the same and yet not the same. The clean, soft, springy tatami; the immaculate neatness of the light, unpainted

[2] This anecdote was told in 1939 to the biographer by Atomi Gyukushi, then eighty-two, head of an art school in Tokyo. In 1882 Morse, Fenollosa, and Okakura visited the Kyoto art school in which she was a pupil, and, in response to Fenollosa's invitation, she transferred to the new Tokyo Academy of Fine Arts, of which he had just been named director. She became noted as a painter of cherry blossoms. Annually, for fifty years, she received an imperial command to view the cherry blossoms with Her Majesty the Empress and paint the day's scene for the palace album. In 1904 Atomi Gyukushi visited the St. Louis Exposition and afterwards, by invitation, visited Morse in Salem. "I remember it was November 3, the Emperor Meiji's birthday, when I called on him at the Boston Museum [of Fine Arts]," she said. "He was very kind and sponsored an exhibition of my paintings, of which I had one hundred and fifty with me. Afterwards I went to Worcester. I remember a German scientist there had raised a bird, a cross between a turkey and a quail, and asked me to paint it, and I did for the Clark University museum. I made many sketches in America which afterward were acquired by Her Majesty. Professor Morse asked me to paint on silk for him a picture of bamboos and swallows. This I did, and he kept it."

woodwork with its satiny feel . . . , the shoji whose tightly stretched paper sounds, like a drum, at every touch; . . . the thin broad boards of the ceiling over which an occasional rat kicks up a row which it is hard to account for on any basis of rat civilization . . . ; the fat, little smiling servant girl with her bright eyes and white teeth, her gentle manners and imperturbable good nature — even Fenollosa stretched out flat on his back reading Emerson. Everything is just as it was a year ago. But we miss the old Woolly. We want to see the look that he used to wear when he came in from a raid round a new town, accompanied by two or three grinning, astonished, and obsequious dealers bearing a pile of boxes and loose pots. And then the triumph of the O. W. [Old Woolly] when he displayed his treasures — got them all out on the floor, finally producing from his pocket as a thing too precious to let the other men bring home, some particularly demoralized-looking old coprolite. "A Koyashi, by God! 650 years old; a genuine Unko Koyashi with a stamp! Never saw but two others, and one of those Ninagawa asked 30 yen for, and I got this for 15 sen!" . . .

A bronze vase that stood for years in the crowded clutter of Edward Morse's library in Salem epitomized his influence upon the history of Japanese art, hinted at in Bigelow's letter above. Morse himself never told the story of the vase, never realized, perhaps, how far-reaching his influence had been. The vase was sent to Salem in 1878, with an accompanying letter from Baron Kato, president of Imperial University, Tokyo, in appreciation of Morse's help in securing two professors for the University. The two professors were T. C. Mendenhall, physicist, and Ernest F. Fenollosa, but they were only the first arrivals in Japan of a group of Americans, attracted by Morse, who laid the foundation for western understanding and appreciation of the art of the Orient.

Now that the tale has been gathered — from faded letters

and archives in Tokyo, Boston, or Salem, from stately museums in Uyeno Park on the banks of the Sumida, or the groves of Nara where the spotted deer flock about the temples, or the white-columned museum in Boston's Fenway — the bronze vase becomes a symbol. If anything can give meaning and encouragement to the often meaningless and discouraging process of living, it is a tale like this of the great by-products of one man's existence.

Edward Morse, without aesthetic culture, without wealth, without a fully conceived purpose, accomplished through his enthusiasm and appreciation the building of an artistic heritage for all humanity. Through a chance remark overheard by Kakuzo Okakura in 1883, he was directly responsible for the intelligent preservation of Japan's ancient art treasures. Through his influence on the life of Ernest Fenollosa, he was indirectly responsible for the West's introduction to Oriental art history. Through his immediate influence upon four wealthy connoisseurs and his indoctrination of thousands of American and British hearers at lectures, he fostered appreciation of the art of the Orient.

Yet the beginning was casual enough. During his first years of teaching in Japan — 1877–1878 — Morse commented with his customary candor to President Baron Kato on the fact that the Japanese were being victimized by self-styled authorities. Sea captains and merchants with no pedagogical training had presented themselves as candidates for high-salaried posts in Japan's new schools and universities. Morse, whose courage outran his tact, bluntly exposed the deficiencies of these candidates. Baron Kato, after a discreet investigation, was so impressed with Morse's unselfish thought for the welfare of the university, that he placed implicit trust in Morse and asked him to name, from his own acquaintance, desirable nominees for vacant posts.

Again Morse won the admiration of the Japanese by his frank simplicity. In the field of science, he knew men whom he could recommend, like T. C. Mendenhall and Charles O. Whitman, former pupils and associates of his; but in an alien field the amateur personnel-director refused to suggest names. "I don't know any philosophers," he told Baron Kato, "but I know how to find out."

Thereupon Morse wrote to Professor Charles Eliot Norton of Harvard, and Norton sent to Tokyo his most brilliant young student, Ernest Fenollosa of Salem. By 1882, when Morse brought Bigelow to Japan, Fenollosa had already deserted philosophy to delve into a lifelong study of Japanese painting and sculpture.

Lowell, to whom Bigelow also refers in the letter of 1883, was Percival Lowell, whose interest in Japan had first been kindled at Morse's Lowell Institute lectures in 1880, and who as a result made a visit to the country, with introductions from Morse to leading personages. Morse had had some share in suggesting Lowell's official appointment as advisor to the Korean government, and in the attachment of young Miyaoka to his staff as interpreter for the journey Lowell undertook to conduct the Korean embassy to Washington. Two other Bostonians interested in Japan by Morse's influence and enthusiasm were Dr. Charles Goddard Weld and Denman Waldo Ross.

Interest in the culture of Japan became a common bond with this group of men, who had all originally been at Harvard in the period when Morse was an instructor there. Fenollosa had been graduated from Harvard in 1874; Bigelow in 1871, but he remained to take his medical degree in 1875, the year that Denman Ross was graduated. Weld was younger, having been graduated with the class of 1879 and taking his medical degree in 1881. Bigelow and Weld were physicians, Ross a student of the fine arts, Fenollosa a philosopher, Morse a zoölogist. Each found in Japanese culture an individual line of taste and study, but all ultimately deposited the noteworthy collections they amassed in the same institution, the Boston Museum of Fine Arts. When its new building flowered in white marble in the Fenway in 1909, the Asiatic art department required a special wing, designed in subtle suggestion of the Orient, to house the greatest single collection of Oriental art in the world. Edward Morse had played a vitalizing role in the inception of that collection.

Letters from Ernest Fenollosa show clearly how Morse's influence in a summer journey in old Japan went on operating, after Morse's departure, to change a professor of philosophy into a great authority on and interpreter of Oriental art history:

Tokio, August 2, 1882

My dear Morse, I received your two letters of June 25th and 30th by last mail, and was surprised and pleased that you had already found occasion to speak so much of my work. It will indeed be the making of me, if I can get those lectures assured beforehand, and I shall never be able to thank you enough for your kind interest. But I don't doubt that I shall find some way eventually to return you the very great favor. . . .

The doctor is well and happy, and I see him most every day. What a magnificent fellow he is to have for a friend. I have also seen a good deal of young Mr. [Percival] Lowell, and also some of Mr. and Mrs. J[ack] Gardner of Boston. All are interested in paintings, as soon as they see them. Wakai of the *Kosho Kwaisha* is buying up everything he can lay hands on to send to Paris for sale.

What do you think of my publishing some preliminary papers on Eastern painting in Harper's Monthly during the next year? Thanks for you speaking to Gonse at Paris. He sent me a specimen of his reproductions. I shall subscribe for the book. . . . I hope you will get your house built and pottery displayed soon. Isn't it fun to take treasures out in America which you haven't seen for months? . . .

Yours sincerely,
E. F. FENOLLOSA

Tokio, April 26, 1884

My dear old Morse, I haven't heard from you directly for a long time. . . . Thanks for your efforts in behalf of my lecturing. . . . I have decided not to come home till 1886. From this year, a Japanese lately returned from Europe is to take my courses in Political Economy; and I shall have nothing but Philosophy to teach after next summer. This will leave me lots of time to write my book. Isn't it a job to write a book? You have perhaps heard from the Doctor [Bigelow], about the recent formation of our new club, Kangwakwai. Three artists, including the present head of the Kano and Tosa schools and myself, compose the criticising committee of four. We hold meetings at which we exhibit old paintings, I lecture on art, and we undertake to criticize any paintings brought, and give certificates, if desired. At the last meeting more than three hundred people attended. You will find a notice of our meetings in the paper I send you by this mail. Our next move is to get together the more original of living Tokyo artists, make them paint their best according to my suggestions; and we will try to sell for them such of their work as we approve. In case we succeed in this, may I call on you to help us, and be a sort of Agent in the vicinity of Boston? We talked about this, you remember, last year. Now the thing is going to begin. We intend also to publish the proceedings of the Club in English, in which I shall have opportunity to let out many interesting and important facts. In fact we are a live Art Club right out here in the East, and we have a mission to discharge both to Japanese and foreigners. . . .

You will see in the "Mail" that Brinkley declares me to be the greatest living critic of Japanese pictorial art. That's a good deal for an Englishman to admit. He is much interested in the Club, and will write for our Journal if we want him. I don't know that I told you that I have received letters patent from Kano Yeitoku, the present head of the Kano house, adopting me (artistically, of course, not legally, though under Tokugawa it would have been legal) into the Kano family, and authorizing me to use the name Kano Yei (Yei being the first character of the artistic names of all past men of the line) and I have taken the name Kano Yeitan, the character Tan being that in the name Tanyu. My *nobu* name is Masanobu so that my full name is Kano Yeitan Masanobu. This I write in Chinese characters and have special seals. I give certificates by myself for old paintings in a novel form (I send you a sample) and I sign my name together with the other three critics on the certificate of the Club. It is quite a thing to have the three greatest critics of Japan admitting me to equality in certifying; and again that Japanese are demanding these certificates. . . . You will hear more of this business before the end of two years. It ought certainly to give me some prestige. . . . Mrs. F. sends regards. With love from us all to all your family. . . .

Tokio, September 27, 1884

My dear old Morse: I've had an H— of a time this summer; been off 2 months and a half, with the Doctor [William Sturgis Bigelow] and an officer of Mombusho, who was sent on a special commission to travel with me and study art. We have been through all the principal temples in Yamashiro and Yamato armed with government letters and orders, have ransacked godowns, and brought to light pieces of statues from the the lowest stratum of debris in the top stories of pagodas 1300 years old. We may say in brief that we have made the first accurate list of the great art treasures kept in the central temples of Japan, we have overturned the traditional criticism attached to these individual specimens for ages, the Dr. has taken 200 photographs and I innumerable sketches of art objects (paintings and statues); and, more than all, I have recovered the history of Japanese art from the 6th to the 9th centuries A.D. which has been completely lost. I have done it by a chain of close reasonings based on a comparative study of all the specimens, many of which have never been shown to anyone before. The priests were everywhere greedy for my certificates, and I have issued more than 100 for things which were absolutely unidentified before. I have found Chinese things called Japanese, and vice versa, many Japanese called Corean, new things called old, and even some old ones called new; and as to names of individual artists hopelessly mixed up. Yet this is the result of native criticism for centuries. Of course it is natural that some of the old fogy Japanese are suspicious and unwilling to trust me. I am proving that some of their supposed treasures are relatively worthless and bringing forth the real gems from unknown holes. But it is significant that the Mombusho, hearing that I was to travel this summer for the purpose, should have appropriated money for a special commissioner to travel with me, to study according to my methods and to find out for the government where the really good things are. It is a big thing that the Mombusho is beginning to take art matters in hand. I think I shall be able to write you more about it in a month or two. I did not return until the middle of September, and the Doctor is down there yet. I bought several pictures dating from 700 to 900 A.D. Already people here are saying that my collection must be kept here in Japan for the Japanese. I have bought a number of the very greatest treasures secretly. The Japanese as yet don't know that I have them. I wish I could see them all safely housed forever in the Boston Art Museum. And yet, if the Emperor or the Mombusho should want to buy my collection, wouldn't it be my duty to humanity, all things considered, to let them have it? What do you think?

While I was away, my little review of Gonse's chapter on painting in *L'Art Japonaise* came out, first in the Japan *Weekly Mail*, afterward in pamphlet form. I sent word to the office to have a number of copies sent to your address. . . . You can see from this a little what sort of a book I am going to write. Of course, in it, I didn't give away any unnecessary points of my own; I confined myself to those where Gonse was in serious error, and on which it was most important to set people right at once. I intend to follow this up with other small publications as opportunity may offer. One has to prepare people a little bit, I suppose, to appreciate the worth of his work. For myself, I must confess to you, who will hear me in private without thinking me conceited, that I cannot see why my work this summer was not just as important at bottom as much of that which the world's archaeologists are doing in Greece and Turkey. Of course people don't see the practical importance of Eastern civilization for the world with the same vividness as they do that of Greek culture. Perhaps it is not so important. But from the point of view of human history as a whole, it is absolutely indispensable. I expect the time will come when it will be considered as necessary for a liberally educated man to know the names and deeds of man's great benefactors in the East, and the steps of advance in their culture, as it is now to know Greek and Latin dates and the flavor of their production.

Now, as to my book. I am at work on it all the time, but have not given any part its final shape. I am sure I can get a publisher for the text; but as you know my ambition is to have it illustrated with the actual masterpieces of all Eastern art, in a way which will force my conclusions home to people's hearts. I want to fix it so that there will be no question in the mind of an intelligent reader when he has finished. It will not be dry; it will present entirely new views of the origin and nature of culture in China and Japan; it will try to make the great artists real living human beings, but radiant with genius, to the imagination. For this, the illustrations must be absolutely the best that modern methods can supply. . . . Lots of things in the air for the future. . . . Love to all, EFF

Indeed "lots of things were in the air," as Morse ended his third and last visit to Japan and set out for his first visit to Europe, via the Indian Ocean and the Mediterranean.

November 11, 1882

Ho! Gould: . . . Am in the throes of packing 2,900 pots. When that is accomplished am off to China, Java, India, Europe, and home. Am accumulating material that will last me for a lifetime; that is, it will if I can ever leave this angelic country.

He had invested borrowed capital of five thousand dollars in pottery—more than he had or could earn. To finance his travels and purchases, he had borrowed twenty thousand dollars, a loan arranged by John Gould through a Portland bank, putting up his pottery as collateral. For the next ten years he was to wade through a morass of debt on account of the pottery.

For the Peabody Museum he had completed an exhaustive ethnological collection of materials relating to old Japan, making it at a moment when the feudal culture was being superseded by western modernization. He was the only person to perceive the exigency of making such a collection. Today it is unique in the world, unrivaled even in Japan. He had sketches by the thousand and interminable scrawled notes for books on Japanese customs and architecture, which he had discussed in Japan with Henry Greenough, the Boston architect, who was traveling in the Orient at the same time. He had likewise gathered a nucleus for studies on tiles, a natural off-shoot of his interest in pottery, and on arrow release, having a theory that primitive cultures might be classified by their bow and arrow technique. These objects and manuscripts he packed and sent home; then he set out alone to voyage westward.

With characteristic energy, he managed to make enough notes and sketches for a book, *Glimpses of China and Chinese Homes,* during the few days' stay that his steamer made in Shanghai and Hongkong, with a brief excursion up the Pearl River to Canton.

Ignoring factors with which he was not conversant, such as history, art, or sociology, he concentrated on fields in which his knowledge gave him background, and produced twenty years later a work which is unique in its lore of Chinese kitchens and privies. He already had in the back of his head a brochure which ultimately attained a large private circulation, *Latrines of the East.* Unfortunately the conditions which inspired his researches also led to his contracting what was, presumably, amoebic dysentery. He noted in his journal:

Canton, March 7, 1883: At the hotel here are two bright well-dressed in light silk Chinese boys to wait on me. Last night I was writing my Journal at the table in the dining room, the only room where there is a fireplace and fire. The Spitoon, one of the big, old-fashioned kind, was out of my reach and I asked the boy to move it nearer. I expected, of course, he would push it with his foot, or if he took it in his hand, to catch hold of the outside. To my surprise, he put his hand inside the hole and lifted it that way and brought it to me! [Sketch.] And only just before he had cut bread for my dinner. It was dirty beyond endurance.

Morse made an excursion into the country to study pottery at first hand, and found the journey far different from his expeditions in Japan.

March 9: I was amazed when I got to China to find that pottery making was confined to a few towns and porcelain still more limited. At Shanghai I should have to take a trip of 400 miles into the interior to see the process of porcelain-making and from this place, Fukiang, I believe it is carried hundreds of miles to Canton to be decorated. . . . The nearest place [from Canton] was a village 24 miles up the river called Shiucwan.

The landlord said I might need a passport, so I went to the American Consul and he said that passports were only issued at Pekin. He would give me a letter if that would do any good and then proceeded to tell me how a short time ago two Americans were stoned, mobbed and driven to their boats. . . . I was bound to see pottery-making before I left China and so, engaging a boat and a crew of six hands with my funds, I started, and this time I put my pistol in my pocket.

In going up the river I noticed how rough the public were. Many would shout at me in a way that was not friendly. When I landed at the pottery town, the place was covered with pots, and I hoped that the pottery-making was in the immediate vicinity, for I confess I did not relish going through the streets with a hooting mob after me. However, there was no help for it.

The guide started ahead and I after him. A troop of dirty ragged boys followed after me, ran ahead in front yelling "foreign devil." Men insolently peered under my hat rim, ground their teeth at me. Indeed, the look of withering scorn and hatred can be better indicated by a Chinese than any other man in the world. It was useless to make any friendly advances and so I looked as savage as I could under the circumstances, stepped backward on any bare toe within reach when crowded too hard. . . .

At last a pottery was reached after going over a mile through stinking narrow streets and a hostile set of cusses. We went through a narrow alley way followed by a hooting mob of boys and roughs. The potters left their work as they heard the racket coming up the alley and so when I got to the place work had ceased and all gathered about me. However, I had come far to see it, and so I pushed through the crowd and into a shed and found a lathe there. I offered ten cents to a man to make a pot. The sum was evidently equal to two days' work for him, yet without an indication of thanks he shouted out coarsely to an attendant to turn the lathe. It was difficult to get the sketch I did with the mob about me. I have not represented their number, as it is too much work to draw figures. One helper kicks the wheel with his foot while the potter models the pot. He just puts sand on the turntable and when the pot is lifted from the table he does not separate the bottom with a cord or wire but pulls it bodily from its rest. In so doing the pot is moved out of shape, which is adjusted again. The ovens are after the manner of the Japanese, larger and more substantially built. Brick posts support the wooden roofing above. The ovens are small at the bottom and widen as they go up. Pots are used for saggers, thus killing two birds at one shot. At least I infer that the pots were to be used again.

Not a sign of a flower or garden in the vicinity of this pottery. It was as barren as a brickyard at home. I could not help contrasting my reception here and the receptions I met in every part of Japan from the potters there. Fancy a desolate backyard, covered with pots instead of bricks, a number of workmen, few at their duties, the rest shouting at each other or to me. A horde of ragged, barefooted boys howling and calling me Fan Kwai, which means foreign devil, and then I thought of my walk back to the boat again amid this howling and hostile crowd. The boys kept up such a yelling that my approach was signalized down the street, so that I passed through a serried array of frowning and angry faces or contemptuous-looking ones. . . .

A pleasant country this! It made me mad, and yet considering the horrible way in which they have been universally treated by Christian nations I don't blame them in the slightest. Indeed, if I was a Chinese, I would do the same and much worse.

Although he might be "mad" at jeers and hostile gestures, it never occurred to Morse to fear attack. The next day he

rode along the top of the wall for a considerable distance and then got out and walked, picking a few wild flowers. . . . The guide took me to a great, square five-storied pagoda which was built I believe in 1300. . . . When I came down from this five-story tower, a boy, who had followed me and had observed me pick up a snail shell, met me with a big handful of them which he gave me and for which he stood ready for his pay. The yard in front of the tower was full of them. It is a large light brown sinistral *Helix* that I knew as a boy as *Helix Senegalensis* and looked upon as one of the rarest in my collection. I never expected to find them in numbers among a lot of dead leaves and rubbish.

Morse bought the snails from a ragamuffin or poked among rubbish to find them himself, blissfully ignorant of the risk he ran. It was nearly ten years more before Japanese scientists, trained in the laboratory technique Morse had introduced, discovered that snails were carriers of the deadly blood and liver flukes of the Orient. Yamagiwa, Fujinami, Mayagawa, and Suzuki — with what glow of pride Morse in later years read the names of the men who had traced the life-cycle of *Schistosoma japonicum* from egg in human feces to free-swimming larva in water, to snail on the bank, and to cercaria returning again to fresh water in rice-paddy or river to penetrate the skin of human laborers. It was a circle completed, from the *Helices* of his boyhood in Maine to the brilliant research of men of the far land to which he had carried the light of science.

At the foot of the five-storied pagoda in Canton, in March of 1883, however, Morse knew only that he had found the habitat of *Helix senegalensis*; and did not guess the lingering souvenir he was carrying with him from Canton.

By the time his steamer reached Singapore, he was sick with a dysentery. He made no sketches, only fragmentary notes. His chief memory, in after years, of Singapore was that people there had never heard of Boston but all knew instantly where

Salem was. They were familiar with it through the numbers of clipper ships and sea captains in the Orient trade hailing from Salem. There were reports of cholera in India, so he gave up the idea of visiting that country.

Alone, unheralded, speaking no Continental language, Morse landed at Marseilles late in April 1883. He wrote home that the buildings looked very lofty after a year in the Orient surrounded by one-story native dwellings. He had his mind set on examining the Oriental pottery which had been forwarded to collectors in Paris, and he tarried not, but entrained at once for the city on the Seine. Someone had recommended to him the Hotel d'Orient, but either his destination or his accent betrayed him to the cab-driver as a greenhorn. Morse, unwitting, reported that "it took my hackman two hours to find this hotel."

The journal notes that his first preoccupation was with pottery.

May 2, 1883: After breakfast . . . I started for a Japanese exhibition of Art Objects. . . . Met Mr. Bing, the great dealer in Japanese Art and M. Louis Gonse, director of the *Gazette des Beaux Arts.* The united collections of pottery seemed poverty itself in contrast with my accumulations. . . . From there [the Louvre] to Bing's store and saw the largest stock of Japanese pottery I have yet seen, but such prices! If my collection could be retailed for half what Bing charges I should make Vanderbilt jealous.

May 6: To the Museum Cluny. . . . Back to lunch with Mr. Munro at his club and who should come in but the unique Oscar Wilde, wearing long trousers and with his hair cut short and slightly curly. Quite an improvement in his appearance. He knew me and . . . resumed the discussion I had with him in Boston on cheap editions. He was very cordial and if we did not agree in regard to cheap editions, we did agree on Japanese art. In spite of the fact that he is an unquestionable damned fool, there is genius about the fellow that some day will be recognized.

May 7: M. Gonse and Mr. Gay came for me and carried me to the house of Henri Cernuschi, 7 ave Velasquez, a house of great magnificence, filled with a marvelous collection of bronzes, Chinese and Japanese, with nearly two thousand specimens of Japanese pottery. . . . I named some for him and amazed him by picking out as the gem of his collection a jar he and no one else had noticed. It was an old Seto four hundred years old and probably made by Hanobuchi [?]. I have two specimens by him and should have liked this. Cernuschi is an old bachelor and has left his house and collection to the City of Paris when he dies. Breakfast for fifteen or twenty is served at 12; certain of his friends have a standing invitation to come in any Sunday. . . . A royal breakfast it was, topping off with strawberries which Consul General Walker on my right said must have cost a franc apiece . . . , that is, 18 cents. . . . The Prefect of Police was there and I expressed a desire to see the degradation places [*sic*] of the city. He said he would send an officer to escort me Tuesday night.

May 9: On getting through Mr. Bing's dinner, with a Japanese, Mr. Hayashi, I went back to the hotel and there found waiting for me two detectives . . . ; one was Martini, secretary of the chief of detectives, the other was Houlier. The last men in the world one would imagine to be detectives. Mr. Martini was very short, redheaded, and wore a little hat like a country apothecary's clerk; the other was medium height, pale faced, and looked like a country lawyer or perhaps a drygoods clerk, and their very success as detectives was based on the fact that they looked like anything but detectives. . . . We left the hotel in a cab at 9 and never got back until 2, and without sketches I could do no justice to the remarkable scenes I saw. The dark and narrow passage ways, the underground dungeons with gamblers, drunkards and thieves, dissolute women, every variety of vice. I was amazed to find, however, so little dirt and smell. Paris is clean even in her dens of infamy.

Morse spent but a week in Paris, crossing to London to keep appointments.

May 10: To the British Museum. . . . Mr. Franks was very cordial and pleasant; we lunched at his house and he brought out a number of *cha-ires* [pottery] in drawers, and I named a number of them which were wrongly labelled. He invited me to a dinner of the Royal Society . . . , twenty or more men. I was introduced to Galton, the Hereditary Genius man, a very pleasant cordial person.

A result of meeting Galton was correspondence on heredity over many years and an offer by Morse to copy from the directory of every city he subsequently visited long lists of names as data on the transmission of family traits.

During the five years 1877–1882, spent chiefly in Japan, Morse added to his absorption in marine biology a keen interest in Oriental art and ethnology. No one subject submerged the others. He maintained parallel channels in his mind into which poured continual streams of observations, ideas, and analyses. With Dr. William Sturgis Bigelow, it had been otherwise. Going to Japan, as a vacation from a medical career, Bigelow became so deeply interested in Buddhist philosophy that he allowed Morse to sail alone for Europe while he remained to study Buddhism.[3] Morse, however, traveled through life in the spirit of Tamburlaine, the Mongol emperor who, on his deathbed, could not bring himself to think of naming his successor but dreamed still of beckoning projects unfulfilled:

[3] In later years he taught Buddhist philosophy at Harvard, but his heart was still in Japan, and by his direction, his burial took place at Midera temple Kyoto, Japan. Tradition in Boston still maintains that, although his family arranged for funeral services in Trinity Church after his death, the coffin was not open to view because Bigelow had insisted that he should be dressed for his sepulture in the Oriental robes of a Buddhist neophyte.

> . . . See what a world of ground
> Lies westward from the midst of Cancer's line
> Unto the rising of this earthly globe;
> And shall I die, and this unconquered?

Now, turning from pottery in Paris, Morse plunged with equal enthusiasm into zoölogy in Britain, with excursions into sociology and society.

Saturday, May 12, 1883, London: Went to the opening of the Fisheries Exhibition in which all the world has an exhibit. . . .

Soon came the procession . . . , the Prince of Wales and his wife, Duke of Edinburgh and the Duchess. I was interested in watching the Prince as he bowed pleasantly on both sides and the Princess beside him was very beautiful and bowed slightly here and there; but the Prince appeared as if he looked upon it as a good-natured joke that if all the people wanted to keep up this flummery he would do the best he could. The brothers behind were all good-looking fellows. Hearty English cheers went up. The Prince is behaving very well in coming out so often, opening the Royal Academy, Fisheries and making good speeches, presumably his. The bustle and confusion was more than I could stand so I came away and back to my hotel.

Morse seems to have forgotten that on October 18, 1860, he had seen the Prince before, when the future Edward VII, a boy of nineteen, visited the Museum of Comparative Zoölogy in Cambridge.

Sunday. May 13, 1883: Went down to St. Paul's Cathedral, a wonderful pile. Inside a big service was going on; I noticed however that though the service was in the interests of the Prince of Peace the monuments about the building were to famous generals with inscriptions about the victories over their enemies. . . . After the general service a High Church flummery was being prepared at the end of the nave which seemed to excite the curiosity of many.

On my way back I passed a Hall where the Salvation Army were having services. Two large colored lanterns hung at the entrance, "Salvation full and free." Inside the place was full, at one end seats arranged like our oration nights at Mechanics Hall and the seats filled with men and women many of the men having a bright red shirt with big buttons on the breast marked "Salvation Army." They were singing when I went in and in this crowd were big trumpets, big bass drums, violins, bugles and a number of tambourines played by girls after nigger concert fashion. In front of this motley crowd were five or six people, the General himself among them, Booth by name; long beard, big nose and demonstrative to the last degree. Experiences were told, he approving, emphatically in every way, as much as to say "I told you so. . . . Isn't that so." It was the most marked exhibition of emotional insanity I ever witnessed and the calm cool way in which young men faced this audience and then wildly harangued them, getting up without the slightest embarassment, reminded me of the way people in insane asylums in the same way address a crowd. I don't think this has ever been alluded to before. . . . The praying bands of the West are not a circumstance to this thing and if it ever comes across the Atlantic it will have a great run in such places as New York and the West.

May 17, 1883, Liverpool: . . . the dirty slovenly women on the street, big bare-footed girls selling papers, little beggar boys with matches, old women coming out of ginshops. Tonight I saw a detachment of the Salvation Army, a young man with a flag, red; another with a bass drum; four young girls looking like factory girls, and one or two other young men marching down the street singing at the top of their lungs, followed by a ragged mob of males and females. Four horrid-looking wretches, men, marching in the ranks as scoffers. It was a pathetic, frightful delusion that animated them but to a thoughtful ignorant person, if ignorant persons can be thoughtful, a very impressive sight. The simple admission that it is God's work makes the whole thing so absurdly inadequate to what they intend to do. If God can do such things, what might be achieved if He ordered the royal family and all the Princes and Dukes to be seized with mania with a battalion of Bishops and regiments of vicars and curates parading with thundering chorus down the street, having Huxley, Owen, Tyndall and the rest in the brass band!

May 18, 1883: Liverpool. Visited the Free Museum and found the whereabouts of Rev. W. W. Higgins with whom I corresponded nearly 30 years ago. Sent him shells and he sent me a large collection of English shells. . . . Mr. Higgins came to the hotel and we spent an hour or two at the Museum. The collection indicates the intelligence and liberality of the Liverpool people. For the support of the Museum a tax of one penny on every pound of income is collected. At 4:30 Mr. Higgins took me to his house in the country about 10 miles from Liverpool; explained to me that he was a clergyman but of the broadest type, that he could not subscribe to the articles without hypocrisy; he therefore chose a position in which he would not be bound and took the position at a Pauper Lunatic Asylum which he had occupied for 30 years. . . .

May 19, 1883: Spent the entire day at the British Museum of Natural History, South Kensington. Dr. Henry Woodward kindly unlocked the cases for me and got together all the specimens of *Mya arenaria* he could find from the Pliocene beds. Many were broken and I could only find a few safe to measure. While there Prof. Owen came in, to whom Dr. Woodward introduced me. Prof. Owen was extremely cordial; shook hands heartily and immediately carried me to his room where he was working away on *Echidna,* an investigation that he didn't want me to say anything about.

Sir Richard Owen (1804–1892) was a noted English biologist and paleontologist. After he became head of the natural history department of the British Museum in 1856, he originated and carried through the movement for removing the natural history collections to a new building at South Kensington. Knighted at the completion of this work in 1884, he retired to his home, Sheen Lodge, Richmond Park. Among his special studies were the brachiopods and mollusks.

It was interesting to watch the old gentleman as he carefully made a dissection. He wore a big hat, a thick and big waistcoat buttoned up to his chin and shoes with holes in them. I asked him if I should dress to come out to his dinner and he laughed at the idea. . . .

Sunday, May 20, 1883: This morning I went to Dr. Woodward's to breakfast. At 2 o'clock he kindly went with me to Richmond Park where Prof. Owen lives, a most charming place, a park 3 miles long in which are a few houses owned by the Crown. One of these houses was given to Prof. Owen by the Queen. . . . The house is of stone with roof heavily thatched with a kind of reed brought from some other part of England and men from the same region have to come to put it on. Prof. Owen told me it had been mended once in 30 years. The house is buried in a mass of shrubbery and flowers, and the garden is nearly wild. I heard a great many English songbirds.

While we were in the garden Mrs. Lankester, her son and a Scotch lady came down the path; they were to be at dinner, too. A good old English dinner we had, beef; the ale was delicious. Mrs. Lankester asked me to dinner Tuesday night to meet her son Dr. E. Ray Lankester and Mr. Moncure Conway. Should like to have the time to write out all the interesting things she told me about Carlyle, Emerson and others. A story about a famous landscape artist whose name I do not remember being bothered by a want of something in his picture. A green meadow; it was too green and he wondered what he should do to it. The picture being hung up in the Gallery, the artist had a few days to touch it up previous to the public opening. Turner, who was nearby, was asked what it needed and, looking at it a moment, made five daubs of brown in the deepest part of the picture, to the dismay of the artist and his friends. They asked Turner what it was and he, looking at it a moment, called it a dog. Shortly after Edwin Landseer came in, and they lugged him to look at it, and he with a few touches converted it into a remarkably good pointer! . . .

May 21, 1883: This forenoon I spent with Prof. [Thomas Henry] Huxley who showed me his class room, working rooms, museum and introduced me to his assistants [at the South Kensington Museum]. We had a good long talk over the Dinosaurian Reptiles and kindred subjects; he gave me his book on the Crayfish and a number of pamphlets and said if I could only be here a little longer he would invite Spencer to meet me at his house to dinner. Nothing could be more cordial and hearty than his welcome. His assistant hunted up a lot of clams which I measured. Huxley had to go off to the Fisheries Exhibition, but while I was at work Prof. E. Ray Lankester came in and we were introduced. He said I was right about the Brachiopods [not being Mollusca] and he confessed he had been wrong in regard to their affinities. We had a little brush in *Nature* several years ago. Huxley also told me I was quite right about them. It is agreeable to have them come around one after another this way. I lunched with Lankester.

Huxley, in his "cordial and hearty" reception of Morse, doubtless had referred to Morse's efforts in 1879 to have Huxley named as his successor in the chair of zoölogy at Imperial University, Tokyo. The proposal had not been realized because of the state of Huxley's health at the time.

The invitation to meet Spencer was tempting, but Morse's leave of absence from the Salem museum was nearly ended and his passage for America already booked.

Margaretta Brooks

Morse in his Study at Salem

CHAPTER TWELVE

THE MORE I THINK OF MINE
1883—1887

MORSE, completing a year's journey around the world, arrived in New York on the *Servia*, June 5, 1883. The accumulation of notes, sketches, collections of pottery, and ethnological objects which he had sent ahead to Salem kept him closely confined to his study at Linden Street and his office at the Peabody Museum for the next four years.

With an obduracy that annoyed family and friends, he guarded his pottery collection as a unit, plunging deeper into debt to build a wing on his house to make room for its display. In London he had written in his journal: "The more I see of these various collections, the more I think of the wonderful character of mine. All the collections here and in Paris will not equal mine in value and hardly in number." He realized that his collection was unique — unique because of his own limitations. It had been made by a scientist rather than an aesthete. It included examples of every period and style, serried as conscientiously as a pupil of Agassiz's might have ranged a series of brachiopods to illustrate the marine fauna of the various geologic ages.

Museums in New York, Cleveland, Chicago, showed interest. Wealthy connoisseurs offered tempting sums for single specimens, but Morse wanted Boston to have it as a unit. He nearly lost the Salem house through the burden of debt. He paid off an overdue note in the nick of time by reluctantly selling some duplicate specimens to the Rookwood Potteries in Cincinnati. Mrs. Nichols, their owner (later to marry Bellamy Storer), had heard Morse lecture on Japan. She arranged to send several of the Rookwood potters to Salem to study and sketch designs from the Morse collection. When her check for thirty five hundred dollars for the duplicate specimens arrived, Morse was enabled to shuffle and redeal his debts. John Gould had arranged a loan at the Portland bank of twenty thousand dollars to finance the home expenses, travel, and pottery purchases. Morse wrote him now:

March 21, 1884

What an intense delight it must be to a man in debt to have left him by some old aunt a sum of money sufficient to pay his debts and a good chunk over! . . . For two days I have been squaring up and I breathe freely. . . . Well, old boy, I feel an eternal gratitude for your securing the loan and Oh, John, if ever that slow craft gets in from sea, — well, well!

The labor of supervising the unpacking, cataloguing, and display of the Japanese ethnological exhibits at the Peabody Museum in Salem was immense. At the same time Morse was having a two-story addition built to the Linden Street house, to afford shelves for the five thousand specimens of his own pottery collection. In the ethnological collection for the Peabody Museum, Morse had endeavored to re-create for Americans the contemporary civilization of Japan. No household article was too small or mean for his purpose. Half a century later, Dr. Jiro Harada of the Imperial Household Museum, Uyeno, Tokyo, appraised the Peabody Museum collection as the best in the world. Morse, he declared, had seized precisely the right moment to assemble types of the feudal Tokugawa culture, about to disappear under the wave of westernization. It would never again be possible, even in Japan itself, said Dr. Harada, to assemble a duplicate of Morse's collection.

With all his other occupations, Morse managed to write the text and make hundreds of careful drawings from his notebook of sketches for a delightful, and technically accurate, book on Japanese architecture, *Japanese Homes and Their Surroundings*. It was published in installments in the *American Architect* before being brought out in book form in 1886 by Ticknor and Company of Boston. A later edition was published in 1895 by Harper & Brothers.

In Japan, Morse had discovered the beauty of tranquillity. His eye was charmed by the soft-toned harmony of natural wood, translucent white paper shoji, and light-green *tatami* on the floor. After seeing the exquisite bareness of a Japanese room, focused on a single hanging scroll, a lone vase with one spray of plum blossom in the *tokonoma*, his mind revolted at the remembrance of American interiors of the seventies which he had seen in lecture tours from Maine across the continent to California. In *Japanese Homes* he summed it up with characteristic candor:

> Such a maze of vases, pictures, plaques, bronzes, with shelves, brackets, cabinets, and tables loaded down with bric-à-brac, is quite enough to drive a Japanese frantic. We parade in the most unreasoning manner every object of this nature in our possession. . . .[1]

Anthropology engrossed him while he worked in the Salem museum at rearranging the odd assortment of ethnological curiosities brought back to Salem by the sea captains of the East India Society. He delivered a paper on "Man in the Ter-

[1] Page 309.

tiaries" before the American Association for the Advancement of Science in September 1884. He still dreamed of writing his second book on zoölogy and his work on evolution. All he achieved was a memoir, *Notes on the Condition of Zoölogy Fifty Years Ago and Today*, for the Essex Institute's annual meeting in 1884, which evolved into *What American Zoölogists Have Done for Evolution*, delivered at New York in 1887, when he retired as president of the American Association for the Advancement of Science.

The success and the never-flagging personal interest and delight in research that made life exciting and fruitful to Edward Sylvester Morse over the long span of seventy years rested on the tenacity with which he seized and pursued a new idea. In 1856 at Bethel, observing differences between tiny earth-colored land shells, he began the long course of study of Mollusca that led him to Agassiz, to Darwin, to the far island of Japan, to Virchow in Berlin, to Sir Richard Owen, Huxley, and Haeckel in London. In 1878, in Tokyo, he initiated profound observations of differences in pottery that made him the world authority of his time on Oriental ceramics. About that same time an idle afternoon's sport started him on another line of inquiry that made him a leader of his day in ethnology.

> My interest in the matter [of arrow release] was first aroused by having a Japanese friend shoot with me. . . . Being familiar with the usual rules of shooting . . . , it was somewhat surprising to find that the Japanese practice was . . . totally unlike ours. . . . [This] led me to inquire further into the use of the bow in various parts of the world. . . .[2]

[2] "Ancient and Modern Methods of Arrow-Release," *Bulletin of the Essex Institute*, XVII, 146–147.

Ardent evolutionist that he was, burning apostle of a new faith, Morse could study nothing without tracing its evolution. If he found a common quohaug shell on Marblehead beach in the walks he invariably undertook the day after a storm, seeking deep-sea treasures in the wrack at the high-tide line, he must trudge inland to an ancient shell heap for comparison between old and new. His Salem study had a table thirty feet long, marked off in millimeters, on which he could range three hundred shells at a swoop for comparative measurements. A saucer in a Tokyo china shop could set him off on thirty years of ranging from prehistoric pottery in the Omori shell mounds to the latest "Yokohama muki" fresh from the kiln for export trade. So with arrow release. His afternoon's sport began fifty years of collecting data on the use of the bow, gleaned from ancient Assyrian sculpture and Egyptian paintings, from the Bayeux tapestries and carvings in mediaeval cathedrals, from engravings by Dürer and oils by Rembrandt, equally with photographs and observations of living primitive peoples.

No sooner had Morse begun to study arrow release than perforce all his friends and acquaintances were drafted into the work. Dr. Thomas Barbour, visiting New Guinea, must photograph for Morse a Papuan in the act of shooting arrows. Professor F. W. Williams of Yale must ransack Rangoon for a photograph of a Burmese archer. Mrs. Zelia Nuttall in Mexico City had to copy ancient Mexican codices in heat and dust. Dr. S. J. Mixter, on a vacation at the Bay of Chaleur, must get the Micmac Indians at Cascapedia in Canada to perform. "An East Indian officer, Captain John Johnstone, visiting Boston, promised to secure for me a release of primitive tribes." [3] Morse

[3] *Additional Notes on Arrow Release* (Salem, 1922).

wrote, his magnetism never better illustrated than by the trouble comparative strangers took to carry out his requests. Captain Johnstone, on arriving in India, persuaded a Captain Luard who was compiling the gazetteer of Central India for the government to interrupt his official labors while he had sketches and photographs of arrow releases of the Bhil tribe prepared, to be forwarded by Johnstone to Salem.

Any odd half hour served Morse for a bit of a go at arrow releases. Frank Hamilton Cushing, back from years in the Southwest, bringing Zuñi Indians to visit his patron, Mrs. Mary Hemenway, at her summer home in Manchester, Massachusetts, was promptly "invaded" by Morse from near-by Salem and queried as to Zuñi methods. Cushing didn't know, never had noticed such everyday acts. Eagerly Morse expounded to Cushing and Mrs. Hemenway the theory he had evolved of five types of arrow release, indicative of the progress of civilization in a race. There was the primary, used by children and the rudest races, grasping a knobbed arrow between thumb and bent forefinger; the secondary, adding the second and third fingers to the grasp of the bowstring; the tertiary, needing more strength and precision, with the tips of the thumb and forefinger; the Mediterranean, using three fingers only, thumb and little finger withdrawn; and the Mongolian, employing a jade, ivory, or brass thumb ring to draw the string powerfully, launching the arrow from the notch between thumb and forefinger. As Morse talked, the Arizonan explorer and the Boston matron kindled to his enthusiasm. Cushing hastily constructed a bow and arrow from sticks and cord, and called a Zuñi chief to shoot. Morse had predicted, from Cushing's accounts of the tribe, that the release would be secondary. It was. Amazed, Cushing burst out, "Spl — endid!"

On his way from Japan to Europe in 1883, despite the weakness and misery produced by dysentery, Morse in the few hours' stopover at Singapore hunted down a philologist of Malacca, D. F. A. Hervey, and got him to interpret while a Temiang Malay from Sumatra demonstrated arrow release. Forty years later, learning that a Siamese embassy was visiting Washington, Morse in no time had Nai Tuan with bow and arrow in his hand demonstrating Siamese methods. The St. Louis exposition in 1904 was Morse's chance to see in the flesh members of the Bagoba tribe from the newly acquired Philippines, and no savage was too young to contribute his mite to science. "A negrito child among them, nineteen months old, shot for me in the most vigorous manner," using the tertiary release, Morse recorded.[4]

The afternoon's sport in Japan in 1878 acquired scientific standing at Philadelphia, September 10, 1885, when Morse read two short papers, one on Eastern archery, the other on the use of the plough in Japan, before the American Association for the Advancement of Science. A reporter on a New York newspaper, frankly bored with the learned papers that had gone before Morse's, wrote that these "proved almost as interesting as the notes regarding Corea he presented yesterday" and understood that "the distinctive manner of drawing the bow was a fact which he [Morse] thought tended to show that the Assyrians had no commerce with the Japanese, as was asserted by someone at the British Association in 1883."

The multiplicity of his lines of research drove Morse to anticipate filing systems later to be developed commercially, and on June 8, 1886, his patent for "Pamphlet Jackets" was granted.

[4] *Additional Notes on Arrow Release.*

Margaretta Brooks was the sole person to benefit by the invention, which was never exploited; but in the Salem library now appeared manila paper folders in which she could locate the information on brachiopods, Tertiary man, arrow release, Japanese pottery, or Chinese kitchens which Morse carried unerringly in his head.

From the jumble of facts, observations, and deductions in his own mind, Morse evolved in these years in Salem a social philosophy far in advance of his day. He expressed it in his president's address to the American Association for the Advancement of Science in August 1887:

> A third period in American zoölogical science, and by far the most important awakening, dates from the publication of Darwin's "Origin of Species." . . . Man now becomes an object of rigid scientific scrutiny from the new position which has shed such a flood of light upon the animals below him. His habits, behavior, the physical influences of his environment and their effects upon him, transmission of peculiarities through the laws of heredity,—all these factors are directly implicated in the burning questions and problems which agitate him today. Questions of labor, temperance, prison reform, distribution of charities, religious agitations are questions immediately concerning the mammal man and are now to be seriously studied from the solid standpoint of observation and experiment. . . .
>
> [Science] offers an unassailable argument today for a wider, broader and deeper education for the masses. The untold misery and suffering of the working classes . . . would be avoided had the rudiments of social science . . . been appreciated by them.[5]

A perquisite of the presidency of the American Association for the Advancement of Science was a voyage abroad as delegate

[5] *Proceedings of the American Association for the Advancement of Science*, XXXVI (Salem, 1888), 1–40.

to the British association of the same name. Morse sailed for Europe again, landing at Liverpool on August 17, 1887. His mind turned to sociology, for the first entry in his journal gives a vivid picture of English life:

Great number of people out of employment, groups sitting on steps of public buildings or plodding along wearily and hopelessly. It is a curious fact that a man gets in a state of utter dejection and despair. The squalor seems dreadful, after leaving America. . . . Women barefooted . . . , tramps, bummers and dissolute women in clothing that once belonged to the upper classes. Fancy an elaborate spring bonnet crushed and filthy on a frowzy head; or a swallowtail coat, one tail missing, sodden with grease, on the back of a dirty tramp. . . .

Respectable English society seems to care most for outward demeanour of respectability, submission and piety. I saw a boy on his knees, blacking a man's boot, a rent in the boy's trousers exposing skin so dirty that at first I thought he had on grey underdrawers. The man with one boot blacked . . . and a policeman stonily sending them away to some alley. Somewhat indignant at this interference with the boy's occupation, I stepped up to the policeman and asked if the boy had committed any offence. . . . "Oh, no, sir; his trousers is torn, sir, and I sent him away."

On August 27 Morse paid a shilling to see the home of the Duke of Westminster: "fine old English park six miles around. . . . [Hall] enormous affair with huge belltower, chapel with a beautiful organ on which a blind organist was playing; expanse of carved wood, polished stone columns, tessellated floors . . . more like a gaudy barroom in San Francisco than anything else. . . ."

When he and his companion drove back to Liverpool that day, they had tea with his "old friend, Mr. Higgins."

He lives in the city now, having given up his place in the country. . . . I got a glimpse of the woes of landlords and the great agricultural depression in England from a confession he made to me. . . . His brother died, leaving him an estate of over two thousand acres within sixty miles of London. A village on this estate has one thousand inhabitants. In ordinary times the estate ought to yield him clear the sum of fifteen thousand dollars a year. With the great depression in agriculture he had in reality not received a single penny. It had been rather a burden than otherwise, and a few months ago a mill burned down which had been built in company with another man by his brother, with whom an agreement had been made that if the estate changed hands, the sum of seventeen hundred pounds would be paid. This claim was made and he had to pay down the amount in cash.

He attended the sessions of the British Association for the Advancement of Science, opening August 30, 1887, at Owens College, Victoria University, Manchester, and sat on the platform to hear Sir Henry Roscoe, the president, deliver the annual address, with the mayor of Manchester moving a vote of thanks that was seconded in international accord by Professor Asa Gray, Boston botanist. He delivered a paper on arrow release before the ethnological section, "illustrated by a bow and arrow, and the president, Prof. Sayce and Mr. Pitt-Rivers and Prof. Leher Evans remarked on the value and importance of my work."

A sidelight on the social habits of British scientists was obtained when Morse attended

The dinner of the Red Lions, members of the BAAA who meet to make great fun. . . . [I] was seated near the King Lion, Sir Robert Ball. . . . I may say here that during the entire dinner at intervals a roar would start, each roaring as near like a lion as possible. No allusion to members is permitted; it must always be "lion." . . . I was told, if called on, I must not say "Mr. Chairman" but "may it please your majesty." The applause of the lions is shown by standing up and shaking or wagging their coattails. Now it may be faintly imagined how ridiculous . . . and how really comical [it is] to see men whose fame is known throughout the world standing up wagging their coattails and growling like lions.

Just as Morse had kept the governor of Satsuma waiting until the chrysanthemums had to be viewed by lantern light, so he was late to the Manchester Corporation dinner for the British Association for the Advancement of Science.

September 7th was a notable day. [I had] a pass to ride on the engine of the fast express to York and return. . . . I took the fireman's seat as he was busy most of the time piling in coal. . . . When it is considered that 570 trains pass in and out of Manchester station daily and we were travelling at a rate of 63 miles, you may imagine the excitement of the ride. . . . It was terrific when we tore by a station with tracks and switches filled with cars, but the most thrilling was plunging into a black tunnel two and a half miles long. . . . Riding so high, had a fine view of the cities and towns, every building of brick or stone, and all with slate or tiled roofs. . . . At a distance hundreds of huge chimneys indicate the wealth of the manufacturing area; a forest of chimneys, the landscape obscured by smoke, and the streams running the blackest water. . . . The engineer spoke only once . . . : "We put in new valves this morning." . . . I was covered with soot and dust. . . . [Got] to the Town Hall [in Manchester] at six o'clock in seedy-looking trousers, disreputable overcoat and a depraved slouch hat. . . . Made a change of clothes and was ushered in to the Mayor with a massive gold chain of huge links over his neck. . . .

Turning, with his customary facility, from zoölogical conferences and observations on the English scullery to the field of ceramics, Morse devoted the next week to the examination of Japanese pottery in English museums. On his way up to London he stopped off at Worcester to visit the Royal Worcester Porcelain works, where he re-labelled some of the Japanese pottery.

September 13 [London]: Today I have been at the British Museum. The Japanese and Chinese collections which were at Bethnal Green are now brought together here with many other specimens which Mr. Franks has added since I was here last [in 1883].

Mr. Franks was away for a month but Mr. [C. H.] Read, his assistant, was on hand and kindly unlocked the cases, and I have been busy all day correcting wrong labels and writing new ones. There are about 800 specimens, many of them very good, and among them I recognized several types of Ninagawa's *Kwan Kodzu Setzu*. How I wish I had them. . . . My hands were absolutely black with soot by the time I had got through examining the specimens. In the whole lot of eight hundred I saw but very few I did not have; in most cases I had identically the same specimens with very few exceptions, the only ones I envied.

With his characteristic interest in anything that fell under his glance, Morse filled the next page of his notebook with sketches of "various shapes of chimney pots I can see from my hotel window" and his indefatigable mind raced to a new project. Oddly enough, this plan of setting up a tile-factory in Salem is the only memorandum never carried into execution that is found in Morse's notebooks.

I have given a few of the many forms of earthern pipes and cement pipes that ornament the tops of the chimneys in London and in all England, in fact. There are doubtless many others and must be of some use in assisting the draft. What I am struck with now is the great use made of pottery. There is, to be sure, at every turn some pottery appliances that we do not see in our country. The coping of stone and brick walls is finished with a tile not unlike the Iwami tile which I have figured in *Japanese Homes.* [Sketch from memory.] It is shaped like this and forms a most useful coping. I cannot now recall the immense use made of this rough pottery work, but it would be a most useful industry if it could be introduced into our country.

September 14: Today I went to the British Museum again and finished up their pottery, making many discoveries in their specimens that they hadn't dreamed of. Mr. C. H. Read was delighted with the light I threw on the things. . . . He gave me a beautiful Karatsu bowl shaped like this [sketch].

He spent a quiet evening at the hotel reading the London City Directory:

I was first surprised to find so few Morses in London; just fourteen Morses in the commercial part and but eight in the court directory; of Becketts there are nineteen commercial and twelve in the court directory. It occurred to me that here would be an interesting line of inquiry for Mr. Galton to undertake, the percentage of names of a family as represented by their occupation, grading them from the lowest as bootblack or chimney sweep to lawyer, doctor, etc. An index might be proved and a comparison between the figures for England and the United States would show greater chances for advancement in a new country. As soon as I see Galton again, shall suggest the inquiry to him.

Hardships encountered by the Salem traveler in foreign lands are indicated by the next entry: "called on Mr. James Rosford and he is going to get me a bottle of American whiskey, a commodity I have much wanted." Thus fortified, on September 16 he set sail for Christiania (Oslo), Norway, where he observes that the "fiords look like the region around Mt. Desert." On the twentieth he went to the university:

Prof. Sass . . . asked all about my work and showed me . . . a lot of *Cypris* developing in mud sent from Australia. . . . Looked through the Zoölogical Museum. . . . Then called on the professor of northern antiquities, Prof. Rygh, who was a blunt old cuss . . . evidently ashamed to show me the collection. . . . The early pottery of Norway was very interesting, made by hand but ornamented with elaborate figures like rough lace work. The Stone Age showed some beautiful knives, axes, etc. . . . , piled into poor cases, no light, no labels, and a mess generally. An old woman unlocked the building in which was preserved the Viking ship . . . ; the captain of the boat had been buried with it, and everything belonging to it, shields, weapons, oars, furniture. . . . In the afternoon called on Prof. Collett of the University, and he showed me specially the collection of birds. . . . The land and fresh-water shells interested me much. . . . The Ethnological Museum was . . . a few small rooms with the most ridiculous boxes for cases, evidently made by a ship's carpenter. . . . In one case were some fifty to sixty specimens of Japanese pottery, so I offered to name them. . . .

Thus Morse, going through Christiania in two days like a whirlwind, and as irresistibly inspiring Norwegian curators to give him gifts of pottery as he had Japanese noblemen. His name was widely known in scientific circles now. At Stockholm it was the same: "Just one wild jump accumulating material, making sketches, and seeing things." He went to the National Museum and then to the Academy of Sciences, where Professor Levren, "a sweet, earnest, delightful old man," shook him warmly by both hands and asked incredulously, "Is this Professor Morse?" He took Morse to his study and showed him his work on echinoderms. In the library Morse saw various Essex Institute publications containing his own papers. He was also invited to a reception for Henrik Ibsen, where there were "four or five hundred Swedes and a few Norwegians, such fine-looking men and such solid, rational-looking women." Ibsen was "a short, husky, greyheaded man, bushy side-whiskers, high forehead, reddish in the face."

Morse was more interested in *Mya arenaria,* which he discussed for four days at Copenhagen with Johannes Steenstrup, the Danish archaeologist, than in the Norwegian dramatist. Steenstrup "said that in fifty shell-heaps on the Baltic not over ten *Mya arenaria* had been found; . . . evidently not eaten by prehistoric or historic tribes." To a Maine boy whose pet expression was "happy as a *Mya arenaria* at high tide," this dietetic omission was amazing. He liked Copenhagen, though, for even the mayor was a malacologist: "called on Mayor Jonas Collin . . . whose name I have been familiar with as a scientific writer on shells, etc., of Denmark."

At Kiel he first became acquainted with German scientists and called the museum "thus far the best I have seen." At the mineralogical building he found Professor Haas, a paleontologist specializing in brachiopods.

It was delightful to find an enthusiast on the Brachs. He told me he had collected and studied them since he was a boy. . . . He called my attention [in a French volume on brachiopods lately published] to figures copied from my papers on the subject. He told me incidentally that he went some distance from Paris to see D. [a paleontologist], and the Frenchman came to the door and said he would not see a German and shut the door in his face. A similar incident was told me in Copenhagen, that a Danish professor wished his son to learn the German language but, hating Germany so much on account of the Schleswig-Holstein war, he sent his son to Russia instead. . . . It would seem that the new alliance of Italy with Germany and Austria was to get a compact center to oppose the hostile elements which surround the center of Europe.

Morse was writing in 1887. Fifty years later the old entry in faded ink might have been from a morning newspaper.

His first impression of Germany recalled certain aspects of Japan:

October 1887 [Berlin]: . . . Must speak of . . . Hamburg; . . . such beautiful bridges, buildings, parks; long narrow alleys . . . , yet everything neat and clean; flowers in the windows, curtains, door-plates and respectable-looking people. The point I make of these cleanly features among these people and the Japanese is that a set of individuals, eating, sleeping, dying, having the same desires and demands, may be taught or compelled by law to keep clean and orderly. When one looks at England with its reeking mass of pauperism, there is some mistake, something that ought to be rectified; and rectification can come only when the social conditions of the people are studied. England with its piety and prayers is the tyrant of the world and is cursed at home with drunkenness and poverty.

In Hamburg he drove to the museum, where he was received by the director, Dr. Justus Brinckmann:

Told him briefly I would like to see the Japanese pottery, whereupon he became quite delighted and showed me proof sheets of a work he is writing on the art industries of Japan and said the second volume had been delayed by his embarrassment in regard to pottery. . . . He went off in most enthusiastic praise of *Japanese Homes*, said it was the greatest work on the subject and he frequently had to quote from it; I saw he did not know me, so interrupted him by telling him it was my book. It was amusing to see him open his eyes, grasp me again by the hand and express his delight at meeting me. He dropped everything and devoted the day to me, introduced me to several as "the greatest authority on the Japanese nation" and, indeed, nearly turned my head. At Brinckmann's at breakfast was the art director, his wife, children and sister — fresh lobster served in the shell, beefsteak with mushrooms and oysters, and finally bread, cheese, etc., and delicious white wine.

Morse seems to have had correspondence with Dr. Felix von Luschan, preliminary to visiting Berlin, where von Luschan was director of the Ethnological Museum.

October 8, 1887: Dr. Luschan said that my paper on Arrow Release and Cushing's work had made Ethnology a science. Of course I disclaimed any such praise but he insisted and said he had shown my paper to a number of friends and said, "This is the kind of work that must come from such studies."

October 10, 1887, Berlin: Dr. Luschan had invited me to his house to dinner at five and I found he had invited a number of ethnologists to meet me. . . . Dr. Luschan got out his photographs of Asia Minor. . . . Showed me an indistinct arrow release and although it was hardly to be made out, I told him it was certainly Greek and was identical with one shown on a frieze at Phygalia. Immediately Luschan and Dr. [Gustav] Hirschfeld entered into an animated discussion in German, and Luschan turned to me and said the most eminent scholars in their discussions as to the age and relation had finally referred it to the same time and character as the temple of Phygalia. This led him to hunt up an old thumb ring worn by archers which he had dug out the ground about four hours from Damascus. I asked permission to make a drawing of it and, in passing it to me, he said regretfully: "I cannot give it to you because it is absolutely unique." I assured him I only wanted a drawing and went into the library to make the drawing. . . . When I came back, Dr. Luschan said, "So you are the author of that delightful book on Japanese homes! Now I am going to give you the ring," and he made me take it, in spite of all I could do. It seems that Dr. Pleyte had told the company I was the *Japanese Homes* author, and though Dr. Luschan had read the book, he had not identified me with it.

October 11, 1887, Berlin: At the Ethnological Museum . . . Dr. Luschan, the director, took me over the University and introduced me to Van Martens [whom we] found . . . buried in his shells. He showed me a lot of Japanese land shells and was very pleasant. . . .

[Van Martens] was very pleasant, took me in and introduced me to Professor Schultze, who was hard at work at the microscope over his favorite shrimps. He showed me a wonderful glass model made by Blaschka of Dresden, illustrating the anatomy of a *Calcis spongia.* It was a marvelous piece of work.

On October 15 he went to a meeting of the Anthropological Society:

Professor Virchow,[6] the president, . . . opening by stating that there were two guests, Professor Rudolph Baier, director of Stalsund Ethnological Museum, and me. . . . One hundred and fifty members present, all of middle age, strong, intellectual-looking Germans. Professor Virchow gave a talk on the Saxon house, illustrating on the blackboard. . . . He is not a tall or stout man, in profile somewhat resembling Dr. James Freeman Clarke; wears square-toed shoes, dresses simply and looked very much the scholar. . . . He lectures on pathology at the University and is a great ethnologist and also a member of the Reichstag, and a radical, having opposed von Moltke in his new army bill. . . . [He] was very pleasant, spoke of some early prehistoric pottery from the Trans Caucasus and offered to show it to me on Sunday. . . . Afterwards we went to a lager bier saloon nearby. . . . I had a delicious beefsteak and beer with Dr. Virchow, who spoke very good English and talked to me all the time.

The next day, a Sunday, Morse with Dr. Monroe, Scotch author of a work on lake dwellings, visited Virchow's home:

I saw his library, contained in two rooms; books and pamphlets were piled on the floor, tables, and shelves, with narrow lanes to walk through. His table was like a child's school desk with space

[6] Rudolf Virchow (1821–1902), German pathologist who also was noted as an anthropologist and ethnologist and found time for an active career in politics from 1862. He was constantly inimical to Bismarck's policies and achieved great advances in social administration, water supply, sanitation, and budget affairs. He was a member of the Reichstag from 1880 to 1893 and a member of the Berlin municipal council. At the time Morse met Virchow, the latter had already published archaeological works based on explorations with Schliemann at Troy in 1879 and in the Caucasus in 1881.

about a foot square to write in, and that was the only clear space in the two rooms except the narrow lanes and one chair! He then took us upstairs and showed us the Caucasus pottery and then bronze instruments about 1000 B. C. . . . From the house Virchow took us to his laboratory to show us a large collection of pottery from the Trans Caucasus piled on top of cases containing his collection of skulls . . . ; he showed us a Japanese skull containing extra sutures in front of the zygomatic arch, a peculiarity of the Japanese. . . . Indeed it was misery to leave all this richness, rush to my hotel, dash off a few letters, pack my trunk just in time to take the 5:30 train for Dresden. . . . Arrived nine o'clock . . . writing [the journal] ever since, and it is now two. Were it not for the Japanese pottery that must be worked up I would go into a monograph on archery. Dr. Luschan expressed a wish that I would come to Berlin and make a special study of archery implements, offering me every facility for study and examination. The offer was certainly tempting, but it is useless for me to think of it now.

Mark Twain might have made a couple of excellent chapters for *Innocents Abroad* from Morse's naïve peregrination of southern Germany and Switzerland. Getting through museums and literary or historical shrines "in a murderous hurry," Morse pursued his objectives of pottery, ethnology, and geology, as oblivious to landscape as a diver tracing the bright glitter of a coin sinking amid rocks and seaweed.

October 17, 1887, Dresden: The Porcelain Museum being closed until October thirty-first . . . , nevertheless I rang the bell. . . . A man came and . . . I gave out a series of monstrous lies; told him I had come from America expressly to see the Dresden collection; that I was going to Paris immediately, that I didn't care a damn for his European porcelain. . . . Little by little his severe attitude relaxed. Then I told him I had a great collection of Japanese pottery. The way I laid it down convinced him that I was not to be denied, and soon he made a despairing gesture with his arms, smiling, . . . and I went over the whole collection. After knowing Japanese pottery and Chinese porcelain, it is utterly impossible to arouse the slightest interest in European work and the more celebrated these works are in the eyes of European connoisseurs, the more atrocious they are to me.

A tourist's sense of duty took Morse to the Dresden Gallery and the Museum of Armor:

After getting through these museums in a murderous hurry, I got a hasty mutton chop and bottle of beer and started to hunt up Leopold Blaschka, the maker of the famous zoölogical models in glass. I found him a very pleasant old man, and with his younger son, who has an eye to business, passed a pleasant half hour. They are very fond of music, violin and piano, and the old gentleman showed me with some pride oil paintings hanging on the wall as his productions. . . . The father and son make the most perfect and wonderful reproductions of animals that have been conceived of. . . . The most intricate and difficult animals, like *Physalia* [etc.], depicted with the most wonderful accuracy. I asked to see their workroom and found it an ordinary room with a table covered with glass tubes of various sizes, lamps, bellows with treadle below, unfinished acephalids, bits of . . . hydroids, drawings of animals; in another room an alcoholic collection of invertebrate animals. I was shown also some glass models of orchids being made for Dr. Goodale. These were exquisite, and it would seem that their art of modeling might compass the most delicate objects. They lived in a large block of buildings. . . . The buildings here and in Berlin are great substantial blocks, people living in flats.

Morse thus was probably the first Bostonian to see the exquisite botanical models of flowers in glass which generations of Harvard students would dutifully escort mothers and aunts to view in years to come. With the death of the son, Rudolph Blaschka, in 1940, this art died.

October 19, 1887, Leipzig: Called first on publishers who had made some beautiful colored plates of Chinese porcelain for Professor Meyer in Dresden; . . . got their estimate for expense of such plates in anticipation of my pottery book.[7] . . . Then to the Ethnological Museum. Into a dirty gateway, up two flights of stairs, where I found a locked door and a bellknob. Rattled the one and pulled the other several times and had started to go in disgust when two pale, dirty children opened the door from within. They might have been museum specimens come to life. . . . I went into the dirtiest, coldest den imaginable. Atmosphere as antique as the specimens, not only had absorbed odors from Peruvian mummies, Indian scalps and other nitrogenous specimens, but sauerkraut, limburger cheese. Through this tainted atmosphere with handkerchief over my mouth I hastily moved. . . . Judge my astonishment when at the end of the room I found quite a number of specimens of Japanese pottery; indeed a more interesting collection than all the others I had seen in Germany put together. . . . Off I went to hunt up the Director, Professor Obst. . . .

October 20, 1887, Leipzig: This morning Professor Obst came to the hotel, very cordial, apologised for the state of his house, saying he and his family had just got back from Russia and would soon have a new building; expressed great shame that the accomodations were so bad. At the hall found that yesterday I had only seen a portion of the collection. Dr. Obst told me they had hundreds of cases packed away for want of room. Set to work on the Japanese pottery and while there was a lot of poor stuff in it, there were some excellent pieces. I labelled nearly all of them for him, much to his delight. Japanese objects besides the pottery are numerous and some of them extremely rare. . . . I worked in his pottery until eleven,

[7] Morse was already planning his *Catalogue . . . of Japanese Pottery*, published in 1901. The plates were made by a Boston firm in the end.

and then Dr. Obst went with me to the third story of the great University building to see Professor Leuckart.[8] It was delightful to renew a warm welcome and we talked quite a while. . . .

Morse scored off Dr. Leuckart by finding the brachiopods among the Mollusca, and the German retorted, "So many books still have them there that it is a matter of public convenience, but in my library you shall see a pamphlet on brachiopods arranged with the worms."

At Weimar, Morse did not succeed in seeing the collection of Japanese pottery belonging to the Duke of Weimar, and turned instead to other matters less interesting to him. "Resolved myself into a globetrotter and went to see the houses in which lived Goethe, Schiller and Kant." His only noteworthy observation in these literary shrines was that in Schiller's house the stove was of terra cotta; but the night before had been a bad one, and that may explain his indifference:

After freezing to death in my room, though kept warm after a fashion by my exercise in trying to make an infernal German stove give out heat, went to bed. A more exasperating and ridiculous device I never saw. Fancy a bedstead like a box containing a hard, tight mattress just coming to a level with the sides of the box. Upon this is placed a huge ethereal featherbed and a low slouchy pillow . . . and this great fluffy affair undertakes to perform the functions of clean sheets, a blanket or two, and coverlet. It cannot be tucked in because there is no place to tuck it. It makes you think of a partially collapsed balloon. It is simply infernal and damnable, and I understand for the first time the reason the French hate the Germans. Here I have struck the regions of idiocy as regards bedclothes. I yield much to social ways and habits of a people, but it is incredible that they should be happy with such a device, provocative of colds, sweats, and profanity.

[8] Rudolf Leuckart (1823–1898), German biologist and anatomist.

October 25, 1887, Nuremburg: Woke to see the city moat and, rising above, the city walls and quaint towers all covered with dark red tile. . . . [Went] to see the house of Albrecht Dürer and engravings. I examined all the engravings for ethnological details and found some archery . . . ; in a Cupid, the release is Mediaeval Mediterranean.

October 27, 1887, Munich: Professor Zittel . . . went with me to the Ethnological Museum, and I had the pleasure of seeing the collections of old Dr. Siebold,[9] and certainly they were very interesting. Among the pottery were a number I could not identify. The collection represents things as they were fifty years ago in Japan.

At the Munich Academy of Sciences, he was recalled from his new enthusiasm for arrows and tiles by biology:

Professor Hertwig a delightful man who knew me at once from my Brachiopods work and spoke of his paper on the *Coelum* theory in

[9] Dr. Philipp Franz von Siebold (1796–1866) was a German commissioned by the Dutch government to make a secret survey of Japan, which from 1637 had been hermetically sealed against intercourse with the rest of the world, except for a Dutch trading post on the island of Deshima, near Nagasaki. By the terms of the Dutch trade, one ship a year was allowed to touch at Deshima, and once in four years the Dutch trade commissioners, strictly guarded, were escorted to Tokyo to pay tribute to the Shogun. Von Siebold left Europe in 1823 to accompany the next Tokyo party, which left Deshima in 1826. Von Siebold spent three years in study of the language, secret intercourse with scholarly Japanese, and observations around Deshima. He took to Tokyo with him a Japanese physician, a Japanese artist, and several "servants" who assisted him in collecting data. He made large collections in natural history, objects of art, and ethnology, which he sent via Batavia to Europe. His notes and sketches were the basis of his *Fauna Japonica*, *Flora Japonica*, and a miscellany of observations published in a series called Archives for the Description of Japan. All his work during five years in Japan was carried on secretly under pain of death. After discovery at the close of his stay, the Japanese who had assisted him were put to death. Their object in assisting him had been to obtain in return instruction in European medicine and science and the Dutch language. Von Siebold taught them so thoroughly and faithfully that for the next fifty years Nagasaki was the principal school of medicine in Japan. Morse's

which he had quoted me. . . . In his private laboratory special students, . . . one an American from Newark, were working away on the organ of Bojanus[9a] in *Unio*. . . . Told him of the valve I had found at the posterior end of the ventricle in Unio which I had never seen described, and he gave me a slide containing a section showing these valves. The sight of these preparations made me yearn to get into the *Unionidae* again and complete the first work I ever did under Agassiz.

October 30, 1887, Munich: To the Royal Bavarian National Museum, the armor collection particularly fine; . . . walls of the rooms frescoed with battles to illustrate various implements. . . . Could not help noticing a curious blunder in which an archer on horseback has just discharged his arrow at a foe who is also on horseback and protects himself with a shield. The arrow has passed halfway through the shield, but unfortunately the artist has taken a crossbow for his model, being ignorant of archery. The arrow, having no nock, could not be discharged from a long bow! Furthermore, an arrow as short as that could not have been drawn far enough to send it through a pasteboard shield!

Very occasionally in Morse's life, for a moment, he had a glimpse of his own inadequacies; this happened now in Zurich: "I never expected to be here . . . [but finding myself] within half a day or so of veritable glaciers, . . . the chance of seeing a phenomenon I have lectured about for years was too much."

He was distracted by tiles, which kept popping up like King Charles's head:

visit to von Siebold's collection indicates his wide collateral reading, otherwise not mentioned in the journal.

[9a] Details of the journal entries from February 8, 1861, when Morse, at Agassiz's behest, began making dissections, sketch a picture of a slip of the scalpel, an infection apparently of streptococcal origin, and some six weeks of pain and swelling in the right hand, which forced Morse to cultivate the faculty of using his left to write or sketch. Thus, in a sense, his work on the organ of Bojanus was the origin of Morse's "native" ambidexterity.

To Museum of Antiquarian Society [in Zurich]. Collection of Roman tiles was of great interest . . . often had impressed on them the number of the Roman legion who made and used them. . . . While busy taking these notes had but little time to see Zurich.

November 2, 1887, Berne, Switzerland: Saw the Armory. . . . The attendant allowed me to shoot [a crossbow] at the side of a board wall. It did not go anywhere near where I shot at and the force was certainly not equal to my forty-pound bow. . . . Found a most perfect representation of the Mediaeval Mediterranean release in a 15th century Burgundian tapestry at the Historical Museum.

Sternly eschewing tiles and arrows, he finally managed to get aboard the boat, crossed the lake to Interlaken, and took a carriage to the Grindelwald glacier, riding up the valley on November 2 with autumn foliage, exquisite yellows and reds, blazing on the slopes above the green meadows. Then came a stage on horseback: "The horse, being young, kicked up and acted like the devil as a big blister at the termination of my vertebral column proves." At length he stood before an actual glacier. He crossed the snow-covered moraine "with no rubbers, but coming so far I was bound to see a glacier and see one I did." He continues:

Passed in an ice-tunnel in the glacier and realized the ethereal transcendent blue of glacial ice; examined it minutely with a glass and made tracings of the curious erosions and became more thoroughly convinced than ever that Croll's and Moseley's idea of molecular movement of glaciers was correct. The snow was too deep and time too short for me to ascend and get above the glacier.

He returned to Interlaken that night, hurried by train back to Germany, and found that Professor Weismann and his own old pupil, Chiomatsu Ishikawa, whom he called on in Freiburg, were week-ending at Lake Constance, which he had just left.

Chiomatsu Ishikawa, one of Morse's first pupils in zoölogy at the Imperial University, who became one of Japan's most distinguished scientists, after Morse's death published reminiscences of his old master's visit to Freiburg:

> I was out of town for a few days on a trip to Bodensee with Professor Weismann when Dr. Morse called. Professor Wiederscheim, in our absence, took Morse around the city in a carriage. They were passing along Kaiserstrasse when Dr. Morse drew Wiederscheim's attention to the roofs of the houses, pointing out to him that they were covered with tiles which, Morse declared, were Roman and more than a thousand years old. After Morse left, Wiederscheim told archaeologists; they made investigations and to their wonder found that the tiles were just what Morse had said they were.[10]

Morse pushed on to Heidelberg and Cologne, where in the church of St. Ursula he was solaced by finding an "interesting picture painted by Jan von Scheiven in 1456 showing two-fingered Mediterranean release." Later he found "a St. Sebastian showing three-fingered Mediterranean release distinctly, also some Raphaels with two-fingered."

On November 11, in Leyden, he went to the Ethnological Museum, of which Dr. Leverrier was director.

> Had an opportunity of making a drawing of a Siam bow with which they shoot clay balls . . . , beautifully made, and on trial I found it could be an effective weapon for small birds. I am sure the Japanese must have got the idea of the ancient bow in Nara temple [Shoso-in] from this. Dr. Leverrier took me . . . to the Japanese museum in a building by itself and a very interesting col-

[10] Preface to Japanese edition of *Japan Day by Day.*

lection it makes. Many large bronze *daibutsu*, lacquer, dolls, dishes, etc. The collection of pottery is very poor, a few objects of interest only. After leaving the Museum, I was put into the hands of Dr. Shmetze. . . . On the way to the station he told me about my *Japanese Homes* book and how he liked such and such descriptions. It seemed odd to be walking the streets of a Dutch city hearing quotations cited from my book.

November 12, 1887: Reached Paris at noon . . . and rushed immediately for Bing, the great dealer in Japanese and Chinese objects. His place is simply magnificent. . . . Met there Louis Gonse, who immediately invited me to breakfast with him next day; and Bing invited me to dinner afterward. . . . Spent the rest of the day picking out from among the thousands, a few that will fill important gaps in my collection. . . . Sunday had a delightful time at M. Gonse's, examining his collection of pottery, and at three he went with me to Bing's, where I revelled in pottery again. Saw with great envy a number of Ninagawa's earlier types which Mr. Bing bought in Japan before I became interested in the subject. Nothing could induce him to part with them. Thursday — forenoon and part of afternoon going through the Bing store collection.

November 15, 1887: To M. Gonse for dinner and sat up until midnight looking at pottery. Next morning Gonse took me to Hayashi's Japanese store . . . ; found four interesting specimens of pottery. Afterwards Bing took me to Palais of Industry, large exhibition of decorative art, then to Trocadero Museum. Since I was there four years ago they have added a large ethnological collection, a number of objects brought home by Nordenskiöld.[11] . . . From the Trocadero saw the beginning of the huge structure [the Eiffel Tower] to be completed for the great Exposition in '89, to be one thousand

[11] Baron Nils Adolf Erik Nordenskiöld (1832–1901), mineralogist and Arctic explorer, is most widely known for his navigation of the Northeast Passage in the ship *Vega*, 1878–79. Morse had felt a "Japanese" atmosphere about the monument to the *Vega* voyage in Stockholm and evidently admired Nordenskiöld greatly.

feet high; structure is wrought iron; four legs finished up to seventy feet now.

November 16, 1887: . . . Met Bing by appointment and to Mr. Guimet's office where he has brought together a large collection for an Oriental Museum now building which he is to present to the city . . . ; a large collection of pottery but only a few good pieces. [He] gave me his works on Japan, chiefly travelling experiences as he went to Japan accompanied by Régamey,[12] the French artist.

November 20, 1887: Have given three solid days to Bing's collection. It is simply superb, twenty-four provinces represented and eighty six kinds of pottery. In the evening he showed me his Ninagawa types and here I revelled for three hours. I recognized every one of them, showing how characteristic the figures are in *Kwanko dzu Settsu.* If I can possibly secure them for my collection, shall do so.

November 24, 1887, London: Thursday I spent at the British Museum. Mr. Franks courteously opened all the cases I wished to examine. . . . I pointed out the bad arrangement of the Japanese pottery. He admitted all I said but seemed reluctant to change them.

November 25, 1887: To Oxford to visit Professor E. B. Tyler . . . ; delightful day and night. Examined the bows and arrows in the Pitt-Rivers collection, made drawings at the University Museum. Met Mr. Henry Balfour, a very bright fellow, attaché of the museum and much interested in my arrow paper. From the museum Professor Tyler took me to the Bodleian Library and to the Ashmolean Museum . . . collection whose first catalogue is printed in 1656, and here I found to my delight three archers' thumb rings of the Turks, all of carnelian. In this museum was shown a skin ornamented with shells of *Marginella* said to have belonged to Powhatan; Pocahontas may have worn it. In one case the gun of Alexander Selkirk ["Robinson Crusoe"]; in another the lantern of Guy Fox, Faulks . . . give up the spelling of it. Dr. Tyler then took me to

[12] Felix Régamey was the author of a work published in English in 1893 as *Japan in Art and Industry.*

dine in the large hall of Balliol College, professors' table on a raised platform facing the students below; coffee served in a small room and some delicious port.

November 27, 1887: Called on Dr. Anderson, author of *Pictorial Arts of Japan*. Had a delightful time talking over Japanese art, the merits of Hokusai and early Japanese wood engraving. A Japanese student, Mr. Suzuki, came in; . . . has just taken his degree at the College of Surgeons.

November 29, 1887: To Paris this evening, reaching there next morning at six; spent the entire day at Bing's and left Paris eight P.M., reaching London next morning at six, having been in my clothes two days, but have made the greatest addition yet to the collection.

CHAPTER THIRTEEN

THE BAND PLAYED YANKEE DOODLE 1888–1901

THERE WAS nothing of the snob about Morse. He returned to Salem without a regret for the banquets and bouquets from European bigwigs, and busied himself at the Peabody Museum. He worked up from journal sketches and notes a long monograph on ancient arrows and a new method of arrow release, a paper which he read at the Cleveland meeting of the American Association for the Advancement of Science. Within a few months he was hotly involved in a fight with the American government which, he felt, had failed to display the European approach to research.

Morse's nominee in 1878 for the chair of physics at Tokyo Imperial University, T. C. Mendenhall, had returned to America and become chief of the United States Signal Corps. In 1886, with characteristic informality and blending of public interest with official function, Morse had induced Mendenhall to lend him government apparatus. John Gould and young John Morse, enlisted as amateur observers, kept faithful records of ocean temperatures at Salem and Portland, while Morse toured Europe in 1887 and 1888. Home once more, Morse himself carried the apparatus wherever he went in order to take the ocean's temperature whenever he found himself within sniffing distance of salt water.

Mendenhall, meanwhile, had been vainly beseeching Morse unofficially to return the government property, which, as the years rolled around, needed to be accounted for in inventories. Morse countered with an impatient proposal to buy the apparatus. Mendenhall, needing an Act of Congress to sell a government thermometer to a private citizen, urged the return of the apparatus in friendly, personal appeals. Morse finally realized that this was a governmental question and that, given time, any government can wear down any individual. He exploded characteristically to John Gould:

January 19, 1889

Have been so exasperated with that office as I wanted to buy the thermometer but the red tape was too much for me, and if I have anything belonging to them, they will have to send a man and find it. Not another word will I exchange with the poor devils.

Morse's irascibility reflected the strain of thirty years of work on brachiopods, pottery, arrow release, museum arrangement, comparative anatomy, under the lash of perpetual debt. His growing intolerance of extraneous noise indicated the state of his nerves.

Railroad whistles from the freight yards near his Salem home annoyed him. The constant clatter of iron-bound cartwheels and carriages over stone pavements drove him to frenzied outbursts. He engaged in a battle that for years strewed America from Atlantic to Pacific coast with pamphlets, addresses, and letters-to-the-editor. Recruits to the cause included Francis Wilson, the actor, Bishop William Lawrence of the Diocese of Massachusetts, and Charles Eliot Norton of Shady Hill, Cambridge.

Morse's pamphlet, *The Suppression of Unnecessary Noise,*

printed in quantities at his own expense, was mailed broadcast with a slip addressed "to those who wish to aid in suppressing the worst of unnecessary noises." It suggested the distribution of the pamphlet "to local boards of health, selectmen and others in authority . . . single copies, 5 cents; twelve copies 50 cents. Address Civics, 60 Ocean Ave., Salem." Margarette Brooks, Morse's faithful secretary, of course, was "Civics" in person.

As early as 1883, Morse had instigated a hearing of the Railroad Commissioners of Massachusetts, testifying before them:

> The whistling of locomotives in the city of Salem has been of such extraordinary frequency as not only to be a source of annoyance and discomfort but a menace to health. . . . The man at the rear of a train conveys his signals by waving of arms in the day and swinging of lanterns at night. Has the Boston and Maine railroad ever delegated anyone to investigate the undeniable fact that the distance from the engine to the rear of the train is precisely the same as the distance from the rear of the train to the engine?

However, this Einsteinian doctrine failed to convert the guilty engineers, and Morse subsequently communicated to the Railroad Commissioners that he was "compelled to sleep even in summer with the windows closed on account of the awful and wholly unnecessary noise accompanying freightyard work."

He wrote to John Gould on May 27, 1904:

> Hardly a day passes that a horny-handed son of toil doesn't thank me in the street for my efforts in the direction of stopping the Steam Whistle. [He had extended the field from locomotives to factory whistles.] The smaller noises affect only the immediate neighborhood; the hand organ I would not suppress, as the children enjoy it and it don't begin at midnight. But the whistles . . . wake up invalids in the next county, and these I propose to stop.

He tolerated no tolerance of noise by his friends. He added in the same letter: "Your attitude in regard to Steam Whistles is the attitude of all hoodlums who live and have their being in noise. The lower a savage race, the more noise is made at funerals, weddings and other festivities."

Morse's dislike of noise was such common knowledge that in 1916, when the Tavern Club feted him, a poem written for the occasion referred slyly to it:

St. Peter oils the hinges
To prevent Morse having twinges.

Besides Bishop Lawrence, Morse enlisted Charles Eliot Norton in his Suppression of Noise campaign. Norton, professor emeritus of the history of art at Harvard, wrote from Shady Hill, Cambridge, on May 27, 1904, of "the evil of needless noise, not only in our cities but in many country districts."

Morse presented his views forcibly before the Massachusetts Association of Boards of Health at Boston, January 27, 1905, and scattered reprints widely in response to requests from all over the country for lots of a hundred at a time. The noise continued, however. On August 14, 1912, the peppery grey-bearded professor harangued an otological congress, meeting in Boston, on the suppression of unnecessary noise. Fortunately for his overwrought nerves, radios with loudspeakers and continuous broadcasts of programs were not perfected during his lifetime.

Morse's next trip to Europe was made when his lifelong comrade-in-science, Frederic W. Putnam, nominated him to represent American ethnology at a European conference. "Have got to run off to Berlin seminar for a few weeks to represent the

Hemenway Expedition at the Congress of Americanists," wrote Morse in September 1888.

His absorption in science resulted in a negation of self-consciousness that created for the public in his vicinity a continuous drama. Perhaps, fundamentally, it was an egoistic assertion born of his lack of formal education and cultural background. He was in competition with men of university training and family fortunes or with institutional endowments as resources. With Yankee ingenuity, Morse made every opportunity serve his scientific purposes. Dropping anchor on a mail steamer off Southampton to send a tender ashore was a routine, but with Morse aboard it became an episode:

> The stop at Southampton off which lies the Isle of Wight gave me a moment to make an observation. . . . [I] asked the Captain to save the mud which he drew up on the flukes of the anchor. . . . Examining a bucket-full, found masses of peat in which were little fresh-water shells such as *Planorbus, Lymnaea,* etc., and associated with this from a depth of six fathoms were dead shells of *Mya truncata,* a *Buccinum* and a *Cardium,* all evidence of glacial submergence, probably well-known to the English geologist.

It is easy to imagine the scene: weatherbeaten sailors uneasy at the unusual presence of the captain, in brass buttons and blue braid, presiding over the weighing of the anchor; passengers attracted by the bustle and commotion; the bearded professor, magnifying lens in hand, peering at a bucketful of mud with rapturous exclamations and delivering an impromptu lecture on the deck, building from a tiny shell caught in an anchor fluke a panorama of geologic ages.

Landing at Bremen, Morse popped into the local museum before catching the train for Berlin. "Was surprised to find my patent bracket used and the janitor told me he got them in Hamburg." The next sentence in the journal reveals not only Morse's natural pleasure in finding his invention appreciated, but also his abysmal ignorance of world-history. It was not one of the subjects he had studied at Bethel Academy. "Felt in many respects it would have been better if our country had not only been discovered but settled by the Germans." The Italians, French, and British had not adopted the Morse patent bracket for their museums, in other words; and Morse was ready to swap the cultural lot of them for a race who appreciated his bracket.

Even the Germans only acquired merit in his eyes as they contributed to his own lines of inquiry. Virchow might be the dean of international pathologists, but to Morse he was the owner of bones that might be useful to Harvard investigators at the Museum of Comparative Zoölogy.

September 27, 1888: At noon to see Prof. Virchow who was out at the [Berlin] Institute, so, waiting in a room to which I had been shown, employed my time looking for the hyoid bones. . . . Not a trace, though floor, table and window casements were crowded with skeletons and skulls.

On October 2, 1888, the Congress of Americanists opened at the Berlin City Hall.

We all had to go in full dress and white gloves; how I hated to put them on. . . . I was elected one of the vice-presidents and shall preside at a meeting. Tonight the Municipality of Berlin gave the Congress a dinner at the city hall. Virchow made a speech, toasting America; made me sit beside him at a small table and in his allusions to America referred to me. . . . Band played "Yankee Doodle," and I was called on from various parts of the hall and had to respond. I did, ending up by proposing the health of Prof. Virchow; many Germans came up and touched glasses with me. . . . The whole week is packed with engagements; impossible to find a single private minute . . . to make memoranda.

October 31: Today at the Congress I gave my paper on Mr. Cushing's work, stating that these great explorations [in the American Southwest] were carried on by the munificence of an American lady, Mrs. Mary Hemenway. . . .

October 5: A great dinner to the Congress in the banqueting hall of the hotel. Dr. Luschan made a speech on the Hemenway Expedition and spoke in the highest terms of Mrs. Mary Hemenway. Delightful and inspiring to see three hundred German savants jump to their feet, shouting Hoch Hemenway. Prof. Virchow called on me and I made my third dinner speech this week.

The King of Prussia gave a party for the foreign scientists at Potsdam, and Morse, for a sobering moment, turned his thoughts from shellfish and museum brackets to another aspect of Germany. Of the palace of the late Kaiser Wilhelm he writes (October 6):

The walls of the Emperor's bedroom and library covered with pictures of battle-scenes and portraits of officers. Everything about the room suggested bloody war. A map on the desk as covering; inkstand, a mortar, thermometer on the back of a cannon; paperweight, a pile of cannonballs; candlestick, a cavalry saddle. We descended a few steps to another room lined with deer antlers, boars' tusks, all marked with the date of their murder. . . . With all this war and murder, I observed fastened to the head of the bed a small effigy of Christ in bronze. The Garrison church, having the tomb of Frederick the Great, had the walls hung with hundreds of French flags, worn and faded. . . . It seems impossible that the French could ever outlive the ignominy of their defeat.

A day at the Zoölogical Institute roused in him immense envy of the resources available to German scientists.

Two halls devoted to Entomology alone, each as large as the East India Museum hall at Salem. Prof. Movius told me he had an appropriation of $25,000 for cases alone. Were it not for my family and friends and a sense of duty I feel to build up our own institutions, I would come to this city, the most perfect city in the world for the student; the happiest way of living; the most natural enjoyments, and such museums, such meetings for studying everything conceivable. Everyone esteems the value of study and importance of facts. How long will it be before our American people become fully conscious of the importance of these things?

October 17: Here in Brussels have not spoken to a man except Dr. Dulls at the Museum. . . . In the evenings write my Journal and go to a bright beer saloon where I sit drinking and smoking my cigar and contemplating the crowds of middle-aged men and women with their children drinking beer, playing dominoes, cards. Such sights I have never seen in our country and the result is that our people break down at a much earlier age.

It was vain to attempt to reconstruct the American social scene by arranging to take his wife and children out to drink beer of an evening in Salem, he knew; and his mind reverted to what might be done in the museum field.

The struggle for a good case, the variety of methods employed for displaying specimens, convinces me that the time must come soon when there shall be two museums for each institution; one for the public and the other for the student. It is expensive as well as difficult to blend the two.

Arrow releases continued to arouse his enthusiasm.[1]

[1] Anyone Morse knew, of course, was pressed into service in collecting data on arrow releases. Years later Robert E. Peary, Arctic explorer of the United States navy, dutifully delayed his search for the North Pole to report (Au-

October 20, London: Called on Mr. Longman and he invited me to dinner and the Toxophile club rooms in Regent Park; went there at four in a cab. . . . They had finished shooting, as it was late, but asked if I would like to see a little shooting. He and Mr. Erskine stood at 100 yards and each put three arrows at that distance, one of them being a gold. This, of course, was unusual shooting and they said they had better stop, but walked across to the target and shot back, got five out of six in the target. It made me simply crazy to get at archery again. . . .

In one case were the bows and arrows, thumb ring and arrow-rest presented to the Toxophilite Club in 1794 by Mahmoud Effendi, secretary to the Embassy from the Sublime Porte. Case was opened and I made a sketch; walked to Mr. Longman's house and spent delightful evening examining his collection of bows of the world. Andrew Lang[2] had been invited to dinner also and soon came, a thin, tall man with gray hair and rather an effeminate air; and, it appeared to me, an affected air, though doubtless it was perfectly natural. Both he and Mr. Longman talked so indistinctly that I had to concentrate to get what they said.

Morse missed the cheerful German *bierstube* on an English Sunday in London. "In this damned of all damned cities I was blue and thoroughly miserable. Air is filled with misery of hundreds of thousands hopelessly poor and degraded, flaunting their wretchedness; loud-talking bullies, snobs, pious policemen. Cant and affectation; English that is jargon."

---

gust 13, 1894) that "the arrow release of the Arctic Highlanders is the Mediterranean," and to promise Morse a bow and arrows.

[2] Andrew Lang (1844–1912), Scottish man of letters and historian. Morse had been born too soon to feel the heart-warming association younger generations were to have with the name of Lang, familiar on the cover of the *Blue Fairy Tale Book* and other much-loved collections of fairy stories. The *Blue Fairy Tale Book* was published in 1889, the year after Morse supped with Lang in London. Morse's acquaintance with him was, undoubtedly, based on those solid works *Custom and Myth* (1884) and *Myth, Ritual, and Religion* (1887).

He found a link with Boston, visiting the American artist F. D. Millet.

October 20, 1888: . . . to Broadway, Worcestershire. Millet's house is at the entrance to the village. Mr. and Mrs. Millet greeted us heartily and a hot chop and baked potatoes were ready. . . . Quaint buildings two hundred years old. Millet's studio is a grand affair, an old barn fitted up, oddest old wrought-iron appliances, enormous fireplace. Village . . . roofs thatched or of slate. At his house are stopping Mrs. Williams, a sister of Alma Tadema, and a Miss Mead whom I met at Mr. Havemeyer's. In the evening Mr. Abbey came from town.

In the journal Morse made a sketch of the artists at work; perhaps Edwin A. Abbey was even then making studies for the Galahad murals now in the Boston Public Library.

His trip to the meetings in Berlin lasted less than six weeks, but he brought home ideas enough to last for six years. Morse's diaries contain many passages illustrating his ability to maintain several lines of thought and research simultaneously. In his European travels he could snatch an hour from conferences on Brachiopoda with biologists, or on Japanese pottery with connoisseurs, to collect arrow releases, an aesthetic sacrilege that would have horrified devotees of art for art's sake, like Oscar Wilde or Morse's artist friends, Millet and Turner. This was not due to lack of artistic taste on Morse's part, but to his knack for making "common and homely things" serve for research. As a matter of fact, Morse's perception was keen, and in a time when some of his fellow Bostonians were diligently collecting Landseer stags or Bouguereau beauties Morse pounced on the Ruysdaels.

In 1887 he had written:

October 15, 1887, Berlin: Went back to see the gallery of pictures at the Royal Museum of Art. Must say that a large hall full of Crucifixions and Holy Ghosts was a little too much of a good thing. . . . The fact is that one gets a wrong impression of the old masters by the lifeless attempts at reproduction that . . . do injustice to the originals. . . . I was amazed and delighted at the portraits of Dürer and particularly interested in the delicious landscapes of Ruysdael.

October 17, 1887, Dresden: I must confess that with the exception of the Sistine Madonna I took more interest in the ethnological details than I did in the pictures; Japanese pictorial art has ruined my appreciation for this stuff. I examined chiefly for [ethnological] points and got quite a number. St. Sebastian has been of some use in the world, as we get through him as a target some points on arrow release. Sandro Botticelli, Old Cat. No. 34, 15th century. Here was a new form of tile represented . . . ; it was not on a roof but on an overhanging portico. . . .

A line of enquiry allied to the arrow releases was Morse's study over seven years of a curious bronze object which he first saw in European museums identified as a *Bogenspanner*, or bow-puller. Being a practical archer himself, Morse was certain that the device, a double ring of bronze, shaped roughly like spectacles or a yoke, with three spikes on the upper surface, could in no way be utilized for pulling a bowstring. At the Washington meeting of the American Association for the Advancement of Science, Morse presented his studies and fifty sketches, reprinted in the *Bulletin* of the Essex Institute, Volume XXVI. He demolished all the explanations that had been advanced, such as that the object in question was a bow-puller, a lamp wick, a chariot-rein holder, and posed the question of its use as a mystery.

Scores of letters from classicists and archaeologists of America and Europe poured in upon him. Even the now venerable

Charles J. Sprague, who had started Morse collecting fungi in 1854 and who had remained, as was customary with Morse, a lifelong friend, wrote: "It will haunt me now, as the famous Sir Hilary enigma did years ago. If your fertile mind has not solved the riddle, my weakening one surely cannot." C. Howard Walker, Boston architect, dashed off a delightful sketch, of a Greek youth employing the bronze article as bottle opener for a jar of wine, inscribed: "Of course! What could be more appropriate for a warrior, Etruscan or Greek. Shades of Anacreon, why was this idea so long delayed? In vino veritas." Dr. Charles G. Weld, sojourning in Paris, devoted hours of a vacation to wandering through the Museum of Antiquities, seeking to solve the riddle. "I believe that those prongs go into holes punched in leather and the eyes are used for thongs to pass through to be used on harness or armor. How is that for a theory?" Weld was closer than the experts. Dr. Daniel G. Brinton, professor of American ethnology and archaeology at the University of Pennsylvania, after five years of correspondence with Morse on the puzzle, devoted a month in Paris in 1897 to solving it. He wrote to Morse:

> In Paris in September I went carefully through the museums of the Louvre. My labor was rewarded in the most satisfactory manner by finding an admirable representation . . . a full-length figure in marble, No. 68 in the Salle des Caryatides. . . . "Athlète Vainqueur au Pugilat"; fresh from his victory, naked, each hand covered with a caestus and upon each a myrmex, separate from the caestus, as perfectly defined as the implement could be in stone. . . . This discovery removes all doubt of the correctness of my identification.

In other words, Morse's puzzle turned out to be the ancestor of the twentieth-century "brass knuckles."

Meanwhile Morse had put into shape and published, first as

a paper read before the Essex Institute, December 21, 1891, later in a series of articles in the *American Architect*, the notes he had been accumulating on roofing tiles. In Kyoto, with C. Niwa, Director of Japanese Expositions, he had begun the study of tiles, which he later pursued in Europe. "It is a noteworthy fact that the earliest type of terra-cotta roofing-tile ever exhumed still forms the roof-covering of the greater mass of mankind to-day . . . ," wrote Morse. "Wherever the Romans went, the typical Roman tile may be found, often impressed with the stamp of some Roman Legion."[3]

On Japanese tiles he remarks:

> Fortunately the varied tastes of the Japanese collector have led to the treasuring-up of old roofing-tiles. . . . Ninagawa, the famous Japanese antiquarian, contemplated the publication of an illustrated work on ancient roofing-tiles, to form one of the numbers of his "*Kwan kodzu setsu.*" The lithographic plates were prepared for this number; whether the text was ever published I cannot say. Fortunately securing a set of these plates, I managed to get from the author, some years before his death, the names and dates of the tiles figured.[4]

One of the dozens of letters commenting on Morse's study of tiles was from Marcus B. Huish in London, mentioning the founding of the Japan Society in Great Britain. The letter is undated, but its interleaving in Morse's own copy of "Roofing Tiles" indicates that it was written in 1892.

> You will perhaps have heard that we have started a Japan Society here and that it has so far as numbers are concerned succeeded our

[3] *Bulletin of the Essex Institute*, XXIV, 2.

[4] "On the Older Forms of Terra-Cotta Roofing Tiles," *Bulletin of the Essex Institute*, XXIV (1892), 5.

anticipations, over two hundred members having joined. I think that the transactions will make it well worth anyone's joining, — the first volume will be out in November and will consist of some 120 pages. The subscription for foreign members is only 10/6. We find that the flow of works of art to England from Japan has seriously diminished of late.

Gustave Schlegel, professor of Chinese at the University of Leyden, reviewing the brochure on tiles, neatly defined one of Morse's characteristics: "By Mr. Morse's publication we see once more that even the most common and homely article made by man may throw an unexpected light upon the history of mankind itself." [5] Morse's whole career, in fact, was built on his ability, deprived of wealth and opportunity for exotic research, to interest himself and others in "the most common and homely article," whether the tiny, earth-colored land shells of Maine, neglected fossil shells on a railroad cutting in Japan, or roofing tiles.

Morse had a frank, unabashed interest in anything pertaining to human civilization. Not many men of his day would have put into writing, with sketches, the article published under his name in the *American Architect* for March 18, 1893, "Latrines of the East." Reprinted in pamphlet form, it had a wide circulation among medical men, health officers, architects, and ethnologists. The comment of James Mooney, writing from the Bureau of Ethnology, Washington, D. C., was typical: "I am glad that you have handled the subject. Such matters have been generally dodged by writers, but we can have no knowledge of a people's development without knowing something of its ideas of personal cleanliness." Practical, as always, in his

[5] Unidentified clipping in the possession of Mr. John Morse.

point of view, and quick to synthesize science and human affairs, Morse embodied in his observations on latrines the new bacterial discoveries of Pasteur and Koch in terms of soil pollution and human infections. He was impressed by the Japanese age-old tradition of ablutions, and horrified by the promiscuous filth he had seen in China. He described both vigorously, enlivening his description with sketches. Furthermore, he applied Darwinian principles to make a springboard for a prescient paragraph, a generation ahead of workers who would later demonstrate the production of antibodies in human beings to combat bacterial invasion:

> Among the teeming millions in China, there is every reason to believe that a struggle for existence has gone on for centuries, killing off those who could not stand the microbes and bacilli of certain kinds, and, as a result, the survivors are proof against conditions that would instantly depopulate one of our cities.[6]

Morse never analyzed his own feelings about Japan, but his every reference to the country or comparison of Japanese ways with those of other lands betrays a country-bred boy's nostalgia for the old, simple way of life. In *Latrines of the East* Morse obviously is not suggesting that the introduction of plumbing is necessary or desirable for Japan; he is delighting in the transmogrification of a Maine back-house into a Japanese *benjo* wreathed in morning glories, screened by wood-carving, and equipped with a blue-and-white porcelain basin and bamboo dipper for ablutions.

Japan's impact upon the unromantic, uncultured, realistic nature of Morse is the strongest proof possible of the fairyland

[6] *Latrines of the East,* p. 4.

character of her medieval civilization as it was viewed by the first western visitors. Artist and mystics like John La Farge, Ernest Fenollosa, Sturgis Bigelow, Lafcadio Hearn, might be suspected of having colored their views with their own ideology; but Morse saw things as they were. If the unromantic Morse felt the romance of Old Japan, romance it indubitably was; and Morse characteristically perceived it in the most prosaic of places.

Meanwhile, friends of art waged a vigorous campaign to secure the subscription that would keep Morse's pottery in Boston. In 1888 Mrs. Clara Erskine Clement Waters, a well-known art critic of the day, wrote to the editor of the Boston *Journal*:

> I have written to one of the best authorities we have on this subject, and he writes me thus: "You ask if I know anything about Morse's pottery. I do. It is the best collection in the world, not excepting any in Japan. . . . It covers the ground historically, from the earliest, prehistoric pottery down to the work of living men, by provinces, by makers, by forms, and, of course, by types of work. No one but a born classifier like Morse could have got it together at all. . . . I hear rumors that people are trying to get it for the M.M.F.A. [Metropolitan Museum]. It would be a disgrace to Boston to let it go."[7]

Denman Waldo Ross, great Orientalist and art patron of Cambridge, also urged the acquisition of the Morse pottery by the Boston Museum of Fine Arts. In a letter to Ross Turner, dated at Cambridge, February 28, 1889, he defined his interest:

> . . . The collection illustrates, better than any collection of works of Art which I have ever seen, the principle which underlies all true artistic activity, — the principle that it is not enough to invent new

[7] Boston *Journal*, April 18, 1888.

types of things, but each type must be improved and perfected according to the ideal which it suggests to the imagination.

In our modern life, European and American, types are multiplied beyond all precedent, but very few of them are improved or perfected. We seek, all the time, new things, rather than perfect examples of the things we already have. We differ from the Japanese in this respect. They are more interested in producing perfect and beautiful examples of existing types than in the production of new ones. The Japanese artist takes a typical form of pot, and devotes years, or even his whole life, to the production of exquisite examples of it; and he labors, not for gain, not for the public, not even for the amateur, but to realize the ideal of the imagination, the ideal of his soul.

You remember what the old philosopher said: A little thing makes perfection, perfection which is not a little thing. This is the lesson which Prof. Morse's pots and jars teach us. The differences between one example of a type and another are slight; but every example represents a distinct effort to realize the ideal. While we are satisfied to hand on the type which we have invented to the mechanic, to be reproduced by machinery, the Japanese artist prefers to improve and perfect his type every time he reproduces it. . . .

Morse's last journey to Europe was made in 1889, when General Charles G. Loring's efforts to raise funds to purchase the Japanese pottery collection for the Boston Museum of Fine Arts were nearing the goal. Morse felt justified in plunging more deeply into debt for the sake of making his collection as complete as possible. He sailed on April 27, landing at Bremen.

May 19, 1889, Paris: I have been in the city a week ago yesterday; this is the first moment I have had to make a note in the Journal. . . . Every day have worked at Bing's from morning to night, picking out specimens for Mr. Waggaman. Thursday went to the Exhibition, amazed at the extent and magnificence of it. As to Eiffel's tower, the more one looks, the [more] marvelous does it appear. In all history, the higher the monument, the thicker and wider the base; and it strikes me that the merit of Eiffel's was the conception of an arch supporting a tower. In these grand arches, with a span of four hundred or more feet, beautiful vistas are seen of the Exhibition grounds and buildings. . . . So wonderful and attractive is this tower that throughout the afternoon and evening, wherever I was, I continually turned my eyes to this marvel of the 19th century. In one of the buildings [was] a bridge exhibit, and among the various models and drawings are some high bridges and spans of Mr. Eiffel designed for railroads in Spain and South America. I believe from a study of these one could see how the idea of a tower might originate. Having conceived a bridge across a wide ravine and river, what more natural than to think of placing one of high truss piers on top of the arch, and in some such way I believe it was suggested to him. . . .

Had a most delightful evening with M. Gonse; looked over all his pottery with Mr. Ishida, a Japanese who has been very kind.

Morse evidently had some American acquaintances in Paris, for, after leaving Gonse and Ishida, he joined a group of tourists:

In the evening with Mrs. Webb and her cousin, Miss Genton, went to the studio of Miss Browner, a Bohemian, where was also Miss Rado, a Hungarian, these two ladies dressed each in their native costume and they looked charming.[8] Mr. Lucas was there, also Mr. Dallin, a sculptor from Utah who has the commission to

[8] Louis Gonse, editor of the *Gazette des Beaux-Arts*, "chronique des arts et de la curiosité," at 8 rue Favari, did not know about the Bohemian and Hungarian ladies who caused Morse so speedily to forget a promise he had made. Gonse wrote from Paris on August 19, 1889: "Your cordial friendship will pardon me if, in this connection, I remind you that you offered to give me some specimens in exchange, and that I left the choice to your selection. We collectors, you know, are terrible persons; and the hope of adding, if but a single piece, to our series, keeps us on the alert."

make the statue of Paul Revere.[9] Dallin was a very nice fellow who, the day before, carried me through the art school, Julian's, I believe. It was interesting to see them model from life, actually making sketches in clay with as much ease as an artist does with his brush.

As Morse walked and dined alone in the city, his mind returned to dinners in Japanese inns; translucent *shoji* drawn for seclusion, low tables of scarlet or black lacquer, and bowls and saucers of exquisite pottery, the gentle maid-servants treading noiselessly in white cotton *tabi* on soft straw *tatami*:

May 19, 1889, Paris: The lavish appeal to the senses shows how far the Parisian has departed from enjoyment of simple things. One does not wonder that revolutions and communes break out at times among a people who demand constantly so much excitement. Even the idea of drinking their wine and beer in quiet is not to be heard of, but tables and chairs are trooped out on the sidewalks of crowded boulevardes, and here they enjoy sipping and even eating. A nation cannot long exist when the highest pitch of excitement is demanded every moment.

Later he returns to the subject:

May 27, 1889: . . . Went to Mr. Paul Berly with Mr. Bing this morning and saw his collection, a few specimens of pottery; a large library of Japanese books and I saw many new to me. After this Bing and I took lunch at a place where there was a continual roar of traffic over the pavement and with all this hideous and deafening racket, the Parisian people prefer to sit out on the sidewalk and eat. The nerve wear must be enormous. I called on M. Clemenceau,[10] the Liberal Deputy who was asked to form a Cabinet but was too

[9] Not until January 1940 did the City of Boston, through the George Robert White fund, appropriate $25,000 to pay for the statue Cyrus Dallin had been commissioned to execute fifty years earlier. It was then placed near the Paul Revere house, North End, Boston.

[10] Georges Clemenceau (1841–1929), war premier of France, 1917–1920.

radical for them. He has a large collection of Japanese incense boxes, perhaps two hundred. . . .

Four journeys to Europe in seven years were civilizing Morse unconsciously. He still employed his pet adjectives about London, but his attitude towards Londoners had changed.

June 1, 1888: . . . reached the heavy, dirty, poverty-ridden, piety-stricken city of London. . . . To the Royal Academy and saw old pictures . . . some Ruysdaels which struck me as superb. I observe, too, that the more I see the old masters, the more I enjoy them and for anyone to say he don't care for them at first inspection is hazardous, for I remember my first impression was one of disappointment. Now the pictures . . . have an added charm, for they always interest me from an ethnological standpoint [sketches of arrows and tiles]. . . . While I think of it I must record that I never sit down at a hotel or restaurant table without getting purposely into conversation with anyone near me and I never talk in such a case with an Englishman without recalling what Taine says," An Englishman has a fund of sound and rational ideas." It is certainly so and coming from the continent one feels relief to think that he has sprung from this little island.

June 6, 1889, London: To Thomas Hotel to breakfast with Mr. and Mrs. Havemeyer and Mr. Higgins . . . from Greenwich, Connecticut, travelling with them. Mr. Havemeyer showed me a Japanese place where I bagged three interesting specimens. . . . Then to the Royal Society. . . . Met Dr. E. Ray Lankester, who invited me to University College, and Professor John Milne of Japan, with whom I had a glass of beer after the meeting; and to my amazement, Mr. Tanakadate, who has grown a black beard but I knew him at once.

Aikitsu Tanakadate had been one of Morse's Imperial University students at Tokyo. He became Japan's leading physicist, a pioneer in aeronautics, and, incidentally, the author of the system of romanizing Japanese printed words that was officially adopted in 1939. When data were being gathered for the biography of Morse in Japan in 1939, Dr. Tanakadate had been appointed by the Emperor a member of the House of Peers of the Japanese Diet, representing the Imperial Japanese Academy of Science. He brought out from his library the original notebooks in which he had written down, in English, Morse's lectures on zoölogy, interspersed with accurate copies of the illustrations Morse drew on the blackboard as he talked.[11]

Friday, June 7: With Mr. Huish to see the collection of Mr. Alexander at Kensington . . . ; some very fine pieces but so far as I can learn the collections are few and far between. In private collections, Paris is much better and specimens are infinitely better than any I have seen in London. Mr. Alexander, a man of wealth, has a most comfortable house with large garden, lawn and three tennis courts. . . . Invited me to midday lunch and very hospitable.

A curious incident has occurred in the pottery line. . . . Last evening, Mr. Havemeyer gave me a pottery *inro,* the first one I had ever seen. He didn't know why he bought it but thought it was curious and, on my expressing a desire to make a drawing of it, gave it to me with his usual generosity. I told him I had seen a porcelain *inro,* but never a pottery one, and as I was collecting every object made in pottery, was delighted to get the specimen. Today Mr. Alexander, just as we were getting through, said he had a bit of pottery he would like to give me, as he had found it in the drawer of a lacquer cabinet he had bought several years ago . . . and he gave me a pottery *inro.* Quite different in form from the specimen Mr. Havemeyer gave me. . . . Seems curious I should get two in twenty-four hours and both by gift.

With this coincidental shower of gifts, Morse sailed for America. It was the final turning of the leaf for his European

[11] The notes are reproduced in Dr. Tanakadate's book, *Kuzu no Ne,* of which there is a copy in the Harvard College Library.

journals, and in the forty years of life remaining to him he never found time to edit them for a book, unlike the journals dealing with Japan and China.

Morse's collection of Japanese pottery was acquired in 1890 by the Museum of Fine Arts in Boston, at that time situated in the brick building in Copley Square. Under the enthusiastic stimulus of General Charles G. Loring, many donors made up a subscription fund of $76,000. The negotiations were protracted, and Morse deposited the collection with the museum some months before the money was available. He wrote to John Gould about it at intervals:

December 23, 1889

Wish you a Merry Christmas. When pottery is sold, shall make Christmas presents! It is a fact that the collection is in a fair way to be sold, and you may imagine it is a great relief to me.

March 29, 1891

. . . The subscription for the collection is so far advanced that the Committee have no doubt of completing the sum by July 1. It has reached to over $50,000 now and the circulars prepared by them has [*sic*] not yet been issued. . . . The collection going to the Art Museum has lifted the insurance bill I had to meet and for which I owe Fred.

March 5, 1892

. . . my pottery is squared up and after paying my bills I shall have something left to invest.

Money made no perceptible difference in Morse's mode of life. There were no alterations in the house on Linden Street, no extra bathroom or stable to set up the carriage and horses that were the badge of luxury in Salem in the nineties. There were no voyages to Japan or to Europe for pleasure. Morse paid off the mortgage and then did the thing he had promised himself to do thirty years earlier.

He had kept careful account of the money John Gould had advanced from his first earnings to finance Morse at Harvard with Agassiz, as well as of the thousands that Gould, become a bank president, had later invested in Morse's judgment on pottery. Now Morse not only "squared up" his debts but, in affectionate gratitude, bought for his old friend an annuity. John Gould would have balked at a present in cash; Morse thought of a gift Gould could not refuse.

Transferring the pottery from Linden Street to the Italian Renaissance brick building in Copley Square was the work of months. Before packing, each piece — there were one hundred and five barrels — had to be checked against the card catalogue. For purposes of identification, Morse had sketched each piece on its card beside the notes in Margarette Brooks's clear handwriting. One day a card was missing. Morse hastily sketched the piece on a duplicate card with swift freehand strokes. A few days later Margarette found the original, and as she compared the duplicate with it, she was struck by the similarity in the two drawings. Tracing one, she superimposed the tracing on the other. They coincided to a hair's breadth. It was a striking example of the amazing image-memory and deft coördination that made up half of Morse's genius.

His visual memory would have delighted the Jaensch brothers in their study of "eidetic imagery" at Marburg fifty years later. One of his platform tricks that pleased thousands was to take a piece of chalk in each hand, start the tail of a fish or the wing of a butterfly with his left hand, the head of the fish or the opposite wing with his right hand, and complete a perfect drawing simultaneously. Joseph Lindon Smith and Frank Benson, Boston artists, once asked Morse to demonstrate this ability at the Boston Art Club, and verified the simultaneity of diverse action. Then Lindon Smith took the crayons and tried to emulate Morse. "I felt as if my eyes were rolling out of my head on either side," he confessed.

Morse had to start off on his 1890 lecture tour across the country before the expert packers from French's old china store in Boston completed filling the hundred-odd sugar barrels and carted them to Copley Square. Only a scientist as unscientific and informal as Edward Morse would have entrusted to two young women the task of unpacking $75,000 worth of rare ceramic pieces and arranging them in the museum cases. He calmly left it to his daughter Edith and his secretary Margarette, confident that under his training they were responsible, accurate, and competent to handle the task. Margarette reported:

> Nothing is broken but the tip of an Imbe bottle . . . only fifty years old and not a particularly nice one. Picture to yourself two lone females, big aprons on, heads covered with towels, moving barrels, unpacking, sweeping hay, picking up paper. . . . Of course you would easily pick out the pieces but there are some I have to look up.[12]

For fifteen years, in Japan, Europe, and America, Morse had carried in his mind a picture of his collection, searching for items missing, automatically discarding duplicates offered. In Japan in 1883, said Chiomatsu Ishikawa, Morse bought in a second-hand shop a bowl of which the lid was missing. In 1888, walking in Piccadilly, London, he saw in a shop window a bit

[12] Morse had left diagrams with sketches of each piece to guide the arrangement in the shelves.

of Japanese pottery which he recognized as the missing lid. He bought it, took it home, and found he had not been mistaken.

A letter from Morse to his crony, Francis Wilson, the actor, gives another instance.

> You wanted the story about a rare set of Imbe I got in Japan. Be it known then that the National Museum [in Tokyo] had a set of cups of this kind. I did my best to get one but the authorities would not break the set. I sent drawings to various bric-a-brac dealers with generous offers for an example. I never saw one, either in the shops or in private collections and left Japan in 1879 disconsolate. The third time I was in Japan my notes for the book on Japanese Homes led me to the attic of a building for the purpose of sketching the rafters. Coming down the ladder, I noticed a long rather deep box, too deep for a *kakemono* [scroll painting]. I lifted the lid and to my joy found it contained cups identical with the set in Tokyo. An old Japanese antiquarian, on seeing them repeated a proverb which, translated, says "A man who has his net out all the time must catch big fish!"

The Morse collection of pottery had been valued by appraisers at $100,000, and Morse had refused $60,000 for half of it before the Boston Museum managed to obtain by subscriptions $76,000. In consideration of his accepting the lesser sum, Morse was named Keeper of Japanese Pottery at the Museum of Fine Arts, receiving an annual stipend of one thousand dollars. He spent much time supervising the making of additions to the museum collection, which eventually totaled two thousand items. For ten years he also worked three days a week regularly with his secretary, Margarette, on the monumental task of writing and editing his *Catalogue . . . of Japanese Pottery*, published for the museum in 1901. Printed at the Riverside Press, it was a large quarto of 384 pages, with 68 photogravure plates and 1,545 cuts of potters' marks, drawn by a Japanese. George Ives of the Riverside Press, writing to Margarette Brooks, gives a glimpse into the problems of the staff aiding Morse in the gigantic task. Saying that he is glad to aid in the task of correcting the proofs of Morse's catalogue, he adds, "I sincerely trust that Mr. Morse will attend to his syntax. In places it has given me the cold shivers."

Whatever his syntax, Morse made a minimum of errors in research, and forty years later his work was still the most complete in the field, the Japanese finding it simpler to translate Morse's catalogue than to write a new one. Russell Sturgis called it "one of the most perfect pieces of catalogue making and one of the most scientific and also sumptuous renderings of a collection of all those which the museums of the world have yet produced."

From the British Museum came the most gratifying praise of all, that of his peers in the museum world. R. L. Hobson wrote on October 8, 1908:

> In cataloguing the collection of Japanese pottery in this Museum I have made myself very familiar with your magnificent "Catalogue of the Morse Collection." Indeed I may say that it has been my mainstay. I found it invaluable for the identification of the marks as well as for the information given about the potters and kilns. That the notes you made when you went through our collection personally have been exceedingly useful goes without saying. I am preparing a dictionary of marks and would very much like to supplement the Japanese section which is largely drawn from the collection here, with a few of the marks published in your catalogue. . . . If you would permit me to make some drawings . . . I should be very grateful.

CHAPTER FOURTEEN

NEVER ORDERED THEM

IF YOU are a man, and live in or near Boston, nothing can be pleasanter than to be invited to join the Tavern Club, as Edward Morse was in 1891. Artists, musicians, authors, actors, architects, educators, and men of the world make up the group that meets in the old brick building on Boylston Place. Beyond the intimate pleasures of chat and comradeship with congenial minds, the Tavern Club is famous for the dinners it gives in honor of distinguished men visiting Boston. In the years after Morse was admitted, for instance, honored guests included Vladimir de Pachmann and Ignace Paderewski (1891), Jean and Édouard de Reszke (1892), Rudyard Kipling (1895), the Crown Prince of Siam, Alexander Agassiz, Baron Takahira, and Arturo Toscanini.

The Tavern Club was fun, and to a fun-loving nature like that of Edward Morse it was a perpetual feast of joy. When Henry M. Stanley declined a dinner planned for him in 1891, Curtis Guild, a governor of the Commonwealth of Massachusetts, substituted for the recalcitrant invitee, appearing in white ducks, monocle, and topi, which he shed to lecture on "Darkest Africa" in black cotton tights and woolly wig. When Samuel Clemens did come to dinner, January 15, 1901, no one roared more heartily than Morse at the practical joke of having a member made up as Mark Twain's double and seated in advance in the place of honor to which Major Henry Higginson conducted the guest. But Clemens stole the Tavern Club's thunder, jumping to grasp the jokester and crying out, "There's Mark Twain! All my life I've been hunting for that rascal, that impostor who steals the credit for all I do."[1]

The Tavern Club made the most of Morse's personality. In 1896 they were already cartooning him as a Satsuma teapot whose spout jauntily brandished the inevitable cigar and from whose top grew a bald head, bespectacled and bearded. In 1912 the Christmas poem included a line, "There's Morse with his stories so spicy and warm." Perhaps Morse had been reciting

I sat next to the Duchess at tea.
I knew just how it would be.
Her noises abdominal
Were simply abominable,
And everyone thought it was me!

Of course Morse, after admission, gravitated to the "Hellions Table" opposite the head table, nearest the door, where the maddest pranks and most undignified actions went on. After a dinner to Paderewski ("whose hair was full of ginger; he played the Seventh Symphony with a felon on his finger"), Morse and William Winch at the next meeting were observed bubbling with mischief. "We miss that wonderful Bohemian atmosphere," declared Winch, giving Morse the cue. "Yes, we need to cultivate that atmosphere," cried Morse, darting at the nearest member and ruffling his hair in imitation of the great Pole's. In a twinkling he passed around the table, tousling the

[1] By permission; from *A Partial . . . Semi-centennial History of the Tavern Club* by M. A. DeWolfe Howe, privately printed in 1934.

hair of each man until he reached a Beau Brummel of Boston society who had long been suspected of wearing a toupee. The man's face was a study in horror and fear as he watched Morse dancing towards him like a whirlwind, eyes gleaming with mischief, hands fluttering, clutching . . . and then ostentatiously passing over the victim to seize the hair of the man beyond.

Then there was the time when Joseph Lindon Smith in Egypt and Arlo Bates of the Massachusetts Institute of Technology arranged an elaborate hoax. Bates had printed mock newspaper clippings recounting the landing in San Francisco and the progress across the country of a maharajah coming to make a survey of Harvard. Tickets were sold and guests invited to a Tavern Club dinner for the maharajah, and Major Higginson was given an address at which to call for the guest of honor. Meanwhile Smith, returning unheralded from Egypt, had himself made up at a costumer's with beard, eyebrows, black glasses, turban, and Oriental robes.

Major Higginson ushered the guest into the dining room, to which the members had already been lured by a fake telegram saying that the maharajah's train had been delayed. Members and guests rose, to do him honor. Only Morse, with his sharp eyes and his extraordinary visual memory, recognized Lindon Smith's hands. Delighted with the joke, he kept the secret but disconcerted Smith by murmuring "Maharajah Jo-Jo! By thunder!" *sotto voce,* and by keeping up, from his table, a merciless barrage of bread pellets that found their mark unerringly on the maharajah's face. He so broke down Lindon Smith's composure that the artist could not go through with the speech included in the hoax. When introduced, he rose, began "Gentlemen!" and then broke down, salaaming and pulling off turban and beard before he raised his head.

It was even fun paying bills in the Tavern Club. Morse invariably decorated his bill with a humorous sketch and a pregnant remark before returning it with a check. He paid a bill for $6.10 in February 1900 and then pretended to consider the receipt mailed back as a dun, returning it ornamented with a picture of an harassed man holding his aching head and the legend "Damnation! How many times am I going to pay this bill?" Burnett returned it with a sketch of a smug cat smiling inscrutably above a mouth filled with feathers. Another bill, with a significantly long list of sixty-five-cent items, Morse illustrated with a sketch of himself reading a scroll inscribed Rum, Whiskey, Beer, with an emphatic wording, "Never ordered them." Burnett laconically annotated the receipt, "Thanks just the same." The shell motif was bound to appear in anything connected with Morse. When he received a bill of $12.40 one month, he drew on it a hand, forefinger caught in a scallop shell, "Damn it! How it pinches! Can't you take off a few dollars?"

One Valentine's Day at the Tavern Club, Henry M. Rogers took for his subject the zoölogist from Salem.

To Edward S. Morse

To begin at the end — the end of his birth,
For man has two ends when he comes upon earth,
A first end and last end, of course —
The Doctor stood still and shook his bald pate
As the baby appeared yelling, "Hell, what's the date?
I'm on time and my name's Edward Morse!"

From then till today he speaks very well
The language of Heaven and the language of Hell
In his daily and earthly discourse.
He can draw with both hands the planets and charts
But the best of his work is the drawing of hearts
This wonderful Edward S. Morse.

He will tell you of Herschel and Darwin and Kant
Of the Snoozers who know of every known plant
And have lived with the five-footed horse;
Not Dodo nor ape nor man-eating shark
Nor dog with his tail in the place of his bark
Can puzzle our Edward S. Morse.

So hail to the chief of the mollusk and clam
And hail the Professor who uses his damn
With precision and artistic force;
And hail above all to him from Japan
Who knew all there was ere the others began
My valentine, Edward S. Morse.

The supreme memory of the Tavern Club evenings for Morse, however, was of the dinner given there in his honor on January 14, 1916, when, with delicious food, wonderful music, gay exchange of affectionate and intimate quip and jest, the clever tributes in verse of DeWolfe Howe and C. Howard Walker were read. Walker built the refrain of his poem on an objurgation chosen by Morse in boyhood to circumvent Deacon Johnathan's abhorrence of profane language, which had become a lifelong safety-valve of feelings.

INFANDUM

In come June, or come December
I remember
Long ago 'twas first I met him,

Promptly in a place I set him
Quite apart from all the rest.

Since then, whether East or West
We've played mental battledore,
Give and take, and damn the score.
Be it better, be it worse,
You could wager you'd find Morse
Leering,
Jeering,
Cheering,
Driving home each point he made
With one word he always said—
  'Infandum!'

* * * * *

When (may it be long belated)
He's translated
And St. Peter oils the hinges,
To prevent Morse having twinges
If the gates should creak, perchance,
As between them he shall prance,
Then d'ye think he'll mend his manners?
Not at all, 'midst waving banners
And the shouts of loud Hosannahs,
Angels gathered 'round the throne,
Morse advancing all alone;
What most men would deem a crisis.
Hark, what is that sound arises
First, a murmur,
Clearer, firmer,
Till the tumult empyrean's
Hushed to his ecstatic pæan,
Drowning out all other cries
Morse's climax of surprise,
  'Infandum!'

William Sturgis Bigelow, attending the dinner, was "more impressed than ever by your [Morse's] amazing vitality, from the fact that you said you took four cocktails before dinner. I took one, of half strength, and dined on dip-toast, a little thin smoked beef, and tea, and did not come to the club till half-past eight." In the same letter (January 17, 1916) he said:

> I have been kicking myself for not having summed your work up in the obvious nutshell by saying, "When Morse took up the study of Japanese pottery he found that almost nothing had been done by anybody else. When he finished, there was almost nothing left for anybody else to do." You had the "jack of all trades" criticism on your mind at one time, but I think that was squashed that night as flat as a bedbug under a family Bible. You were extolled by everybody as a past master in everything you have undertaken. . . . Altogether, it was a great evening.

Younger men, who knew Morse only in his later years, were always impressed at his vigor and agility in tennis, which he played with ardor until his seventy-fifth birthday. On June 18, 1923, Morse played three sets on Russell Robb's court in Concord. Next day he read in the newspapers that "the noted professor spent his 75th birthday tranquilly at his home receiving congratulations" and declared he was through playing tennis if that was all the credit he got for it. M. A. DeWolfe Howe, in his Tavern Club poem, speaks of tennis:

> 'You are old, Father Edward,' the young man said,
> 'And your beard is becomingly hoary,
> And how do you manage to pack in your head
> Such a treasure of ancient story?'
>
> 'I carry my scissors wherever I go,'
> Said the sage, 'And make many a clipping!

Now here is my latest — I got it from Joe
  (Not Millet, but Miller) — it's ripping!'

'You are old,' said the cub, 'yet at tennis, they say,
  There are none who would dub you a vet!
It's all very well like a Peabo to play —
  But why keep on jumping the net?'

'In my youth,' Father Edward replied with a wink,
  'They thought me a witch — or a wizard —
For Salem was full of them then — and I think
  'Twas my jumping that saved me my gizzard.'

'You are old,' said the youth, 'and I have not a doubt
  That your words are the truth all unvarnished;
But how, at your age, do you still bring them out
  With oaths so enchantingly garnished?'

'In my youth,' said the graybeard, 'to science I took,
  To all learning I made my salaams,
But the rivers of knowledge from book after book
  Had drowned me long since but for Damns!'

'You are old, as I've mentioned before,' said the youth,
  'Yet excel in all juvenile games;
But the crown of your skill is the accurate truth
  With which you remember our names!'

'Get out!' — the Professor replied in a rage,
  'I've answered conundrums in plenty.
You're Jim, Bill, or Harry, whatever your age —
  I'm Morse — and some day I'll be twenty!'

On May 12, 1916, the Tavern Club awarded Morse a silver medal in token of his quarter-century of membership. W. R. Thayer wrote the citation:

You must born him down East, in the Dirigo State,
Deserted by sons who expect to be great, —
Then bring him to Cambridge, to Agassiz's care,
To study the bones of the Pliocene Bear,
Count the gills and the scales of preposterous fishes
And all the queer critters in Agassiz's dishes;
Let him ogle the oyster, make love to the clam,
Have his pet Lyopoma, as Mary her lamb.
Perfect him as draughtsman, upon wood or from it,
Make him hunt like a sleuth on the tail of each comet,
Not to mention Old Sol, Lady Moon and the stars,
Or that little affair between Venus and Mars.
When you think he has vanish'd in nebulous spaces,
Behold him enamor'd of Japanese vases!
Don't forget that to round out this versatile man
You must baste with ceramics and lard with Japan!
Add delicate morsels — nine ends and ten odds;
Stuff well with sea-urchins and gasteropods;
Throw in malacology, beetles and botany,
And everything else, excepting monotony:
Endow him with arteries supple as silk,
So that cocktails to him are as harmless as milk:
Give him eye like a hawk, and at three-score-and ten
He will tucker at tennis the strongest young men —
Proving equally ready to volley and lob —
For whatever he's doing, he's on to his job:
Have him settle in Salem, a privileg'd wizard,
Cheerful alike amid Dog Days or Blizzard:
Garnish with wit, in hilarity steep him,
With plenty of commonsense wholesome to keep him:
Then serve at all hours: — the product, of course,
Is our savory Taverner, Edward S. Morse.

In the year 1923 Morse, who was seeing much of a young Japanese student, K. Mizuno, a guest of the Adams family in

John Gould in 1913

Salem, took the Japanese to dinner at the Tavern Club and introduced him to Professor Charles Townsend Copeland, teacher of English literature at Harvard. Mizuno wrote:

> Both Dr. Morse and Professor Copeland were born in Maine. The two were quite extraordinary in their manners and character. Probably Professor Copeland was more eccentric than was Dr. Morse. Once this Professor [Copeland] said to me: "Morse says funny things. We people of Maine are all eccentric, so all crazy people are called Maniacs."
>
> In the Spring of 1925 I went to bid farewell to Dr. Morse for I was to leave Salem then. He pointed at the piles of books lining his study and said: "I'm going to give all these books to the University of Tokyo after my death." Then, at our parting, he said in solemn tones: "I think this is going to be our last meeting. Next time when you come to America you won't find me here but in hell. But you will find my brain in Philadelphia, so I hope you'll go there and take a look at it if you have time."[2]

Morse was never introspective. He had no time to "dissect" his own entity and perceived no contrast between what he preached and that which he practised. Invited to deliver a Founder's Day address at Vassar College in 1895, he chose as his title "The Importance of Good Manners."

Perhaps some of the petulant truculence he had exhibited in his campaign for the suppression of unnecessary noise had arisen from nostalgic memories of Japan of the old days, when rikishas spun noiselessly over dirt roads, telephones were not known, railroads not yet built, automobiles not yet invented. The Japanese, moreover, living for a thousand years behind walls of paper, had developed a technique of silence, an elaborate convention of courtesy. Morse apparently never reasoned

[2] *Japan Times and Mail*, July 10, 1938.

it out consciously, never himself adopted an Oriental imperturbability and restraint, but he was homesick for it, and mentioned specifically in his address his impression of Japanese *mores*.

His address, printed as a pamphlet, was widely circulated. Ironically, "The Importance of Good Manners" led Morse into ill-mannered controversy. The peppery Salem professor was so nearly an "institution" of New England that the Boston Sunday *Globe's* "Uncle Dudley" took a sly poke at his incongruities. Morse fulminated, in a reply printed in the *Globe* May 27, 1894:

> Your editorial assumes that because I make insidious comparison between Japanese manners and ours I necessarily concede to the Japanese superiority. This assumption is ridiculous. It is because I appreciate the intellectual strength of our people, their independence, their nerve that I would have them equally great in refinement and manners. . . . I would have them make a sufficient distinction between the polish of their behaviour and the polish of their boots and hats.

There was never — well, hardly ever — any polish on Morse's boots, and it was because his friends appreciated his "intellectual strength, independence and nerve" that they put up with his manners. He was apt to flare up savagely if addressed as "Doctor" or "Professor"; he had little use for physicians and, as for the term "professor," he characterized it as the favorite title of barbers, bootblacks, and dancing masters!

Morse scrawled — and Margarette Brooks copied for the convenience of editors and compositors — his credo as a museum director in his article, "If Public Libraries, Why Not Public Museums," published in the *Atlantic Monthly* for July

1893: "To bring up young children in . . . a town [that can show nothing but the mere elements of material existence] is to stunt their intellectual powers, and to narrow persistently the horizon of their life."[3] About the nature of the public museum he was explicit:

> First and foremost . . . the town museum should illustrate the natural products of the immediate region, . . . animals, plants, rocks, and minerals. . . . Second, a general collection of similar material from elsewhere, to show the relation of the county to the rest of the world. . . . And finally, a series of forms to show the phylogenetic development of the animal kingdom. . . . [What a public museum should be] has been attained in part at the Peabody Academy of Science in Salem. The collections comprise, first, a remarkable series of the animals and plants, rocks, minerals, and archaeological specimens collected in the county of Essex. . . . Besides this there is an epitome collection of the animal kingdom, brought from all parts of the world . . . ; and finally, an ethnological collection, arranged by countries. . . . Special colored labels [call] attention, by title and shelf number, to books in the public library referring to the immediate group. . . . Courses of lectures are given in the Academy Hall every year, which are practically free to the public. The city librarian supplements these lectures by printed lists of books treating of the subject matter of the lecture. . . . A like service is often done for the free courses of lectures given by the Essex Institute. In this manner three institutions coöperate with one another in utilizing the collections in their possession in an educational way, and for the good of the general public.[4]

Morse was fifty-three, middle-aged enough to pursue collecting for "rest and recreation," but he still managed to get into mischief over it like a boy. He had tried, on several spring

[3] *Atlantic Monthly,* July 1893, p. 119.
[4] *Ibid.*, pp. 114 ff.

trips to Maine, to find a loon's egg for his collection. In 1893 he went to Rangeley with Frederick W. Dixon of Philadelphia and secured three, packing them in the center of the clothing in his trunk for safety. He forgot about them; and when his wife finally undertook to unpack for him, she found three loon squabs.

Morse could readily forget small details. On his lecture tours it was customary for the chairman of lecture committees to entertain him overnight, to save paying his hotel expenses. Once in Cincinnati, he was allotted a guest chamber encumbered with bric-a-brac, including a French gilt clock, adorned with Cupids entwining their arms about the dial and showering rosebuds over the case. Its fussy tick-tick, tick-tick, maddened Morse. He tried to silence it, but it was a tamper-proof marvel of jewellers' craft. Eventually he wrapped it in all the clean towels in his room, wadded his clean shirts and underwear around it, and stuffed it into his suitcase. He did not think of it again until he looked for his nightshirt in the guest room of the next city on his itinerary. Then he returned it, but wrote no apology; he felt that he was the aggrieved party.

Details not connected with brachiopods or pottery lay lightly on his mind. N. F. Abbott, cashier of the bank Morse used in Salem, met him one day in the post office, scanning a postcard just taken from his lock-box. "My errands," Morse explained. "Can't remember 'em. Write myself notes on postcards and mail 'em. When I come uptown, memoranda all here in post office box."

Morse's mail was heavy, for he was a good correspondent and kept in constant touch with friends on three continents.[5] An

[5] He once even acted as amanuensis for his cat in a correspondence with the cat of Robert Louis Stevenson. A delightful letter postmarked January 4,

instance is a French newspaper clipping about Morse's sun-heater, sent him from Rome by an American friend. In 1883, during the poverty-ridden days of building his pottery collection, when the Linden Street furnace would not work, Morse had invented an ingenious sun-heater, installed in his library windows at home and in his office windows at the museum.

It was a box of black corrugated iron, several inches deep, covered with window glass. The outer air passed behind the sun-warmed iron, ascended, and entered the room at the top. It was pioneer air-conditioning, for it changed the air in a room every forty-five minutes. Installation, at home in Salem, cost only eighteen dollars, but the Boston Athenaeum adopted it on a scale that cost three hundred. Morse, reporting it to the Society of Arts at Massachusetts Institute of Technology, observed conscientiously, "This method of heating can only be auxiliary, as it will be inoperative on days when the sun is obscured."

However, like all pioneers and eccentric characters, Morse was fast becoming fabulous and a source of legends. By 1893, the New York *Sun* had transformed this modest device into "a house with all its rooms fronting southward . . . the southern front of glass, and by means of reflectors Prof. Morse is enabled on sunny days to heat the whole house by sun alone." The Salem *Observer* endeavored to correct this exaggeration: "The real house is of the cottage type, with a large addition at the rear for the professor's library. It is heated by a hot-air furnace

1888, was addressed to "Jamaica Ginger, Box 204, Salem." (The Stevenson cat, by a coincidence, also bore the name of Ginger.) It was evidently written by Mrs. Stevenson when the Stevensons were at Saranac and contains references to Henley and George Meredith, as well as to the various Stevenson pets. The letter is in the possession of the Peabody Museum, Salem, Mass.

and open fireplaces with wood fires. Mr. Morse naturally feels disgusted at the ridiculous accounts of his old invention, especially as he is pestered at each mail by numerous letters with requests for plans, figures, details of the wonderful heater." [6]

Still the legend grew and the fame spread. Elihu Vedder, living in Rome, clipped and sent to Morse an item from *L'Italie* in French: "Cette puissance des rayons solaires, le professeur americain Edward S. Morse de Salem a réussi a l'utiliser pour la ventilation et le chauffage de deux bâtiments. C'est le résultat de chaleur artificielle le plus curieux que l'on ait obtenu jusqu'à ce jour"! ". . . it makes me mad way through to see how that sun-heater is roared by the papers of the country at intervals, each time adding a bigger lie to the one preceding it," raged Morse in a letter to John Gould.[7]

He never needed *chaleur artificielle* to make him hot over even mildly controversial subjects. He had been appointed a judge of Oriental pottery for the Columbian Exposition at Chicago in 1893 and discovered that the Daughters of the American Revolution were sponsoring a replica of the Liberty Bell to be made of contributions of money, old jewelry, and historic metal. "Articles of historic interest will be particularly appreciated. Historic mementoes of Washington, Jefferson, Franklin have been promised," said a circular that Morse found in his mailbox. It mentioned one unique contribution: "steel rims of eyeglasses belonging to Hon. Daniel J. Merrill, chairman of the executive committee of the Centennial anniversary in Philadelphia in 1876 who died of softening of the brain and these were found in his pocket after death"!

[6] Undated clipping, Peabody Museum, Salem.
[7] January 22, 1893.

This was too much for Morse. He raged, in a letter to the Boston *Herald:*

> One hardly knows which to condemn the most, the promoters of such a project or the weak-minded enthusiasts. To melt down rare historic mementoes . . . is an idea so monstrous that an injunction should be issued till the public conscience is aroused. . . . So long as such idiots are in the land, we shall next hear of a demand for autograph letters of Washington or folio editions of Shakspere to be boiled into pulp from which tickets are to be made to admit one to the first tolling of this "Columbian Bell." Prehistoric fabrics from Peru and Egypt, rare tapestries and laces might be ground up into shoddy to make a carpet on which the bell-ringer shall stand to toll a knell on the obliteration of rare things. Surely it is time to protest against such criminal imbecility.

The letter to the *Herald* gives a faint impression of the vivacity of Morse's ordinary conversation, made up of bubbling profanity, extravagances, innuendoes, and slightly off-color jokes. A typical note to a friend, dashed off on a postcard read:

> The idea of your stopping at a Hotel is simply monstrous! Nellie and I will repudiate you and yours and your primogenitors as well as descendants forever if you even breathe such a brutal act. I will meet you at noon. We will have dinner here and a good smoke and you may go to Bolles then, or any other place with double "l's" in it.

His own conduct was forthright to the point of brutality. Mrs. Jack Gardner (with whom he had enjoyed flashing exchanges of wit, mitigated by a common love of museum objects) invited him to a dinner in her newly opened Fenway Court in Boston. The dinner was served on antique refectory tables in a lofty gallery, paved with flagstones, walled with still damp plaster, and feebly heated by a single fireplace. It was a brilliant company, and Mrs. Gardner scintillated in black velvet *décolleté* with her famous ropes of pearls. Morse shivered through the soup and fish, then left the table, returning with his overcoat and overshoes on. Mrs. Gardner was furious; Morse triumphant. She never invited him again.

Morse had not a single inhibition as a result of what other people might think of his actions or his interests. Generally he swept both new acquaintances and old friends impetuously along in the current of his enthusiasms. In Boston, about 1900, he was introduced to a man of the same patronymic, George W. Morse, street railway corporation counsel who lived in Newtonville. They quickly discovered a common ancestor in "Anthony Morse, the Settler," and George Morse invited Edward Morse to dinner.

The first time I remember Cousin Edward was a Sunday [recalled Genevieve Morse Christian]. I was about seven years old. I had cut a mask of a goat's face from the Sunday paper. He wondered how many of the masks were being played with all over the country. He became so fascinated that he and Father got Michael the coachman to drive them around to see how many they could count in the streets. He came back very pleased; in an hour he had seen twelve.

Morse never gave a thought to his host's Sunday dinner getting cold and a coachman nettled at hitching up the horses on Sunday morning without warning; but he himself would not wait two minutes for the Salem hackman ordered to drive him to the Salem "depot" when he was taking the train. If the hack was not at the door when Morse came downstairs, Morse would start walking up the street. "Damn the man!" he would splutter.

He grudged the time and effort spent in dressing or undressing, took no interest or pride in the cut or material of his clothes, but when his tailor sent home a new suit with the pockets more shallow than the old ones he kept stuffed with clippings, specimens, and notes, Morse devoted an hour to measuring the new suit and an old one, comparing them as exactly as though they were two specimens of *Arca granosa*. Then he calculated to the fraction of a square inch how much material had been eliminated by the tailor's shears, and wrote a stinging letter of complaint, specifying the cash value of the saving effected and its disproportionate relation to his inconvenience.

He was a good trencherman all his life but grudged a moment of waiting, when called from his study, if a meal were delayed when he appeared. To work off this irritation, it became an invariable habit of his to sit down at the piano and bang out the Pilgrims' Chorus from *Tannhäuser* as a vent to his feelings.

Yet everyone loved him and spoiled him, from his wife to his secretary and servants. There was no malice, no surliness, in his manner. "Jean, where's my coffee?" he would roar, waking early in the Concord house of his daughter, when he visited there. "Ye'll just have to wait till it's brewed," would answer the Scotch waitress, disappearing into the kitchen to hurry matters up. And Morse, chuckling, would sit down at the piano and play "The Campbells Are Coming" over and over until the coffee appeared.

The days just were not long enough, time lacking for all the ideas and the interests and the plans that streamed through his mind.

CHAPTER FIFTEEN

I WILL, BY THUNDER
1899–1910

CHANGES had come with the years. Grandma Owen, who spent her winters in Salem, her summers with her son George Owen, and a visit always to Jane Beckett Morse in Gorham, was gone. Jane Morse died in 1896, during one of the flying visits her boy Ned paid her as often as he could manage. He was standing in the kitchen, at the foot of the steep stairs put in when the cottage was enlarged for his marriage in 1863, and Jane Morse, coming down to him, was stricken and died in his arms. John and Edith Morse had grown up, the boy to enter business in an insurance office, the girl to use her artistic talents in designs for tapestry and needlework in a little shop with Jennie Brooks.

In his peregrinations over the country Morse, by chance, brought home matrimonial partners for both his children. In New York he had met David Blakely, a dynamic former editor of the Chicago *Evening Post,* the St. Paul *Pioneer Press,* and the Minnesota *Tribune.* A passion for military bands drew them together. In 1890, after hearing Strauss music on a European visit, Blakely had signed up Edouard Strauss for an American tour of seventy-three cities. In 1892 he heard John Philip Sousa with the Marine Corps band and, with the high-handed imperious procedure of a veteran newspaperman, went over the brass hats of the army, direct to the President, to buy Sousa out of the service and put him under contract. Americans went wild over Sousa and his band at the Chicago World's Fair in 1893. "The Stars and Stripes Forever," a march familiar to every American, was Sousa's tribute to his long association with Blakely; it was written when the news came of Blakely's sudden death.

Blakely invited Morse to stop off sometime at the old homestead in Bradford, Vermont, where the Blakelys spent their summers, and Morse saw at once that his friend had three most charming daughters. He insisted that they return with him for a visit in Salem. Handsome young John Morse had always said it was against his principles to fall in love at first sight. Meeting Miss Julie Blakely unexpectedly, as he returned to Salem, Saturday night from the insurance office in Boston, he adhered to his principles only by postponing his surrender until Sunday morning. On June 21, 1899, Edward and Ellen Morse went to Bradford to see their son married.

Edith Morse had been married the year before to Russell Robb, a young electrical engineer who was later to be a partner in the firm of Stone & Webster in Boston. Morse had met Russell Robb in Dubuque, Iowa, before the first trip to Japan. Robb was an orphan, in a home where Morse stayed as a lecturer. Impressed with the boy's personality, Morse asked why he didn't insist that his father send him to college, instead of letting him work, at fourteen, in a grocery store. It wasn't his father, only the brother-in-law of his dead mother, young Robb explained.

Morse must have thought of his own boyhood and his own struggle for an education. He knew another uncle of the young orphan, Governor Bagley of Michigan. Somehow, by conversation or letters, Morse awakened the aunts and uncles to the necessity of furthering the boy's education. Young Robb went to stay at the Bagleys' in Detroit, with the promise that, if he prepared successfully and passed the examinations for Massachusetts Institute of Technology, Morse would find work near Boston to help finance him, and the Morse home in Salem should be open to him for week ends and vacations.

It is another proof of Morse's gift for quick scrutiny and intuitive analysis that in a few hours' talk he could thus recognize ability in a country lad. Russell Robb never forgot the inspiration and aid Morse had given him. As long as he lived, a poor boy could have from Russell Robb the money for tuition at the Massachusetts Institute of Technology, and it was one of his joys that "his boys" in turn paid back the loans to enable others to be helped. Russell Robb's firm built the white walls and dome of the Massachusetts Institute of Technology on the banks of the Charles River in Cambridge. If the hand that draws the bowstring is responsible for the mark of the arrow, then Edward Morse should be credited with a share in the building of that landmark.

I thank you very much for the kind letter acknowledging receipt of my book. It was dedicated to you mentally, if not formally, for however slight it may be as a piece of work, it is due entirely to the forcing brought about by your generous interest. . . . There is nothing so invigorating as the active interest and intellectual sympathy you have so much of. It has a peculiar effect and one would be very dull not to be aroused to some sort of exertion after feeling it. I can trace back your influence on my life for a great many years and even if circumstances be such that I cannot always have the influence so direct in the future, there is at least always a desire to have you think well of me that would be a stimulus,

Robb wrote in 1895 to his future father-in-law.

Russell Robb planned another book that he did not live to write. Much of the material for this story of Morse's life was carefully amassed and stored in his home at Concord, Massachusetts. Unfortunately he lived only two years longer than Edward Morse, dying suddenly of heart disease on February 15, 1927.

The marriage of his children and the removal of the Japanese pottery to the Museum of Fine Arts left the Salem home more quiet than Morse had known it for thirty years. With the receipt of the money for his pottery, the constant lash of debt was taken from his back.

June 22, 1899

Dear John [Gould]: John was married yesterday at Bradford; all went up and had a great time. . . . I am going [to Rangeley, Maine] tomorrow to be gone a week. . . . The burden of my pottery catalogue is off my hands and I am into Brachiopods again . . . and it is rather amazing to me after being away from the study of them for twenty years, I go back fairly hungry for the subject.

A framed parchment had been hung on his library wall this month by Margarette Brooks, denoting that the Emperor Mutsuhito, under whom Morse had served in the Imperial University at Tokyo, 1877–1879, had conferred on him the Third Class of the Order of the Rising Sun in recognition of his services to education and to archaeology in Japan. Morse kept the jeweled decoration in the vault of a Salem bank. He told a Japanese friend that he valued it too highly to wear except on occasions associated with Japan.

For twenty years, from 1881 to 1901, Morse had published no scientific paper of original research on *Mollusca.* His work as director of the Peabody Museum at Salem, his European journies, his researches in arrow release for ethnology, his preoccupation with his work as vice-president of the National Academy of Science and president of the American Association for the Advancement of Science, his tremendous labors in completing, cataloguing, and arranging the Japanese pottery collection for the Museum of Fine Arts, and his writing of *Japanese Homes* was enough to fill the hours of any one man. Nevertheless, though they were forced to flow as an undercurrent, he continued his observations on the shells which had made the theory of evolution valid for him and led him first to Japan. On November 20, 1901, at a meeting of the Boston Society of Natural History he read a paper, *Observations on Living Brachiopoda,* the reprint of which was widely circulated among biologists and university men.

Forty years had passed since *The Origin of Species* was published, but still the controversy between religion and evolution that parted Agassiz and his pupils found expression. From the Massachusetts Institute of Technology, October 8, 1902, wrote Arlo Bates: "I thank you much for your monograph. I could understand very little of it but I suppose it is all right. I should be sorry to think it could cast any reflections on revealed religion."

The beginning of another scientist's world-famed work in embryology is announced in a letter to Morse from E. G. Conklin of the Zoölogical Laboratory of the University of Pennsylvania, November 16, 1903:

Thank you for your kind note calling my attention to your early observations as to the openings into the mantle chamber of *Lingula*. . . . I intended to devote myself to Brachiopods last summer but was diverted by the discovery of the most wonderful egg ever seen, that of *Cynthia*, an Ascidian in which the substances which go to form the muscular, digestive, and nervous systems are most clearly differentiated in the *unsegmented egg*. I shall have something to say about it at the meeting of the Zoölogists which will be held here during the Christmas recess.

One of the earliest to recognize the importance of Morse's return to marine biology was Sir William Van Horne of Montreal, president of the Canadian Pacific Railroad.

August 11, 1902

I have just now been giving a few hours to your work on the *Brachiopoda*. It is like all you do — founded on the solid rock of absolute knowledge and built up with force and precision, leaving little for followers to do. Your drawings are marvelous. They are so simple that few will recognize their extraordinary merit but to me they are the very perfection of illustration.

W. M. Wheeler of Harvard's Bussey Institution at the Arnold Arboretum, Jamaica Plain, Massachusetts, voiced his contemporaries' view of Morse's dualism. Referring to Morse's "beautiful work on Brachiopoda," he adds that he was

delighted to see it contained so many observations on the living animal. We Zoölogists have every reason to feel deep regret that you have not been able to devote all your time to investigation of our science but, doubtless, the Ethnologists do not look at the matter the same way!

Throughout his life Morse had financed his own laboratory and researches, but publication is a costly affair when an author is also an artist, demanding an extravagant number of illustrations. The memoir on the Brachiopoda was made possible, once it had been read and evaluated, by grants. Glover M. Allen, secretary of the Boston Society of Natural History, wrote on December 20, 1901, a month after hearing the paper read: "I note you now have $1,000 promised to pay for the plates. By all means go ahead with the work and I guess the Society can pay any balance. If not, I will pay it myself rather than lose so valuable a paper for our Publications." The remainder, however, was forthcoming in a few more weeks, when Charles D. Walcott, treasurer of the National Academy of Sciences, sent a check for five hundred dollars, "being the amount of the appropriation from the Bache fund."

Morse perhaps was feeling trepidation at launching into print on a scientific subject after twenty years away from laboratory circles. He asked C. E. Beecher, of the Yale University Museum, and William H. Dall to read the proof on his manuscript.

> Your plates are very beautiful and illustrate in a most lifelike manner the characters you describe [commented Beecher, June 3, 1902]. They will form a classic contribution to Brachiopod anatomy. [And a week later, June 9]. By this mail I return the last batch of proof and congratulate you on the completion of your Memoir. This work will be very useful for me to give to my graduate students to assist them in gaining a knowledge of Brachiopod anatomy. I also think that some of the data you give will help to elucidate some of the vascular and other markings on the interior of some of the fossil forms.

Louis Agassiz's son, whom Morse had seen receive his degree from Harvard in the Civil War days in Cambridge, now director of the Museum of Comparative Zoölogy, where Morse had carried so many jars of "alcoholic specimens" up the long flights of stairs, aided in the publication of Morse's exquisite drawings. On May 17, 1902, he wrote to Morse, "I enclose check for $250 for the Brachiopod plates, and am glad the whole thing has gone through to your satisfaction. It will be a pleasure to see all that published at last."

Nearly twenty years earlier, in Dresden, at Rudolph Blaschka's home, Morse had seen the "glass flowers" being made for Professor George L. Goodale of Harvard; now on September 22, 1902, Goodale paid tribute to Morse:

> Your magnificent work on Brachiopods has reached me in camp here [Shelburne, New Hampshire]. Your papers are a great delight to me always, from the perspicuity and straightforwardness of your text and the luminousness of your drawings. Packard was here for a week and he was in time to see the memoir. He was much pleased with it in every way.

Yale agreed with Harvard. Edgar Roscoe Cumings of the Yale University museum wrote:

> Your Memoir on living Brachiopoda is of the utmost value to the paleontologist. Such work is bound to clear up the meaning of a great many puzzling morphological details of fossil shells. I sincerely hope that you will find the time and strength to give us many more similar studies of this group which your unique knowledge so fits you to deal with.

As the years rolled on, a second generation of Japanese scientists had come into prominence, still venerating Morse as their master in science, so wholeheartedly that even in London, at the Royal College of Science, their relationship to Morse was known. Writing of Morse's work on brachiopods, G. B. Howls said: "Sincerest thanks for your great memoir, which I am proud to possess. I have just been revelling in the observations of Yatsu, who seems to me worthy of your encouragement, and now comes your authoritative work. How Huxley would have rejoiced in it."

Naohide Yatsu, always proud to call himself a "grand-pupil" of Morse, because he had studied at Tokyo Imperial University under the great Mitsukuri, of Morse's first class in 1877, made a name for himself in marine biology and became seventh successor of Morse as head of the zoölogy department at Tokyo Imperial University and director of the Misaki marine biological station founded by Morse.[1] K. Mitsukuri held these posts in 1902 and wrote to his old teacher:

November 29

I have received your observations on Living Brachiopoda and hail your renewed activity in the zoölogical line. Not that I have any reason to complain but feel very much indebted to you for your efforts to make my country known to the outside world during the last 20 years. You have seen Yatsu's papers on Lingula, I suppose. I am just back from Misaki. I hear Dr. Bigelow is in Japan but cannot exactly locate him.

The debt the Japanese felt for Morse's interpretations of their country can be understood from a note of James Ford Rhodes:

Dresden, June 30

We spent a week in Oxford and had one of the most delightful seven days which have ever fallen to my lot in a happy life content with the sphere to which God called me. We saw much of our old

[1] In his "History of the Science Department, Imperial University, Tokyo" (in Japanese; copy in Harvard College Library), Yatsu paid high tribute to Morse's work in Japan.

friends and made some new acquaintances among whom was Professor York Powell, Regius Professor of History at the University. One night when he was dining with us, or perhaps when we were taking luncheon with him at his rooms in Christ Church, he asked "Do you know Morse of Boston?" As we are in the history line I supposed he meant John T., of whom I began to speak. No said he, it is not that Morse; it is Japanese Morse; he is the man of all others there I want to meet. . . . The talk then went on to the great catalogue and I could see that Mr. York Powell panted and thirsted after one. Please send one, in the ordinary binding. Place, *bitte gefälligst*, in the book the enclosed card. Send the bill to me. We go to Karlsbad today. I am going to take the cure so that I may drink lots of champagne next winter without an attack of the gout.

By October, Rhodes was back in Boston, sipping champagne and writing from 392 Beacon Street:

October 22

It is all right about the Catalogue. I received advice from the Museum and sent them one dollar to refund to Professor York Powell the amount of express he had to pay on a gift. The evening the books came, Mrs. Rhodes and I spent a while looking over your book on Brachiopods, admiring the skill and address with which the drawings were made and the thought came to us that we knew the man who knew more about brachiopods and Japanese art than any other man in Europe or the U. S. As a writer on the United States endeavoring to present our country in its true light it is a satisfaction to me to know that in these two things, which have no relation to coal, pig iron or steel, we excel England, Germany and France. This is what pleased me so in Professor York Powell's unconscious tribute to your genius.

Morse's sympathetic interest in Japanese students never waned in the forty-five years after his return from Tokyo. Every student who came to America bore a letter to Morse

Sensei at Salem; they were proud to style themselves "grand-pupils" of his. In October 1904 Morse gave up one of his cherished mountain-climbing trips because, as he wrote to John Gould (September 1, 1904), "Dr. Kitasato, the discoverer of antitoxin for diphtheria, will be in Boston between the 1 and 4 of October . . . and if possible I want the Tavern Club to give him a dinner. At any rate I should be there to show him some attention."

Dr. Emil von Behring and young Kitasato had announced in the *Deutsche medizinische Wochenschrift* for December 4, 1890, experiments in immunization against diphtheria and tetanus by injections of blood serum from subjects experimentally infected who had recovered. The basis of blood serum therapy was thus laid by these brilliant pupils of Robert Koch's. Welch of Johns Hopkins and Parker of New York adapted and carried on researches in the new therapy, and in February 1895 the New York Board of Health announced "a new laboratory at the Willard Parker Hospital in New York," where there was "serum from 20 horses . . . sufficient to treat 1,000 children a week," adding that "American mothers . . . need have no more fear that their children will be strangled to death by the demon diphtheria." In 1902 the American Academy of Arts and Sciences elected the discoverers "Foreign Honorary Members . . . in the Section of Medicine and Surgery." Kitasato was also founder and director of the Institute for Infectious Diseases in Tokyo.

Though Morse gave up the mountain trip, he managed to get in a flying visit to the St. Louis Exposition before hastening back to entertain Kitasato. The fatigue of his dash to the Mississippi was outweighed by the chance of observing the arrow-releases of natives of the Philippines, newly become wards of America. As usual, he poured forth his enthusiasm to John Gould:

August 28, 1904

. . . Could not resist the chance of going to St. Louis [to judge Oriental pottery at the Exhibition], being appointed a Juror with expenses out and $7.50 a day. . . . The only thing that tempted me was the opportunity of getting material for another paper on Arrow Release. Think of the Pygmies from Africa, Tierra del Fuegans, Philippine Island tribes and North American tribes galore! The head of the anthropological department, Prof. McGee, is an old friend of mine and every opportunity will be had for what I want.

In the summer of 1900 all the world was reading in newspapers a serial story, marked by drama and suspense, of the Boxer rebellion in China and the international expedition for the relief of the beleaguered diplomats in Peking. Morse looked up his old journal sheets of 1883 and wrote the book he had been too sick to write on shipboard in the Indian Ocean. *Glimpses of China and Chinese Homes* was published in 1902 by Little, Brown & Company, Boston. It is the shortest and most vivid of his books, illustrated with the original sketches from his journal, packed with terse comments brewed from reflections twenty years after.

If one soberly considers the manner in which the Chinese have been treated by Christian nations, he cannot be surprised at the attitude of the Chinese towards him. In the plainest way, it may be stated that the "Foreign Devils," under the guise of a diplomatic phrase known as "spheres of influence," have stolen thousands upon thousands of square miles of territory, have robbed them of nearly every open port, and have extorted untold millions in indemnities.[2]

[2] *Glimpses of China*, pp. 202–203.

This paragraph was rather detached for an American whose country had just occupied Cuba, taken the Philippines, and was moving towards acquiring the Panama Canal Zone, but Morse was too good an anthropologist to make an idealistic internationalist:

Cleanliness, sanitation, good roads and schools, coinage, postal-system, fire-apparatus, and all the wonderful development of steam and electricity and the thousand instrumentalities of life that we have acquired through persistent study of the behaviour of nature's laws make no more impression on their brains [those of the Chinese] than it would on the brain of an ox. In these respects they represent a savage race, and for these reasons the Chinese Exclusion Act is justified.

The Chinese must certainly lament that they are not strong enough to issue an edict excluding the foreigners from their country, for the foreigners have already wrought untold calamities for their people, just as the free admission of the Chinese to our country would, in the end, work untold miseries for us.[3]

Morse throughout his life deprecated what he regarded as the hypocrisy of preaching to "heathen" nations doctrines that had not succeeded in producing Heaven on earth in Christian countries. John Gould, devout church member and supporter of foreign missions, often argued the question with him. One of Gould's projects for helping some savage tribe elicited a characteristic response:

April 25, 1910

. . . Have read the remarkable record. I don't know which is the most pathetic: the squalor and absolute worthlessness of these buggerers or your attitude in devoting your time, your brain, your talent

[3] *Ibid.*, pp. 208–209.

to such worthless creatures or animals when your city and your country is filled with hoodlums . . . these right at your door, so to speak, the deserving poor, and then to roam the coast for a few half-breeds! It is equivalent to a breeder getting together a few mongrel dogs or hens and caring for these instead of devoting his time to encouraging a good line of descent. I can't see it, dear boy; to me it is unphilosophical, unscientific and ghastly.

Morse's extraordinary capacity for carrying on simultaneously several lines of thought, coupled with appropriate action for each, appears again in the busy years of the first decade of the twentieth century. It would be tedious to recount on each page of his biography Morse's activities as director of the Peabody Museum in Salem from 1880 until 1912, but it should not be forgotten that his daily administrative duties at the museum were for more than thirty years the background to his manifold accomplishments in the fields of research, literature, astronomy, civic campaigns, ceramics, and the never-ending correspondence.

His work as director of the museum had kindled enthusiasm in the community. Trustees and public-minded citizens of Salem were coöperating generously with him to build a fireproof addition to old East India Hall to house the irreplaceable ethnological collection of the Peabody Museum, but Morse was even more deeply preoccupied with the scientific question of whether loons dived deep enough to bring up certain minute shells from the ooze. Gould sent him a clipping about a selectman of Sandwich hooking a loon in a trawl sixty-six feet deep. Frank Benson, Salem artist, known for his paintings and etchings of wild fowl, told Morse, as he wrote to Gould on March 15, 1904, that the coot can swim down several fathoms and his boy opened the stomachs of four coot and in each was a shell looking like a fresh-water mussel. The loon never went down that depth; he was hooked near the surface, possibly ten feet. I can't believe the bird goes deeper than that. The interesting thing about loons is that when they dive they literally fly in the water, making the same graceful turns that they do in the air. I saw this in the aquarium at Brighton, England.

P.S. Have just raised $25,000 for this Museum for important changes I am making in the building.

The "interesting thing" about this letter is that it illustrates two things about Morse: one, his remarkable visual memory, for he had seen loons swimming in the glass-fronted tanks at Brighton in 1887, almost twenty years before; and the other, his relegation of the administrative feat of raising money for a new building to a mere postscript.[4]

In 1904 both money-raising and the determination of the diving propensities of loons were incidental, for Morse was heart and soul with Percival Lowell in a study of the canals of Mars.

For thirty years peppery Percival Lowell and profane, positive Edward Morse had been close friends. Lowell's interest in the Orient had first been aroused by Morse's lectures on Japan at the Lowell Institute. Morse's friendships in Tokyo paved the way to Lowell's appointment as counsellor to the Korean

[4] That Morse approached reluctantly the money-getting part of his duties as museum director is shown by a fragment of a draft of a letter he scratched off in 1915: "A dismal duty of college presidents as well as Directors of Museums is to beg, and the few times I have been compelled to exert this function has [*sic*] been disagreeable to me in the last degree. An important collection has come to us from Japan which I should like to show you. We want at least six hundred dollars for it. . . . Would you be willing to contribute?

"P.S. And my no less respect and esteem whether you help or not."

government, and when Lowell accompanied the first Korean embassy to the United States Morse's young friend Tsunejiro Miyaoka was his secretary and interpreter. Lowell's fine interpretations of the Orient in *Noto* and *The Soul of the Far East* might not have been written, had he never met Morse. Conversely, Morse would not have written *Mars and Its Mystery* had he not known Lowell.

It was Jane Beckett Morse who first kindled her son's interest in astronomy, however. Her own intelligent pleasure in the heavens is shown in her letters to him from her quiet country home in Gorham, Maine:

> December 16, 1882
>
> My dear Child, I don't know where this letter will reach you [in Japan] but I want to write you if only to tell you how dear you are to me and how thankful I shall be when you get through your wanderings and are safe and well at home. I know it must be delightful to see so many new and strange places but it must be fatiguing, too. . . . Mary is with me this Winter which makes it very pleasant for me as I love to be at home more than anywhere else. What a glorious sight the Comet was, you must have had a good view of it. We had to get up at four o'clock to see it. We had a good sight of the transit of Venus, too. I remember well of seeing this Comet in 1842. . . . As you get farther South you will see the glorious Stars at the South Pole, six of the first magnitude, nine or ten of the second and no end of smaller ones. You must enjoy it so much, I know.

Lowell had been studying Mars since 1870, beginning with a two-inch telescope mounted on the roof of his father's house in Brookline and progressing to an observatory built at Flagstaff, Arizona, in 1894, with a telescope equipped with an eighteen-inch refractor.

It was Lowell's custom to summon Morse to Arizona at the seasons when Mars was best to be observed. His investments in railroad stocks entitled him to passes with which he facilitated Morse's journeys, and the invitations were so frequent that they became laconic. "Come Tuesday. R.S.V.P.," he wrote casually from Flagstaff on May 15, 1897.

April 15, 1903

Dear Man: Am I to have the pleasure of your company out here in my eyry within the next few months? I will send you passes or tickets out and back. Mars is showing off well and will continue to do so for some moons yet.

November 6, 1905

Dear Fellow-Martian: Many thanks for the draughts, in your own handwriting, too. In return let me make you a present of yourself, done by Mr. Lampland in the land of the pine and the pined-for. *Banzai.*

March 3, 1914

Dear Morse Sama, *Oki ni ari.* Bravo! Now go right on. Come out at the next opposition [of Mars] and write a book on what you have seen, since one on what you have not seen has taken so well.

Morse visited Flagstaff frequently, at the low mansion on the edges of the mesa, overlooking the pines of Conchino forest and the distant mountains across the desert. Lowell was still a bachelor, and Morse never permitted matrimony to interfere with science, so their companionship was cordial and constant.

In 1906 *Mars and Its Mystery* was published, not because Morse pretended to astronomical learning, but because he wished to defend Lowell's theory of the "canals" as an indication of life on the planet against bitter attacks of other observers and astronomers.

One must insist [he wrote] that the study of planetary markings as well as the interpretation of their meanings comes not only within the province of planetary astronomers, but that any broad-minded man, with an acute eye and familiar with the sciences connected with the surface features of the Earth, is quite competent to make observations of his own and to judge of the merits of the question.[5]

Morse was the despair and exasperation of professional astronomers because he "proved" to his own satisfaction the existence of life on Mars by means of such informal comparisons as the crackling of glaze on old Satsuma pottery from Japan and charts of man-made streets in Montreal and railroads in Illinois. To Morse himself the hours at the telescope were disheartening because he had in them the first physical evidence that he was sixty years old.

It was my good fortune to have the privilege of observing Mars every night at the Lowell Observatory . . . for thirty-four days. . . . Careful to avoid any bias in the matter I rigidly refused to allow either Professor Lowell or his assistants to suggest where I might find a canal or a marking on the disk. . . .

May 16. Occasional flashes of a few lines . . . for the first time I am encouraged to believe that each day an improvement will take place. I saw enough to make my first drawing. . . .

May 24. Am in despair of seeing anything when the others see so much. I must have an old and worn-out retina. . . .

June 5. I find a slow advance in my ability to see the markings though it is exasperating that the janitor of the Observatory talks about seeing plainly certain details . . . and looking at the region I can see no trace of a canal or anything else. . . .

June 12. . . . Every day I notice a very slight improvement in detecting lines. Markings formerly made out with great difficulty are now instantly recognized. . . .

[5] *Mars and Its Mystery*, p. 69.

Mr. Lampland, after many attempts has succeeded in photographing the more conspicuous markings. *The lines do exist essentially as figured by Schiaparelli and Lowell.*[6]

The book was well received and in recognition of it Morse was elected to membership in the Société astronomique de France. One reviewer, however, wrote savagely: "The author is a zoölogist and a student of Japanese porcelain, not an astronomer. His book is carelessly put together, repetitious, decidedly partisan." Morse parried with glee, "Have never studied or collected Japanese porcelains," for his pet aversion was people who confounded pottery with porcelain.

Percival Lowell appreciated Morse's partisanship without overestimating his knowledge of astronomy:

August 30, 1907

. . . now if you will write an article for the *Atlantic*, I can assure you its acceptance and it will net you some money, always a good thing. You can touch lightly on what you are not well posted on and come down heavy on what you are. . . . Let me hear from you with an "I will, by thunder!"

The great Schiaparelli wrote from Milan, November 10, 1906, to draw the sting from criticisms in the reviews: "I have just read the book 'Mars and Its Mystery.' . . . It is the application of common sense to the question of Mars. . . . Chapter VII is very important even for the practical astronomer. . . . We owe a debt of gratitude to the author whose book will . . . contribute powerfully to correct ideas on the subject." James Ford Rhodes said, "Treatment is scientific and sane. . . . I thought the literary style good except in the controversial parts. . . ." The Lynn physicist, Elihu Thomson, was stimulated to write:

[6] *Mars and Its Mystery*, pp. 158–164, 181.

The thin air would make the nights very cold, of course, but that is no detriment. What puzzles me most is the double canal so parallel and a distance away from the other. Can it be that the Martians have put their niggers on a reservation canal off to one side? Can it be that the population is getting too dense and the men and women beginning to live on separate canals by edict of the central executive? Can it be that the parallel canal is a training area for youthful irrigators, or can it be that there is some system of water transport connected with the irrigation scheme and the parallel is the return circuit?

Even after the sale of the pottery had removed the spur of financial necessity, Morse continued to accept invitations to deliver lectures. It had become a habit. He had learned to speak easily, naturally; and he enjoyed the response he elicited from audiences. As his reputation increased, many invitations came from scientific societies and congresses, affording him the pleasure of meeting old friends and co-workers in science, gathered from distant universities or laboratories. Through his detailed studies and sketches of Japanese and Chinese buildings, he had also earned the interest and appreciation of architects. A letter from Cass Gilbert illuminates this side of Morse's life and reveals an unexpected gap in his wide field of acquaintance.

With all Morse's Washington friends in science and art, and all the close friendships Theodore Roosevelt had in Japan, it is strange that these two admirers of Japan were not acquainted with one another, but Cass Gilbert's account of how Theodore Roosevelt disrupted a meeting of the American Institute of Architects and left Morse fuming and cursing on a rainy night in Washington is evidence that the President and the Brachiopodist had never met. Gilbert wrote from New York, December 22, 1908:

I have an explanation and an apology to make you relative to last Wednesday night. Both would have been promptly offered had I been able to find you in Washington, but the whole thing moved so swiftly there that it was impossible for me to keep up with the procession.

The President sent for me Wednesday evening [the night of the meeting] to meet him at the White House at half past nine.[7] I therefore asked Mr. Frank Miles Day of Philadelphia, past president of the Institute, to take charge of the meeting at Hubbard Memorial and gave him a written memorandum of the arrangement I had made with you over the telephone, namely that a taxicab was to meet you at the Masonic Hall at 9:40 and bring you to the Hubbard Memorial at 9:45. . . . I opened the meeting and introduced Mr. Day . . . and I hurried off to the White House. . . . I learned later that he, perhaps not realizing you were to come a little later, although he had the memorandum, adjourned the meeting.

I was detained by the President until twenty minutes of twelve and as I had not had any dinner, I was utterly worn out and nearly starved and did not attempt to hunt you up that night. . . . I want to thank you most heartily for your willingness to assist us in a very embarrassing series of circumstances. I appreciate it most highly, and finally I want to say that the audience was a mighty poor one. I should have been ashamed to have you called on to address so slender an assembly.

Just as Peary must divert his attention from the North Pole to study arrow releases among the Eskimos for Morse, so must

[7] Gilbert, architect and designer of the Woolworth Building and the United States Customhouse in New York, the Minnesota State Capitol, and other notable buildings, had been appointed by Roosevelt a member of the Federal Council of the Fine Arts. He served in the same capacity successively under Presidents Taft and Wilson.

any friend expect to be called on to investigate whatever problem aroused Morse's interest. L. O. Howard, long permanent secretary of the American Association for the Advancement of Science, was chief of the Bureau of Entomology of the United States Department of Agriculture. To him was allotted the friendly duty of preparing the biographical memoir of Morse for the National Academy of Sciences in 1935, in which he wrote: "There should be a book about Morse. One will be written some day." In life, however, Edward Morse, brushing his coat one morning in Salem, was struck by the sight of a minute object which he could not identify and evidently commandeered the chief entomologist of the United States to investigate for him. Dr. Howard wrote on March 24, 1913, "Your letter of the 20th and the enclosed objects attached to the bristles of a clothes brush have come. I have an idea that these are the eggs of some large louse affecting the animal from which the bristles came. . . . I am not familiar with them, however, and this is only a guess of mine. I will send the objects over to the National Museum for the inspection of my corps of systematists."

Pioneer in ideas, student of civilizations, devotee of new techniques in science as he was, Morse remained obstinately conservative in his personal habits. He never learned to use a typewriter or to drive an automobile. He could not bring himself to dictate to his secretary. From the beginning to the end of his long life he wrote all his scientific notes, his voluminous journal, his vast correspondence, and his books in longhand with pen and ink. On November 25, 1904, he told John Gould: "It is one incessant writing of letters. I never sit down to my table that a dozen letters are not staring me in the face, and they must be attended to." Six years later he was in much the same difficulties:

May 1, 1910

I have just come across a letter of yours in a potsdam or some other dam deposit on my table. . . . Life is too short to write an answer so remind me of it when I see you next. . . . I keep my engagements on a blue card eighteen inches long and four inches wide, a sheet for each month, as it can be found among my papers. If I could show you these cards you would see that every day with rare exceptions I have an engagement and sometimes two. Committee meetings, lectures, business, meetings of Boston Society of Natural History, American Academy of Arts and Sciences, Boston Scientific Society, Malacological Society, of which I am president, Salem Civic League, Hawthorne Memorial Association, etc.

In June 1909 he had just completed the extra work of superintending the removal and rearrangement of the collection of Japanese pottery to the new buildings of the Museum of Fine Arts in the Fenway. On June 15 he wrote to Gould, "I am a little overdriven on account of the new Art Museum and the work here." And on June 25, "Have just got back from Boston, working at the Museum of Fine Arts unpacking pottery in a room at 100 deg. It was awful!"

He spent at least one day a week in Boston as Keeper of Pottery at the museum. He was director of the Peabody Museum in Salem, a full-time job for almost anyone. He was called on for various special occasions, such as a memorial to Louis Agassiz in 1907, of which he wrote to Gould (December 27, 1907): "Your kind words about my Agassiz address surprised me somewhat, for it was written as everything is, at odd moments, before breakfast, etc. Do you know, John, that for years, winter and summer, I am through breakfast before seven?"

Ever since the completion of the catalogue of Japanese pottery, and despite his work on the books on China and Mars, Morse had been working with brachiopods, as he had always promised himself he would do when the pottery was off his hands. At the November 1909 meeting of the National Academy of Science at Princeton, New Jersey, when Simon Flexner was making a communication on "the transmission of epidemic poliomyelitis in monkeys" and E. G. Conklin an early study of "formative substances in eggs," Edward S. Morse, a generation older, was reading a paper on "The Early Stages of Acmoea." On February 12, 1909, he had also made an address entitled "Factors of Darwinism" before the Boston Society of Natural History, in honor of the fiftieth anniversary of the publication of *The Origin of Species* and the centenary of the birth of Charles Darwin.

He had no compunctions at sending John Gould, now a Portland bank president, to collect material for his researches.

February 13, 1909

Dear John: What are you really about? Have you retired or are you a soft-handed son of toil? If you have any leisure I wish after some storm you would run over to Saco Beach and collect a few wave-rows of stuff thrown up. You know how the stuff looks, shell sand in little narrow ridges parallel with the beach. The best way is to have a brush and sweep up on a sharp-edged sheet of paper and dump into a cigarbox or contribution box if cigar box is too profane. I am studying the early stages of our shells and find some interesting things. . . .

Gould, risking pneumonia, went obediently but not informedly, for Morse wrote a week later:

February 19, 1909

Dear John: I was greatly amused at your letter. You must be breaking up not to know about the tides. I examined the sample you sent me, nothing but fragments of barnacles and broken pieces of Mytilus. The shell sand I want can only be found between the high and low water mark. The projection of a great sheet of water as the tide goes out pushes along the little shells and leaves them in a ridge [sketch]. I get more sand in scooping up these little ridges of sand and now I have discovered that by using a large baking pan I can imitate the sea by swashing the water back and forth over the sand. The shells are washed up at one end of the pan. . . .

He himself was by no means "breaking up" when at seventy-two he could carry out an itinerary such as this:

March 1, 1910

Just one second to write you a line. . . . Every day and every night is full. (I am not, I would give you to understand.) As as example, last week spoke at a dinner at Economic Club at Manchester, N. H., with Baron Kikuchi . . . Monday night. Up next morning at 6:30 to catch 10 o'clock for New York to make connection with Poughkeepsie, N. Y., Tuesday night; crossed to place near Waterbury, Connecticut, where I lectured Wednesday night; back to Boston to attend Thursday Evening Club in Boston, getting to bed at one o'clock and so on nearly every night. . . . Have been so driven for three months I haven't looked at a shell. So no sand at present. What I need is more sand for myself.

Morse was a magnet. In Japan he had attracted thither Fenollosa, Weld, Bigelow, Percival Lowell, and had joined to them Kakuzo Okakura and Tsunejiro Miyaoka. Now, as he vibrated between Salem and Boston thirty years later, Okakura was curator of the department of Asiatic art and Fenollosa expert on art history at the Museum of Fine Arts; Bigelow was teaching Buddhist philosophy at Harvard; Lowell and Morse had been peeping at Mars together in Arizona; and Dr. Charles Goddard Weld was financing both the art studies of Fenollosa and a new fireproof wing for Morse's Japanese collection at the Peabody Museum in Salem. Salem friends like Mrs. Wheatland, whose husband had been the guiding spirit of the Essex Institute, Judge and Mrs. Alden White, the Misses Ropes, and many another contributed, but the wing and the new auditorium joining it to historic East India Hall were named in honor of Dr. Weld.

Morse, at last able to arrange his ethnological collection of the vanished feudal civilization of Japan in adequate quarters, completed the plan conceived in 1885 when he wrote in the preface to *Japanese Homes*:

> I feel now that the labor [of collecting material for the book] has not been altogether in vain, as it may result in preserving many details . . . which in a few decades of years may be difficult, if not impossible, to obtain. . . . Nothing can be of greater importance [to the ethnological student] than the study of those nations and peoples who are passing through profound changes and readjustments as a result of their compulsory contact with the vigorous, selfish, and mercantile nations of the West. . . .[8]

[8] Pages vii–viii.

CHAPTER SIXTEEN

SOUND AS A TRIVET

1911–1921

SICKNESS for twenty years had more and more isolated Mrs. Morse from her husband's active career outside the Linden Street home. After her children's marriages, she went often for long visits at their homes in Concord. In 1911 she died.

Her death was a shock to Morse, and the task that sickens every heart when its turn comes, of looking through the relics of the dead, turned his mind back to the days at Cambridge in 1861 when he had courted pretty Ellen Owen. He remembered now, too late, his intention of instructing her to keep pace with him in science. He came across the long-preserved journal of 1861 in Cambridge: "It is a good thing that Nellie is going [back to Cape Elizabeth], for I should get no work done." It reminded him poignantly that for forty-seven years his Nellie had loyally endured his constant preoccupations, expenditures, and absences "to get his work done."

Morse lived on alone, by choice, in the Linden Street house. Josephine, his faithful housekeeper attended to domestic details; Margarette Brooks came over daily from her house beyond the garden fence to work with him in his library. Excerpts from letters to Gould illustrate his mode of living:

June 8, 1914

. . . It is ridiculous and preposterous for you to come to Salem and not come a day ahead and spend the night at my house, see the Museum, and my mollusk drawings and hear some good music records in the evening.

May 4, 1919

. . . Shall continue my food economy; I find that half a doughnut satisfies me and they last twice as long. Very little meat will do. Haven't had roast beef in the house for years and never beefsteak unless John's wife is here.

His wife's death drew him no nearer to the comfort of a formal creed. Early in the twentieth century he had put himself on record about religion. When the Berle children wanted to form a "Morse Naturalistic Society," Morse, recalled the Rev. A. A. Berle, pastor of the Crombie Street Congregational Church in Salem, "came to me privately and said that I, being a clergyman, might fear to have him talk freely with the children on scientific matters, as he considered himself an unbeliever."[1]

With the maturing of his children, Morse had lost the opportunities of contact with youngsters that had once meant much to him. During the youth of his own children, he had never neglected a chance to instruct any child in the wonders of nature, taking them on walks in the field or by the seashore, giving them peeps through his microscope. Dr. Harold Colton, director of the Museum of Northern Arizona, was one who as a boy of twelve had his mind turned to a career in science by a visit of Morse to his father's house in Philadelphia. Adolf A. Berle, Jr., prominent in the State Department during the administration of Franklin D. Roosevelt, was president of the zoölogical society, to which his father had opposed no objec-

[1] Letter to the author, 1939.

tion, and of which his brother and two sisters were members. Young Berle achieved fame by qualifying to enter Harvard at the age of thirteen years. On Morse's seventy-fourth birthday he wrote, "We all feel it is a privilege to have known you, and you never had a quartette on whose friendship and loyalty you could more unquestionably count."

Another boy of old Salem, James Duncan Phillips, remembered for forty years the delights of nature study with Morse and described the

> . . . annual Cuvier Club tea over in the little clubhouse in Professor Morse's backyard. The Cuvier Club was a monument to Professor Morse's ability to inspire young men with an interest in Natural History. John Morse, Louis Gavet, Albert Cole, George Webb, Sydney Fenollosa, and six or eight other boys were members. It collapsed as they grew up and left for college but the interest in Natural History and Science never left them.[2]

The trustees of the Peabody Museum, in affectionate solicitude for the effect on Morse's spirit of his wife's death, at the annual meeting of 1912 unanimously voted a year's leave of absence without loss of salary. Morse's first thought was for his beloved shellfish.

> May 25, 1913
>
> Ho, John! Been in the house three days with a cold and have made feeble attempts to clear up [my desk]. An infernal job. Also writing paper on *Solenomya borealis.* Is there any mud-digging going on in Portland harbor?

On the back of the letter John Gould pencilled, retrospectively, "This was the beginning of trouble," for a mollusk

[2] *Salem in the Nineties* (privately printed, 1937), p. 10.

emerged unexpectedly from the mud of the ocean to rouse Morse from the apathy that had followed his wife's death. The government's initiation of a dredging project off Portland had kindled a flash of the old imaginative fire in Morse. He began to feel more energetic at once, and wrote to Gould:

June 9, 1912

Just back from New York where I have presided at a good-sized meeting twice a day (as president of the American Association of Museums), making response to speeches of welcome, putting motions, etc., and got back sound as a trivet. I don't know what a trivet is and haven't time to look it up.

Morse was rapidly recovering from a physical depression that three months earlier had caused younger men, thinking of the inevitable, to send him a grim *memento mori* of which he had written Gould:

April 7, 1913

The Wistar Institute of Anatomy, Phil[adelphia], wanted my brain. I told them to send me on a proper jar for it and I would pay the bill. You should see the jar, with my name etched on it, date of birth, residence, etc. A ripper, and wouldn't let me pay for it. [Sketch.] It's a work of art.

Before his seventy-fifth birthday arrived, Morse, absorbed in shellfish, had forgotten the weight of years and the possibility of death. To his daughter, accompanying her husband on a business trip to England, he scrawled:

June 18, 1913

Delightful to get your telegram, for it assured me you were alive and well. . . . Have finished a scientific paper on a mollusk that is little known, drawings all made. Am enclosing some birthday letters I got. We have having hot weather again; my cold which troubled me for a month has nearly gone.

That same month in which his seventy-fifth birthday occurred, Morse enrolled as a research student in the Marine Biological Laboratory at Woods Hole, successor to the summer school on Penikese Island where he had taught for Agassiz in 1873. He gave Gould no respite in Portland. "If by the merest chance in the world a *Solenomya borealis* is got," he wrote on June 30, "keep it in a large dish of salt water, changing every day, and telegraph me. I should then come to Portland and make the drawings there at once." In August the Brooks girls persuaded him to take a little vacation at Newport, but he interrupted it on receipt of news from Gould.

August 20, 1913

Just got back to my room at nine o'clock P.M. and find . . . two letters from you. . . . Literally I have thought of nothing all day but those great fat fellows in the pan. . . . I must bring down a suit of old clothes and go out [on the dredge] with you. I didn't thank you enough for what you have accomplished. Make observations yourself and I will append them to my paper with your name. You will go thundering down the Zoölogical ages as the first man to study *S. borealis* alive. . . . You are a brick. It is simply great.

He hurried back to Portland, and they had a glorious time, the pair of them seventy-five years old but prowling the beaches and poking through mudflats as eagerly as when they had been sixteen, "playing Schologizeim and acting like a pump-skewking" on the beach.

September 3, 1913

I had a wonderful time at your house; you were all so good to me. Now if you can get a few more *Solenomya* and some of the mud they live in and prove or disprove whether they bury themselves siphon or foot downward, you shall have full credit for it. Kindly cash the enclosed check and fee, bribe or debauch the dredger men in the interest of getting some more live specimens.

September 15, 1913

Can't you buy a new boat, or hire one? I'll pay for the hire. . . . It is contrary to the analogy for a lamellibranch to go down siphon end first. . . . Don't hesitate to tip those men a dollar each now and then.

January 12, 1914

I am crazy over your collecting. . . . Please don't run any risks. If anything should happen to you I could never look at a shell again and, what's more, would destroy all my notes. Am working on my Japanese journal all the time, and attending meetings, lecturing, etc.

Alone in the Linden Street house in Salem, with Bigelow on one hand prodding him to get to work on the Japanese journals and John Gould up in Portland prescribing the time-tried tonic of shells, Morse began filling up the rooms with pans of salt water containing living gastropods and snapping at Josephine, the housekeeper, to tiptoe around lest she startle the creatures he wished to study.

For years I have hunted in vain for a living specimen of *Thracia conradi*. This year my friend John M. Gould has collected specimens alive in Portland. . . . After storms I have repeatedly found the broken shells. . . . The gulls immediately recognize the conspicuous white object on the beach and break the thin and fragile shell, devouring the fat morsel within [wrote Morse, with his chatty way of enlisting the interest of the public in science]. A live specimen was sent to me from Portland carefully packed in seaweed. . . . [It] remained in a vessel of fresh sea water for three days without a sign of life. At one time the occupants of the house were away, . . . and then for the first time the creature timidly thrust out its tubes.[3]

[3] "Notes on *Thracia conradi*," *Nautilus*, XXVII (November 1913), pp. 73–75.

Good physicians are as wise at healing the spirit as the body. Dr. William Sturgis Bigelow, dubious of the wisdom of beach-combing for a septuagenarian, yet aware of the desolation of Morse's house after his wife's death, wrote for his old friend the perfect prescription.

July 1, 1913

The only thing I don't like in your letter is the confession that you are still frittering away your valuable time on the lower forms of animal life . . . instead of devoting it to the highest, about the manners and customs of which no one is so well qualified to speak as you. Honest, now, isn't a Japanese a higher organism than a worm? Drop your damned Brachiopods. They'll always be there and will inevitably be taken care of by somebody as the years go by. . . . For the next generation of observers, the Japanese we knew will be as extinct as the Belemnites.

The shaft went home, like one of the well-aimed arrows Morse loved. Margarette brought out the journals for 1877–1883, thick with the dust of thirty years, and Morse set about transcribing them in his hen-track chirography, weaving them into a connected narrative. The task gripped him, and presently (March 6, 1914) he could write to Gould: "Sixteen hundred pages of Journal already copied. Margarette is making a type-written copy which will be revised. Have been working steadily since September 30 and it is a job."

Work on this manuscript was interrupted on June 25, 1914, by a conflagration that burned over most of the city of Salem. Morse's home on Linden Street was saved only by a fortunate coincidence. The advancing flames were four blocks away and the city water supply had run dry when a new motorized pumping engine, returning from a demonstration in another Essex

County town, passed Salem, and its crew volunteered its services. It was set to pumping water from the old millpond in South Salem, near Linden Street. At the same moment the town of Marblehead completed emergency engineering that diverted its water supply into Salem mains, and a shift in the wind drove back the flames.

Morse had been completely fatalistic about the fire. He resisted all the efforts of Josephine and Margarette Brooks to pack up manuscripts, valuables, or furniture. Dr. Samuel J. Mixter and Dr. Thomas Barbour hastened from Swampscott in two automobiles to salvage the important scientific material they knew Morse had in his home. They found him seated calmly in his study, attempting to play a South Sea Island flute with his nose instead of his lips. "The Solomon Islanders can do it; why can't I?" was his greeting to his frustrated rescuers.

Morse did his share as a citizen in rehabilitation work after the fire.

July 5, 1914

Ho! John, the first help Portland received after the Great Fire [of 1866] came from Salem and consisted of 100 sets of carpenters' tools. It is a pity your Mayor or someone does not awaken some interest in helping [Salem]. . . . The needs are great simply in clothing and feeding the thousands who have lost everything. I shall have blown in two or three hundred before I get through. I have an old man of 83, his wife, and daughter who have lost everything and house not insured. . . . I do nothing but answer letters from all over the country.

All over the world, distinguished men of various nations, reading of the Salem fire, thought of Morse. Typical of the solicitude expressed and reflecting the other great conflagration of 1914 was the letter from Henry Woodward, now retired from the post of director of the South Kensington Museum of Natural History, where, in 1887, he had introduced Morse to Sir Richard Owen. Woodward wrote from London, August 2, 1914:

I read in the papers of the terrible fire which has destroyed half your city and must have brought distress and misery in overwhelming quantities to your door. You seem to have saved your Museum treasures in a miraculous manner and I hope you have not to mourn for personal friends amongst the many who have been rendered homeless by the catastrophe.

And now here are we in our little turbulent overcrowded Island and next door, across the street, the whole of Europe is ablaze, or likely to be in 24 hours, and no one at hand to put the fire out. God alone can help us, for man seems not only helpless but "war-mad."

In the autumn of 1914 Morse was elected president of the Boston Society of Natural History. He had joined it as corresponding member when he was a youngster in Portland, reporting his first discoveries of fungi and land snails. He had become an active member when he went to Boston in 1858 as a draughtsman of pianos and sewing machines at five dollars a week. Only former President Charles Eliot of Harvard had been a member longer than he.

December 20, 1914

This means that I have to go to Boston first and third Wednesdays of every month, snow, rain, hot or cold. That divine afflatus of flying downstairs I used to have but have got over. . . . I go off my front stairs and other stairs two at a time and often go up two at a time. Am grinding away at the Japanese Journal and it is a burden, for I want to get to my scientific work.[4]

[4] Letter to John Gould.

During the World War, Morse read three newspapers daily and felt keen sympathy for the Allies. "Raised $75 for the French sufferers in Cambridge yesterday. . . . This is the third lecture I have given," he wrote on January 27, 1916, to John Gould, for whom a month later he scrawled off a digest of his activities in his seventy-eighth year:

. . . Just for the fun of it I have copied off my engagements for this month. 1 — Malacological Society, Boston; 2 — Boston Society of Natural History; 3 — Thursday Evening Club, Boston; 5 — Peabody Museum with guests; 6 — Museum of Fine Arts with guests; 7 — introduce speaker, Academy Hall, Salem. . . . 9 — Longfellow, Coolidge and Strong at lunch at home; 10 — Meeting of Ropes Trustees, Salem; 11 — Lunch at Union Club, Boston; 12 — Museum Committee, Salem; 13–14 — Dinner at Tavern Club, Boston; 15 — Lecture on Japanese Pottery, Museum of Fine Arts, Boston; . . . 17 — Thursday Evening Club, Boston; . . . 19 — Reception to Gen. Pew; . . . 21 — Introduce lecturer at Academy Hall, Salem; 22 — Concord Historical Society visits museum in Salem; 23 — Pall-bearer at Mrs. Bates' funeral, Trinity Church, Boston, tall hat, black gloves; 24 — two Japanese at Salem museum and to lunch; 28 — Committee meeting at Salem; 29 — Lecture at Museum of Fine Arts and a dinner at 7.

Margarette Brooks wrote the note, "Tall hat, black gloves," on his calendar; left to himself, Morse had a fine disregard for raiment. He escaped her sharp appraisal one morning by appearing in the library with his overcoat already buttoned up, his hat on his head, on his way to catch the train to Boston. It was not until he took off the overcoat, in front of an audience assembled for a lecture on pottery at the Museum of Fine Arts, that he realized he had unthinkingly put on that morning the clothes he had taken off on coming home late from a banquet. He was in full evening dress for a morning lecture.

He was free, in 1916, to attend the social and scientific gatherings he enjoyed so thoroughly, because he had been retired in 1914 as director-emeritus of the Peabody Museum, and young Lawrence Jenkins was doing the work as director. He still went in once a week to the Museum of Fine Arts, running up the marble stairs briskly, tweaking the ear of pretty girl assistants, and telling them some of the milder limericks. He weathered better than his contemporaries:

December 9, 1915

Ho, John! Sorry to hear you are temporarily busted. Why did you break your rib? I am disgustingly healthy. . . . Thanksgiving eat too much at Edith's . . . and the following Saturday eat still more . . . at a Harvest dinner at the Salem Club. Monday no better; dinner of Chamber of Commerce in Boston in honor of Baron Shibusawa, . . . eat one oyster and a little soup. Tuesday, East Asiatic Society at Harvard club, eat five of the six oysters, soup and a cocktail. Wednesday, presided at Boston Society of Natural History, every seat filled; went to a gorgeous dinner of a friend before the meeting. Thursday . . . eat fairly well at big dinner of Thursday Evening Club at ex-Secretary [of the Navy] Henry von L. Meyer [4a] at Hamilton. Friday, dinner at Ends of the Earth Club, New York. . . . Came home Monday sound as a nut. So I must be tough, for I worked all day mentally. In New York catalogued a collection of 150 Japanese potteries. Japanese Journal almost done.

The Ends of the Earth dinners in New York were vastly entertaining, with informal talks from the doers of interesting deeds. Poultney Bigelow suggests their flavor in a letter of October 31, 1911, to Morse:

[4a] Meyer's name was George, not Henry. The mistake is characteristic of Morse.

You will be called on at the "Ends" dinner December 8; also Townsend who has been dredging the bottom of the Pacific; also Col. Thompson, authority on Japanese Navy; also Herb. Bowen of Venezuela diplomacy (and Persia); also Alleyne Ireland of Tropical Colony authorship; also George Kenna; also W. H. Bishop, consul for the Sicily earthquake. There's explosive material enough. . . .

Morse's eccentricities were fondly regarded by his friends, who usually referred to them jocosely in extending invitations. His aversion to noise and the truculence which embroiled him in open hostilities, as when he and his Peabody Museum associate John Robinson ceased speaking and wrote notes from desk to desk when communication was necessary, are referred to in a note (the year is missing) from George Goodale, Harvard professor of botany:

My dear Morse, It would give us pleasure to have you dine with us on Wednesday at six and then you will be all ready to go in your dress coat to Agassiz's reception at eight, returning here for the night. The room you will have is not large, but our clock in it *is*, so that you ought to sleep well! I ought to say that we shall have three or four friends at dinner: you are at sword's points with none of them!

The book compiled from the diaries of his years in Japan was finished late in 1916.

August 25, 1916

Ho John! . . . All through the hot weather I have worked constantly on [the] Japanese Journal. . . . I carried the manuscript and illustrations up this morning in a handbag and it was heavy. Houghton Mifflin are to publish it and they are enthusiastic about it. I am glad to get it out of the house. . . .

The syntax, unlike that of his *Catalogue . . . of Japanese Pottery* twenty years earlier, would not give the printer cold shivers, for his son-in-law, Russell Robb, had read and corrected the typescript page by page. Morse and Margarette Brooks had worked four years continuously, transcribing, editing, and illustrating the old diaries. The book was published under the title *Japan Day by Day*, in two volumes, with seven hundred and seventy-seven illustrations drawn by Morse.

Poultney Bigelow, reviewing it in the *Independent*, employed a happy phrase: "These precious two volumes have the charm of . . . Marco Polo's account of life at the court of Kublai Khan. . . . Were every other book on Japan lost Morse's two volumes would bring the empire of Dai Nippon back to life. . . ."[5]

No book in English furnishes more delight or a better introduction to an understanding of Japan than *Japan Day by Day*. Its objective record, its keen observation, made it as valid when it was reissued in Japanese a generation later as on the day of its original publication.

Dear Professor [wrote Mrs. Isabella Stewart Gardner from Fenway Court, Boston, November 15, 1917], I have this moment finished reading that most delightful book of yours. It has given me so much pleasure that I am bothering you by writing to thank you for writing it. It recalls very long ago days in Japan and also reminds me of very many times when you have talked to me about Japan and drawn your delightful sketches. I wish I could see you oftener and hear more of the enchanting things you can tell. You are much to be envied.

William Elliott Griffis, at Ithaca, New York, himself a well-known writer on Japan, added his praise: "I have just written

[5] January 12, 1918, p. 72.

a booming notice of your *Japan Day by Day*. You have done well to put on permanent record your sights and feelings in Old Japan. Some day, in some form, I may tell more of the old feudal days of '70–'74. . . . Good luck and long life." [6]

"My dear old Sensei," scribbled William Sturgis Bigelow from Beacon Street, Boston, November 15, 1917,

> I have been enjoying the emotions of decomposition for the last ten days — that is, feeling distinctly rotten, which I suppose to be the prelude to disintegration, but before I break up I want to tell you . . . how this book brings back the old days as if they were yesterday. Not only what you write, but what you suggest. It is strange how one thing brings up another, as the man said when he took an emetic. Well, for good or ill, the cruise to Japan was the turning point of my life — whether for better or worse, who knows? What is the difference, anyway, between good and bad? I can only give cases. You, for instance, are a good fellow. Long life and happiness to you, you G.O.M.!

There were warm ties between all Morse's friends on points of common interest, and indeed "one thing brings up another" as is shown in Nathan Haskell Dole's letter from 12 Dane Street, Jamaica Plain, Massachusetts:

> November 15, 1917
>
> *Konichiwa* [good day], you dear old friend! I am enclosing the torso of a much longer article which I wrote for the magazine section of the *Transcript*, forgetting entirely that the collecting part of your book had already been covered in the month of October. If you had seen me playing an *allegro con brioco* on my typewriter at midnight so as to finish it, you would have rejoiced at the music of the click. Well, Edgett ran it in, considerably curtailed but you will understand. Sometime when you are to be at the Museum will you let me bring the first volume and also *Japanese Homes* for your autograph on the flyleaf? Then I shall prize them even more.

[6] Letter of November 5, 1917.

Ernest Fenollosa who, as you know, was my classmate at Harvard and my most intimate friend, invited me to spend a year with him in *Dai Nippon* and I regret deeply that I was then, as I am now and always have been, a human brachiopod fastened to my native rock and I couldn't accept his call. I have had similar invitations to visit Russia and Sweden and Spain and England. With the aspirations of a globe-trotter I have never been anywhere, — except Philadelphia! . . .

I notice that Fenollosa's name appears only once in the Index to your book, but you mention him in several places; once where he and other Professors sang an unappreciated warsong. It was Lizzie (Mrs. F. No. 1) who crost the red lacquer bridge (at Nikko); she herself told me. . . . *Sayonara!*

Had Morse done nothing else in his career, he would still rank high as author of this one book. With the acumen of an anthropologist, the accuracy of a scientist, and the vision of an artist, he faithfully portrayed an alien civilization which he had studied just before it was engulfed by a tidal wave of western modernization.

About the time of his eightieth birthday, June 21, 1918, Yale made him a present — the honorary degree of Doctor of Science. Professor-Emeritus Theodore S. Woolsey of the Law School presented him to President Arthur T. Hadley with the citation:

Edward Sylvester Morse — Born in Portland eighty years ago . . . , a student with Agassiz, in the chair of zoölogy at Bowdoin. The pursuit of Brachiopods led Professor Morse to Japan. Three years in the Orient changed the current of his life. As collector, man of taste and man of letters, he has interpreted Japanese ceramics and Japanese character with loving fidelity. As head of the Peabody Museum in Salem since 1881 he has built up a wonderful institution. As zoölogist and ethnologist he has won an enviable name. A double life is his, the happy union of science and art.

Morse posed for his picture in gown and hood under the Yale elms, but did not let his honors distract him from his latest scientific enthusiasm:

July 2, 1918

Dear John [Gould]: . . . now I hold three honorary degrees from the three oldest colleges in New England. Ph.D. from Bowdoin (1871). M.A. from Harvard, and the Yale Sc.D. or D.Sc. The ceremony was very impressive. I wore a cap and gown and when the parchment was handed me by the President a hood of buff and blue was slipped over my shoulders. . . . Have just sent a third article to *Science* on fireflies flashing in unison.

October 21, 1918

When Yale gave me my degree, Verrill, two years younger, was too feeble to attend. Never touched a drop of liquor nor bit of tobacco in any form. With me rum and tobacco since youth. Feel full of the Devil all the time.

He was becoming retrospective, living alone after his wife's death, with the labor of editing his old journal-pages of the Japanese years emphasizing the passage of time.

January 11, 1916

Dear John: I was never out of debt till I was 54 years old and though I was apparently cheerful, it worried me like the Devil. I have never been in debt since and having no debts and a balance in the bank am sending you fifty dollars on condition that you spend it on personal comforts. If you are inclined to spend it on charity, let me dispose of it, for, God knows, I have in mind those who need it. This European horror has cost me some already.

January 31, 1918

I am gathering up my zoölogical work again. . . . The book "Japan Day by Day" has had a large sale but it cost so like the Devil to make that I probably shan't see a cent of money unless it goes into a second edition. . . . I try to subscribe to all the war funds and last Spring I subscribed $200 to the new Salem Hospital. . . . Did I tell you the conundrum Why is Germany like Holland? Because it is a low lying country damned on all sides.

October, 1918

Haven't we got the Huns on the run?

In the multiplicity of his collections of pottery, shells, and ethnological objects, Morse escaped the contagion of philately, though he never neglected to get a handful of foreign stamps from any scientist he met anywhere for the collection of his son John. Nevertheless, in one of his characteristic explosions over small annoyances, he influenced the science itself. Stamp-collectors who spend hours counting "perforations" will relish a letter signed by the indecipherable scrawl of the Third Assistant Postmaster-General.

Washington, D. C.
February 21, 1917

The Department has received, attached to the letterhead of the Peabody Museum, a newspaper clipping from the Boston *Herald* of Jan. 26 headed "Too Few Holes" and bearing the initials "E. S. M." Some time ago the spaces between perforations in postage stamps were widened to prevent premature separation of the stamps in handling at the factory, in postoffices and by the public. . . . There has, however, been some complaint of tearing of the stamps and the Department is taking steps to increase the number of perforations, without, however, using so many as to cause the stamps to separate prematurely as formerly.

Morse probably did as much as any one citizen to obviate the perennial deficit of the Post Office Department. His correspondence with friends and acquaintances in all parts of the nation and the world was both vast and continuous.

At eighty, with his museum work handed over to the younger men he had trained, his terms finished as president of the Boston Society of Natural History and the American Association of Museums, his book on Japan completed, Edward Morse picked up once more the zoölogy he had always preferred but so long postponed from pressure of more immediate responsibilities. His brief observations on fireflies flashing in unison were published in 1918 and amplified in 1924 in *Science*. To the *Nautilus* he contributed papers and drawings on *Caecum*, *Mya arenaria*, and *Helcion pellucidum*, "a European mollusk never before recorded in America." Working summers at the Marine Biological Laboratory at Woods Hole and winters at Salem, he completed in 1919 his observations on the living lamellibranchs of New England for the *Proceedings* of the Boston Society of Natural History (66 pages, with 143 sketches) and in 1921, for the Peabody Museum at Salem, his account of living gastropods of New England (30 pages, with 119 sketches). His last published work was an article, "Shell Mounds and Changes in the Shells Composing Them," which appeared in *Science Monthly* in October 1925.

During the summer weeks at Woods Hole, he worked in the room of Dr. Gilman A. Drew, who had succeeded him in teaching zoölogy to University of Maine students at Orono. As usual, all his friends were commandeered to hunt specimens. He acknowledged their help gratefully:

While enjoying the hospitality of Mr. and Mrs. Sabin W. Colton, their son, Dr. Harold S. Colton, secured much material for me in the waters around Greenings Island, Mount Desert. Through Dr.

and Mrs. Charles G. Weld I had rare opportunities for dredging in the waters of Penobscot Bay, as well as the use of a summer house for study. While a guest of Mr. and Mrs. Dwight Blaney at Iron Bound Island, Maine, I had the use of their excellent laboratory and a larger part of the drawings were made from material Mr. Blaney dredged in the waters of Frenchman's Bay.[7]

Paying a visit to Margarette Brooks and her sisters, vacationing at Newport, Rhode Island, he "discovered a minute specimen of the beautiful limpet *Helcion pellucidum* of Great Britain. . . . So far as I know this species has never been found on this side of the Atlantic. Miss M. W. Brooks discovered another European species, *Homalogyra atomus*, at Newport." Thinking of the World War the United States had entered in 1917, Morse concluded: "With the tremendous traffic going on for nearly two years in the conveyance of troops and provisions we may confidently look for other introductions of European species."[8]

"Many years ago I resolved to make drawings of the expanded soft parts of New England Mollusca with the intention of ultimately publishing the drawings in some form. One of the drawings used in this memoir dates back to 1855," were the opening words of his monograph on the lamellibranchs.[9] He had only waited sixty-four years for a free moment in which to complete his project. He was eighty-two years old and had never stinted himself of good whiskey or big black cigars; yet his eye was as keen, his hand as true as ever, his scientist's pa-

[7] "Observations on Living Lamellibranchs of New England," *Proceedings of the Boston Society of Natural History*, vol. XXXV, no. 5 (July 1919), p. 144.

[8] "A European Mollusk, *Helcion pellucidum*," *Nautilus*, vol. XXXII, no. 3 (January 1919), pp. 77–78.

[9] "Observations on Living Lamellibranchs," p. 139.

tience as tenacious. He sat for hours waiting until a shy snail would obtrude its "soft parts" to sit for its portrait. "In this memoir I have described and figured the expanded animal of 48 species of New England lamellibranchs."

He did more than that; he launched a characteristic diatribe against "the deplorable condition of the nomenclature [of shells] . . . so appalling that one gives up in despair to fathom the puzzle."[10] He cited the *Buccinum obsoletum* of 1855, which in 1876 had become *Ilyanassa obsoleta,* and by 1915 had been dubbed *Alectrion obsoleta.* He was not poking fun this time, as he had at the "four insane young men" of the Essex Institute who insulted animals with long names in 1870. In 1919 he quoted supporting statements from Charles Sedgwick Minot, Edward L. Rice, Herbert Osborn. Instant support from many quarters showed the force of his criticism.

I fervently add Amen to your sermon [wrote G. H. Parker of the Harvard Zoölogical Laboratory, December 31, 1919]. I can't tell you what a delight it was to meet the good old names, *Modiola modiolus* and *Mya arenaria* and all the old girls as I first knew them around Nahant. The new names make me think of the divorce court; they seem to have been swapping partners.

William H. Dall of the Smithsonian Institution registered a sly hit in his letter of October 15, 1919:

I congratulate you on the valuable information contained therein. In view of it I can bear with complacency the wail about nomenclatorial changes, especially as I remember something of the same kind which arose when you published your very excellent paper on Maine land shells (1864) from our good friends who had been accustomed to call all land snails *Helix*. . . . It is fine to have you putting out

[10] *Ibid.*, p. 140.

good material when most people would be aping Shakespeare's "lean and slippered pantaloon."

David Starr Jordan's letter of May 19, 1922, seems to hold a faint aroma of the old rivalry in which Morse had been offered without effort in 1877 the Tokyo teaching post for which Jordan had been a candidate. There is a trace of a desire to let Morse know that things had worked out very well after all:

Stanford University, California

My dear Morse: I am delighted to know you are getting on so well at an age which few of us can hope to reach. I am only seventy-one and in excellent health but thirteen years more is more than my physicians guarantee.

I read your paper on Lamellibranchs with interest and to a certain degree with sympathy. But in the group in which I am interested there was no tangible conception of a genus except as a more or less convenient pigeonhole until after the days of Darwin. We tried to make each genus represent a line of Evolution. It is necessary always to separate analogies from homologies and to make a genus one of the more or less broken fragments of the huge tree of life. I don't believe that species can be recorded by numbers, for in the groups I worked on it was easier to remember a name than a number. In Ichthyology the old names are now becoming very much better known than those which were current fifty years ago and the fauna of Europe is but a fragment of the huge fish fauna of the world. In the Zoölogical Commission to which I belong we have made an arrangement for the preservation of such names as Amia, Scarus and the like. I am sending you, for the pictures — if for nothing else — a recent paper of mine on fossil fishes of this region.

I have a boy, Eric, who at the age of fifteen published an excellent paper on the Mollusks of Trinidad Head, California, on which he worked with great enthusiasm. This breaks the record as the youngest author whose work has been printed by the Smithsonian Institution. At the age of ten he greatly surprised his father by making a catalogue of fossil shells found in the Santa Monica cañon and he seemed to know what species and genus meant. A year or two later I overheard him tell someone that "I don't like to talk about it but I know more about shells than my father does." I hope he may keep it up and surpass his Daddy in other lines in which his father is more familiar.

Having thus capped Morse's opening paragraph on the lamellibranchs with its reference to a "drawing made in 1855," Jordan, as he signed his name to the typed sheet, thought of another point and added a sentence in ink beneath his signature: "Five years ago Eric had about finished a manual of the shells of California." In other words, Morse might require sixty-four years to get around to a memoir; a Jordan could get the task done before he entered college. President of Stanford, member of the United States Fish Commission, director of the World Peace Foundation, David Starr Jordan had had a brilliant career; yet that youthful disappointment in the hope of visiting the exotic land of Japan had smoldered beneath the ashes for fifty years. Now, loving pride in the next generation's prowess brought the purifying clear flame that extinguished rivalry in amicable acknowledgment.

Simon Flexner of the Rockefeller Institute for Medical Research wrote a benevolent appreciation:

May 31, 1922

How delightful that you go right on year in, year out, without any consciousness of the passage of time to produce and then to produce more. If I could have that kind of luck I should feel I was favored by the gods. I wonder did you enter into some kind of conspiracy with the gods of the East?

Morse was by no means cloistered for his studies.

February 19, 1921

Dear John, Have been here (at the Cosmos Club, Washington, D. C.) since Wednesday. . . . I am here to identify the Japanese pottery in the Freer collection, hundreds of them . . . curious how I remember hundreds of names and recognize them [the pieces of pottery] in many cases before they leave the man's hands who passes them to me. The Cosmos Club . . . where forty years ago I knew everyone, now all are strangers. . . . Most of [the old crowd] are dead. That is the misery of growing old, all your friends die.

Still his own energy flourished and his spirits were sociable.

December 1, 1921

Ho, John! Tuesday morning in a violent storm left Hotel Copley Plaza . . . rode to Salem, visited Naumkeag Mills, our Museum and the great machine works in Beverly, had lunch at the club there, went through the mills, back to Boston, getting there at 6:30, dressed for the dinner at night, made a speech, back home next morning sound as a trapdyke. . . . Rum and tobacco! Nothing like 'em in reasonable quantities.

With the same blithe insouciance that led him to link the crackled glaze of Satsuma pottery with the markings seen through a telescope on Mars, Morse brought Virchow into a dissertation on fireflies flashing in unison. Writing in *Science,* July 26, 1918, he wound up:

While dining with a number of friends at the Parker House, the guests at a neighboring table had been noisy, even boisterous, doubtless we had been somewhat noisy, too. The neighboring table suddenly became quiet and we stopped talking to see if the noisy ones had gone, but they were still there, other tables looked about for the pause and this hush spread rapidly through the hall. Dear old Dr. Virchow had often observed this pause and thought my explanation [transmission of a common impulse] correct. He also told me it was a saying in his country that when this hush occurred an angel was passing through the room; also, that a lieutenant was paying his debts!

Dear old Virchow, whose *gemütlichkeit* in the old days led him to leave new discoveries in pathology pending while he spent hours hunting for a bone Morse wanted, had been dead fourteen years when Morse, at the Parker House, recalled his folklore; but Morse still preserved his letter, dated from Berlin, March 21, 1895, on reading Morse's address to the Vassar girls, *The Importance of Good Manners:* "I think it is like a kind and warmhearted genius and cannot fail to have a good influence."

Virchow, at long range, had made an estimate those close to Morse might not have been able to make at that time. Morse's "warm heart" burned for science and humanity, rather than for those of his immediate circle. As a boy he had defied his father's deepest wishes and sentiments to pursue his study of shells and science. He had allowed his brother Fred to saddle himself for years with the whole burden of supporting their mother and sisters. He had accepted the fourth part of John Gould's small salary to finance his years with Agassiz. He had left his wife farther and farther outside the sphere of his travels and career in science. He abandoned the faculty and students of Imperial University after two years. He calmly accepted the services of Margarette Brooks from her childhood to her death. He was forever dashing away from his post as director of Peabody Museum in one cause or another related to his scientific obsessions. After his wife's death, he obstinately refused to live with his children, remaining alone in the Salem house for fourteen more years to continue his work.

Analyzing him coldly, Morse, like any genius, can be appraised as a monster of single-minded egoism. Yet he never thought that of himself, never was consciously cold or cruel to his family or his friends. People in California or Japan or Germany, France or England, remembered and loved him for decades, nourishing their friendship with the vast stream of letters which he always found time to answer with charm and verve.

Somehow, as an uneducated boy on the long sweep of the Maine beaches or in the green Maine woods, Edward Morse had found in Nature a radiance that shone through his personality with the glow of a kindled taper. He could set sparks glowing in every person with whom he came in contact. He was tuned in on a universal wave-length that found reception in persons of any age, any land, any language. He had within him a quality that evoked in the world, as he passed through it, energies that persisted after him, that led ultimately to the development of museums and collections, the writing of books, and research of permanent value.

CHAPTER SEVENTEEN

HOW MUCH FORMALIN
1922—1925

THOUGH MORSE was noted for forgetting the names of the men who played tennis with him or peered through his microscope or learned from him the difficult potters' hieroglyphs of Japan, the young men did not mind. Still youth thronged about him, though his beard and hair were silvery grey, his step a trifle uncertain. Most of his boyhood friends were gone, but the appreciation of a new generation warmed old bones.

In June 1922 Tufts College awarded the man who had never been graduated from any college his fourth honorary degree:

> Edward Sylvester Morse: Your work as an author and as a scientist and as a connoisseur has been recognized in this country and in far-away Japan. Few there are who have your breadth of scholarship, or who have used their powers in so wide a field. Expressive of the scope of your accomplishments, the Trustees of Tufts College confer on you the degree of Doctor of Humane Letters.

A few weeks later, the Emperor Taisho conferred a second Japanese decoration on Morse, the second-class Order of the Sacred Treasure:

> In recognition of your meritorious services rendered to the cause of learning and culture in our country. With great pleasure and deep sense of gratitude we remember your professorship at our Uni-

versity, 1877–79, when you contributed so much to the introduction of western culture, especially the foundation of modern zoölogy and anthropology in Japan. The present growth of these sciences in our country must be attributed in great measure to your initial efforts.

These things were pleasant, but his rugged health was deserting him, as he confessed to John Gould in a letter of March 23, 1922:

I have had attacks of old age for the past few months . . . indigestion, dizzy head, rheumatism in both arms, and thought I should never write again. Two weeks ago made an after-dinner speech at the Tavern Club . . . and last Saturday afternoon lectured in Cambridge for the benefit of the Margaret Fuller Neighborhood House on the experiences of a collector. I am writing two or three pages daily on a second part of my arrow release.

It was more than old age that ailed him. A few months later, at the Salem Hospital he had helped to finance, he was operated on for strangulated hernia. The surgeons employed local anesthesia because his persistent bronchitis interdicted ether. Morse with bravado smoked a big cigar as they worked, and asked for news of the Dempsey-Firpo fight.

By June 1923 he was well enough again to cope with a surprise party on his eighty-fifth birthday. The Boston Malacological Society, of which he had been the first president in 1910, came to Salem twenty strong, appearing unheralded to find him in his office at the Peabody Museum, working on gastropods as though it were no special day. They brought him a silver pitcher and a tray, presented with a speech by the club president, Arthur F. Grey of Watertown. It then appeared that each member had come equipped with a basket lunch, so they all adjourned to 12 Linden Street after Morse had conducted them through the Japanese collection in Weld Hall. At his home, for several hours Morse indefatigably showed them his shells, his personal pottery collection, and his decorations from the Japanese emperors.

Letters by the score and the hundred poured in, with congratulations, affection, and the friendly raillery he loved so well. Francis Wilson wrote from Lake Mahopac:

I have never objected to anything that was yours except your handwriting! You must have something on God who permits you to go on and on, gliding happily thru the years, the admiration and comfort of your friends! I have never had any skill in unraveling the inscrutable, nor yet His ways, but He knows I love you.

And Poultney Bigelow, at Malden-on-Hudson, addressed an envelope to "The Illustrious Brachiopodarchbishop of Tamasanapolis" and by adding "Morse — Salem, Mass." had his greetings delivered.

My dear damniculous E. S. [wrote Bigelow], you are great and glorious. Long life to you. Bless you and give my love to Margarette for whom all the world cherishes gratitude on account of Japan Day by Day, — to say nothing of other reasons that endear her to connoisseurs of character.

It was exciting and emotionally stirring, the outpouring of affection from so many. Afterwards the old man was tired. His children persuaded him to go to his well-liked White Mountains for a rest and change. He himself realized he had needed the quiet.

August 28, 1923

Ho, John! . . . I have had a fine rest of two weeks in this delightful quiet place [Tamworth, N. H.]. Absolutely not a sound. No whistles or puffing of engines. No crowing roosters or cackling hens. No barking of dogs or squalling of cats. No mooing of cows or bellowing of bulls. No shouting of children or crying of babies. Not even a songbird or cawing of crows. Doubtless the insects chirp but I do not hear their high vibrations so it is simply heavenly.

Three days later, back in Salem, he read with memory and pity of the great earthquake, September 1, 1923, that had laid Tokyo and Yokohama in ruins. He thought of the Imperial University that he had helped to build, and the thought sent him to change his will. He talked it over with his children, Edith and John, who gladly assented to his plan. He bequeathed his entire library to the university at Tokyo, scientific works and pamphlets collected through seventy years, many of them autographed by the great men of science. The books filled sixty-nine large cases when they left Salem after his death.

Morse had keen pleasure in seeing the interest in Maine land shells that had cemented a lifelong friendship between him and John Gould come full circle:

February 6, 1924

Delighted to get the newspaper and congratulate you on being President of the Portland Society of Natural History. It is a curious incident that two boy friends of Portland should become Presidents of the two oldest Natural History Societies in New England. It is a damned shame that the Portland people cannot raise a decent working fund for the Society.

He hated the inexorable inroads of time.

November 19, 1924

Ho, John! Are you still alive? . . . One effect I find of old age I lack initiative. It is difficult to start to write a letter even. Of course I forget things, getting deaf, blind, legs incoherent. Have to go downstairs slowly and am impatient all the time at my old age. Keep busy, though. Attend meetings in Boston, Cambridge and here [Salem]. Read all my magazines, scientific and otherwise. . . . Had twenty-one men [the Salem Literary Club] to a supper.

Even at eighty-six years of age, Morse looked forward to more time in which to carry on his research. He came of a long-lived family. His brother Fred, now ninety, was still vigorous. Years ago his mother, Jane Beckett Morse, had written to Ned, in one of the few letters from her which he preserved:

Gorham, Maine
March 15, 1885

My dear Child: I was quite amused in reading the enclosed letter. It carried me back [to] when I was a little girl and played the same game. I think it was an old English play, for Mother learnt it of her Grandmother Dyer who came from England and she was a hundred years old, lacking three months when she died, and Mother was six or seven at that time and Mother was born in 1788, so it must be quite an ancient affair. Mother was the youngest of ten children, the oldest being thirty when she was born. Mother used to vary the words in the play . . . sometimes she used the word Wizard instead of Witch. "Tripany, Tripany, Trany, Trou; the Cock is awake, the Biddies asleep, it's time to give the chickens some meat; what time is it, Old Wizard?"

The eighty-eighth year is considered a lucky milestone of a man's life in Japan and is signalized by the bestowal of a scarlet cap to keep cold drafts from thinning white hair. Dr. Chiomatsu Ishikawa, a pupil in Morse's first class at Tokyo in 1877 and a successor to Morse in the chair of zoölogy at Imperial University, sent his old professor the scarlet cap on June 18, 1925.

Japanese visitors brought it, Dr. Michinosuke Miyajima, distinguished bacteriologist, and Copeland's one-time student of literature at Harvard, Commander K. Mizuno of the Japanese Imperial navy. Again Morse showed Mizuno the glass jar, waiting on his library table for his brain. To Dr. Miyajima,[1] seeing it for the first time, Morse explained, with scientific detachment, that the Wistar Institute eventually would dissect his brain, looking for evidence of a factor of the ambidexterity he possessed. Morse had genuinely forgotten, in sixty crowded years, the injury to his right hand in Agassiz's laboratory that had caused him to cultivate the use of his left hand.

Morse, dismissing both the glass jar and the scarlet cap of old age from his mind, spent the summer of 1925 over his manuscript on shell mounds and the changes of shells composing them. He completed the paper and seven drawings to accompany it and sent it off to *Science* monthly. For Thanksgiving dinner he went from Salem to Concord, where his children and grandchildren gathered around the table at Russell Robb's home.

On Wednesday, December 16, 1925, Edward Morse took a walk in the early twilight and stopped in at the Brooks house. The women were just having supper. He said he would not eat and sat reading a magazine until they had finished. At nine o'clock he went home, Margarette walking across the garden with him to his door. He was tired and went up to bed. Josephine, bringing his malted milk at ten o'clock, found him stricken. She went to the telephone to call the doctor and, on

[1] At the age of eighty-seven, Morse still exerted a dynamic influence even on strangers. Dr. Miyajima was so affected by Morse's example that in 1930, on receiving a cablegram at Geneva announcing the sudden death of his only son, a medical student in Tokyo, he overcame a deep-rooted racial tabu and cabled back directions that his son's body should continue to serve science by being offered for autopsy.

returning to his room, saw that he had lost consciousness. Edith Robb and John Morse came quickly from Concord, but their father did not speak again. On December 20 his heart stopped beating.

He had told them, when he went to the hospital in 1923, that he would like to be buried from the First Unitarian Church in Salem. The services were held there, in the old stone church with its square tower, next to the Ropes Memorial Garden of which he had been trustee so many years.

At the Peabody Museum, the Essex Institute, the Salem Club, flags drooped at half-staff while they buried Edward Sylvester Morse. The Japanese ambassador sent a wreath from Washington. Delegations and tributes from scientific societies were innumerable.

Fred Morse, ninety-two years old, was told in Portland that his brother Ned was dead.

"A shock?" asked Fred Morse. "Dear me, I didn't suppose Ned was old enough to be in danger of a shock."

In 1930 Commander K. Mizuno of the Japanese navy visited America in the suite of Prince Takamatsu, brother to the Emperor. Mizuno made a pilgrimage to the Wistar Institute and asked to see the brain of Professor Morse.

> "Here I am, Professor," said I in my heart, "to see your brain as you requested." A voice seemed to reply, "Good, my boy, but I am in hell. Are you brave enough to come down here?"
>
> My eyes filled, but I thought even the Wistar Institute where everything is analysed could not tell how much of my tears was sentiment and how much the formalin in which the brain was soaked.
>
> That brain had been weighed, measured, dissected, but nowhere had there been found, so I was told, the source of his ambidexterity.

For us outsiders of science, however, the warm heart was more precious than his cold brain.[2]

It was the warm heart of Edward Sylvester Morse that transmuted his love and genius for science into an ardent and illuminating mission to share knowledge with the "outsiders of science"; to make collections and build museums for the public; to teach children, encourage students, and coöperate with scientists. His notable achievements arose in each instance from his own initiative in utilizing materials that lay ready to his hand, whether it was the land shells of his native Maine or the products of the potter's wheel in Japan.

He was an American, and wherever he traveled, he left respect and affection for that name. He recognized brothers in science in every land, knew no jealousy or prejudice, but promoted coöperation for the good of mankind.

Without formal education, influence, or wealth, solely by his resolution to perform the work in hand to the best of his capacity, this boy from Maine won distinction among his contemporaries as a zoölogist, as a teacher of science, as an authority on ceramics, as a leader in the development of museums, and as a writer and illustrator.

His scientific studies must necessarily be superseded; his books may be forgotten, his name fade from men's memory; but as long as children pass with wondering eyes through museums in Tokyo, Boston, Cambridge, and Salem, the warm heart of Edward Sylvester Morse will fulfill its mission.

[2] *Japan Times and Mail*, July 10, 1938.

APPENDIX

APPENDIX

The memory of Morse was greatly honored in Japan. In the Memorial Hall of the beautiful library of Tokyo Imperial University, built by the Rockefeller Foundation after the 1923 earthquake, there stood on easels the portraits of four men designated as moving spirits in the history of the university. Three were Japanese; the fourth was Edward Sylvester Morse. Each volume of his scientific library, bequeathed to the university at his death, had a bookplate bearing a photograph of Morse. The centenary of his birth, June 18, 1938, was observed by a banquet in Tokyo under the auspices of the Science Museum of Imperial University and the Society for Propagating Scientific Knowledge. Dr. Yatsu, Dr. Hasebe, Commander K. Mizuno of the Imperial Japanese navy, who had known Morse in Massachusetts, and Dr. Kiichi Miyake, professor of botany and marine biologist of Imperial University, who was also acquainted with Morse, gave addresses to an audience of more than three hundred Japanese scientists.

The marine biological laboratory of Imperial University, which Morse first established at Enoshima in the summer of 1877, continued research at Misaki on the mainland, to which it was removed. In the museum, in 1939, were still to be seen specimens collected by Morse. Morse's collection, made in 1878 in the north, was still on display at the Imperial University of the Hokkaido in Hakodate. On August 11, 1939, the Kokusai Bunka Shinkokai society gathered at a dinner in Tokyo nearly a score of distinguished Japanese scientists and men and women who had known Morse in their youth and gladly contributed their recollections for a biography of Morse. It was fourteen years since his death, sixty years since he had sailed away from Japan, but none who had known him had forgotten his radiant personality.

Kojiro Tomita, successor of Okakura as curator of Asiatic art at the Museum of Fine Arts, Boston, was one of a committee composed of Mrs. Charles Goddard Weld, J. B. Millet, John Ellerton Lodge, and Francis Shaw Kershaw which established a fund named for Morse at the museum "where he spent the long, productive term of thirty-three years, the income to be expended to perpetuate his interests and activities in the field of Japanese and allied Far Eastern Art."

On December 25, 1925, the Tokyo *Nichi Nichi* published five columns of eulogies on Morse by his former students in Japan. The issue of *Toyo Gakugei Zasshin* (the publication of Japanese scientific societies founded fifty years earlier by Morse) for February 1925 was dedicated to his memory and filled with articles, in Japanese, on the man and his work.

Chiomatsu Ishikawa, who followed in Morse's footsteps as an authority on brachiopods and as his successor in the chair of zoölogy at Imperial University, had carried his devotion to Morse through the years. After Morse's death, Dr. Ishikawa, with Dr. Michinosuke Miyajima of the Kitasato Institute of Infectious Diseases, Tokyo, made a pilgrimage to Morse's grave in Harmony Grove Cemetery, Salem, and arranged, with the sponsorship of *Keimei-kan,* a Japanese scientific society, to have *Japan Day by Day* translated and published in a Japanese edition (1930). The translation was made by Ishikawa's son, Kiniichi Ishikawa, Princeton graduate, whose undergraduate summers had been spent in Salem as an assistant to Morse at the Peabody Museum. In 1940 the appreciation of Morse's book in Japan led to a popular one-volume edition of Ishikawa's translation, issued by Sodansha, Tokyo.

Three surviving students of zoölogy under Morse on November 3, 1929, participated in the dedication of a monument to him at the Omori shell mounds, near Tokyo, visible from the passing railway trains. The granite base, eight feet wide and five feet tall, with *Omori Kaidzuka* incised in Chinese characters, was surmounted by a granite replica of one of the prehistoric vases unearthed by Morse in 1877. Mr. Motoyama, archaeologist and publisher of *Nichi Nichi,* gave the monument, which was designed by Prince Kashiwa Oyama and Admiral Shozo Arisaka, archaeologists. Dr. Ishikawa composed an inscription in English which appears on a copper plate inserted in the pedestal.

Another monument to Morse rose within a few hundred feet of this at Omori as the result of a touching rivalry of affection. Dr. Chujiro Sasaki, a student who had accompanied Morse on the initial excavation at Omori in 1877, felt that the Motoyama monument was not in the spot where Morse first drove a spade into the ground. Mr. Yonejiro Usui, owner of the land which included the spot as identified by Sasaki, willingly gave the site where, on April 13, 1930, an irregular slab of dark stone from Sendai, in pure Japanese style, was unveiled. Associated with Sasaki in this memorial was Tsunejiro Miyaoka, who added an international touch by incorporating in the fund one hundred dollars placed in his custody by F. S. Kershaw of Boston for any memorial raised to Morse in Japan.

Dr. Naohide Yatsu, seventh successor of Morse in the chair of zoölogy at Imperial University, Tokyo, accumulated an interesting collection of "Morseana" in cases in a small room on the top floor of the new science building constructed after the earthquake of 1923. Morse's portrait was placed in the corridor of this building, and in the department of archaeology in the same building Dr. Kando Hasebe, professor of anthropology, made a special permanent exhibit under glass of the Omori artifacts excavated by Morse. Morse's original drawings for his *Shell Mounds of Omori* were preserved in this department, also. In the summer of 1939, Dr. Hasebe was informed by a custodian that in the *godown* (earthquake and fireproof storehouse) of the department were two boxes marked with Morse's name. They were opened and found to contain aesthetically valueless specimens of pottery, evidently carefully stored after Morse's departure in 1879, "waiting orders." Their interest lay in the fact that they included the first pottery pieces collected by Morse, inexpensive saucers for *o-sashimi* (raw fish) made in the form of shells. Each specimen had a yellowed paper label pasted on it, with its identification in Japanese characters in the handwriting of Ninagawa. This constituted Morse's primer in Japanese pottery.

INDEX

INDEX

Bei Fragen zur Produktsicherheit wenden Sie sich bitte an:
If you have any questions regarding product safety,
please contact:

Walter de Gruyter GmbH
Genthiner Straße 13
10785 Berlin
productsafety@degruyterbrill.com